2023 IEEE Latin American Electron Devices Conference (LAEDC 2023)

Puebla, Mexico
3-5 July 2023

IEEE Catalog Number: CFP23T67-POD
ISBN: 979-8-3503-1191-4

**Copyright © 2023 by the Institute of Electrical and Electronics Engineers, Inc.
All Rights Reserved**

Copyright and Reprint Permissions: Abstracting is permitted with credit to the source. Libraries are permitted to photocopy beyond the limit of U.S. copyright law for private use of patrons those articles in this volume that carry a code at the bottom of the first page, provided the per-copy fee indicated in the code is paid through Copyright Clearance Center, 222 Rosewood Drive, Danvers, MA 01923.

For other copying, reprint or republication permission, write to IEEE Copyrights Manager, IEEE Service Center, 445 Hoes Lane, Piscataway, NJ 08854. All rights reserved.

*** This is a print representation of what appears in the IEEE Digital Library. Some format issues inherent in the e-media version may also appear in this print version.

IEEE Catalog Number: CFP23T67-POD
ISBN (Print-On-Demand): 979-8-3503-1191-4
ISBN (Online):   979-8-3503-1190-7
ISSN: 2835-3463

Additional Copies of This Publication Are Available From:

Curran Associates, Inc
57 Morehouse Lane
Red Hook, NY 12571 USA
Phone: (845) 758-0400
Fax: (845) 758-2633
E-mail: curran@proceedings.com
Web: www.proceedings.com

TABLE OF CONTENTS

Implementation of a ROS Node for Roaming Between APs for an Autonomous Mobile Robot 1
   Luis Calle, Jordi Castel, Marco Amaya

Thermal Effects in Fully-Depleted SOI Devices 5
   Ziyi Wang, Dragica Vasileska, Caroline S. Soares, Gilson Wirth, Jairo Mendez Villanueva,
   Marcelo A. Pavanello, Michael Povolotskyi

A General Toolkit for Advanced Semiconductor Transistors: From Simulation to Machine Learning 9
   Antonio J. García-Loureiro, Natalia Seoane, Julián G. Fernández, Enrique Comesaña

Modeling High Frequency Response of Nanometer SOI Devices Using Monte Carlo Transient
Technique 13
   J. Méndez-V, D. Vasileska, E. A. Gutiérrez, K. Raleva

Improving the Performance of Photovoltaic Laser Power Converters Using Automatic Global
Optimization Techniques 17
   Javier F. Lozano, Natalia Seoane, Enrique Comesaña, Florencia Almonacid, Eduardo F.
   Fernández, Antonio García-Loureiro

Device Parameters Estimation of GFETs at Temperatures Below 300 K 21
   Leslie M. Valdez-Sandoval, Eloy Ramírez-García, Nikolaos Mavredakis, David Jiménez,
   Anibal Pacheco-Sanchez

Experimental Characterization of a Thermoelectric Generator System 26
   Daniel Sanin-Villa, Elkin Henao-Bravo, Juan Pablo Villegas-Ceballos, Farid Chejne

Real-Time Wheelchair Controller Based on POF-Based Pressure Sensors 31
   A. X. González-Cely, C. F. Blanco-Díaz, Díaz Camilo A. R., T Bastos-Filho

Transconductance-To-Current Ratio Based Threshold Voltage Extraction in MOSFETs from Linear
to Saturation Modes.. 36
   Matthias Bucher, Nikolaos Makris, Loukas Chevas

Digital Twin of Electrical Motorcycle Battery Charger as AC Load in a Microgrid Based on
Renewable Energy 40
   Alcides Amado Herrera-Guerra, Elkin Edilberto Henao-Bravo, Juan Pablo Villegas-Ceballos

Non-Iterative Parameter Extraction Method Based on the Single Diode Model (SDM)............................. 45
   Manuel J Heredia-Rios, Luis Hernandez-Matinez, Monico Linares-Aranda, Mario Moreno-
   Moreno

Development of Simplified Lumbar Spine Mechanism Implemented with Tendon-Driven Motion................. 50
   Ruben Florez, Facundo Palomino-Quispe, Thuanne Paixão, Ana Beatriz Alvarez, Lucas
   Angst, Luis Maggi

Precession Motion of a CanSat Rover-Back Prototype During Its Fall-To-Earth Descent............................ 55
   Luis Aiquipa Moreno, Roxana Pastrana Alta, Walter Estrada Lopez

4x1 Patch Antenna Array Using Taper-Based Coupling for 24 GHz Radar Applications 58
   Arlen Gonzalez-Azuara, Gabriela Méndez-Jerónimo, Humberto Lobato-Morales, Germán A.
   Álvarez-Botero

IoT-Based Neonatal Incubator Monitoring System 62
   Alisson Waleska Martínez, Fernanda De Lourdes Cáceres, Kevin Fabricio Martínez

A Fast Transient Full Class AB LDO Regulator 67
   A. Serrano-Reyes, M. T. Sanz-Pascual, B. Calvo-López

A Multi-Stage CTLE Design and Optimization for PCI Express Gen6.0 Link Equalization 71
   Karla G. López-Araiza, Francisco E. Rangel-Patiño, Jorge E. Ascencio-Blancarte, Edgar A. Vega-Ochoa, José E. Rayas-Sánchez, Omar Longoria-Gandara

A New Microstrip Directional Filter Configuration Composed by Hybrid-Mode Resonators 75
   Humberto Lobato-Morales, Gabriela Méndez-Jerónimo, Germán Álvarez-Botero

Fabrication and Characterization of Al-Based Integrated MIS Capacitors 79
   Daniel Rocha-Aguilera, Daniel Ferrusca-Rodríguez, Joel Molina-Reyes

Implementation of Step-Hetero-Oxide & Dual Material Gate Designs on Lateral β-Ga$_2$O$_3$ MOSFET
for Terahertz Applications 83
   Priyanshi Goyal, Harsupreet Kaur

TCAD Investigation of Step-Oxide and Asymmetric Doping Design with Electrode Engineering on
Lateral β-Ga$_2$O$_3$ MOSFET for Terahertz Applications 88
   Priyanshi Goyal, Harsupreet Kaur

Effect of the Thermal Annealing Temperature on the Luminescent and Morphological Properties of
Silicon Rich Oxide Bilayer Structures 93
   J. Juan Avilés Bravo, A. Morales Sánchez, L. Palacios Huerta, J. Federico Ramirez Rios, M. Moreno Moreno

Increasing the Doping Efficiency by Post-Deposition Annealing in a-SiGe: H Films Synthesized by
PECVD 97
   Ernesto Franco, Alfonso Torres, Mario Moreno

Long-Term Potentiation and Depression with Vertically Stacked Nanosheet FET 101
   Nupur Navlakha, Md. Hasan Raza Ansari

Experimental Analysis of HfO$_{2/X}$ ReRAM Devices by the Capacitance Measurements 105
   Fernando J. Costa, Aseel Zeinati, Renan Trevisoli, D. Misra, Rodrigo T. Doria

Thermal Evaluation of 28-nm P-Type FD-SOI MOSFETs 109
   Alan Rossetto, Caroline Soares, Gilson Wirth, Marcelo Pavanello, Ziyi Wang, Dragica Vasileska

Fabrication and Electrical Characterization of Al-Based MIM Capacitors 113
   Daniel Rocha-Aguilera, Joel Molina-Reyes

Simulation and Operational Evaluation of Distributed Storage Devices Connected to a Direct
Current Distribuition Nanogrid 118
   João Paulo De Andrade Machado, Felipe Cabral Reis, Arthur Fonseca Côrrea, Lucas Dos Santos Bulhosa, Wilson Negrão Macêdo, Marcos André Barros Galhardo

Optimization of the Electrical Conductivity and Thermal Coefficient of Temperature (TCR) on
Hydrogenated Amorphous Silicon-Germanium Films Doped with Nitrogen (a-SiGe:H,N) for
Applications on High Performance Infrared Detectors 123
   Oscar Velandia, Mario Moreno, Ricardo Zavala, Alfredo Morales, Alfonso Torres, Luis Hernández

A Proposed STEM Program to Make Institutions More Inclusive for People with Visual and
Physical Disabilities in Panama.. 128
Victoria Serrano, Vladimir Villarreal, Lilia Muñoz, Konstantinos Tsakalis

Development of Drop-On-Demand Inkjet Process for the Fabrication of Thin-Film Printed Devices 132
*Salvador Ivan Garduño, Angel Sacramento-Orduño, Magaly Ramírez-Como, María Isabel
Reyes-Valderrama, Ventura Rodríguez-Lugo, Magali Estrada*

Empirical DC Compact Model for Source-Gated Transistors Using TCAD Simulation Data 137
Patryk Golec, Radu A. Sporea, Eva Bestelink, Benjamin Iñiguez

DC Biased Field Plate RESURF for Further RDSON Reduction of LDMOS Transistors............................. 141
Wendi Wang, Z. John Shen, Ian P. Brown

Multi-Level Operation in Ultra-Scaled MRAM 145
*Viktor Sverdlov, Mario Bendra, Wolfgang Goes, Simone Fiorentini, Abel Garcia-Barrientos,
Siegfried Selberherr*

Perylenediimide-Based Acceptors for OPV Applications 149
*Magaly Ramírez-Como, Desiré Molina, Magali Estrada, Luis Reséndiz, Josep Pallarès,
Ángela Sastre-Santos, Lluis F. Marsal*

Effects of the Compliance Current in the Electroforming Process of HfO$_2$-Based ReRAM Devices 154
Silvana Guitarra, Lionel Trojman, Laurent Raymond

Development of a Biosensor for the Detection of Glucose Levels in Saliva Based on the Oxidation
and Detection of Hydrogen Peroxide 158
*Fernando Sánchez-Hernández, Esteban Martínez-Guerrero, Elsie Evelyn Araujo-Palomo,
Cuauhtémoc R. Aguilera-Galicia, José Luis Chávez-Hurtado, Patricia G. López-Cárdenas,
Jaime Ramírez-Angulo*

Accelerating Engineering Education and Workforce Development in Automation & Control for the
Semiconductor Industry Based on Cognitive Neuroscience... 162
Luis Fernando Cruz, M. Luis Miguel Quevedo, Wilfrido Alejandro Moreno

High-Performance Germanium P-I-N Photodiodes for High-Speed, Hard X-Ray Imaging 168
Ziang Guo, Sergei Mistyuk, Arthur Carpenter, Charles E. Hunt

Development of a Turbine Spirometer Prototype and Signal Digitalization 172
*Heyul Chavez, Carlos Herrera, Felix Llanos Tejada, Jorge Bazán Mayra, Javier M.
Moguerza, Carlos Raymundo*

Author Index

Implementation of a ROS Node for Roaming Between APs for an Autonomous Mobile Robot

1st Luis Calle
GID-STD
Universidad Politécnica Salesiana
Cuenca, Ecuador
lcallea@ups.edu.ec

2nd Jordi Castel
GID-STD
Universidad Politécnica Salesiana
Cuenca, Ecuador
jcastel@ups.edu.ec

3rd Marco Amaya
GID-STD
Universidad Politécnica Salesiana
Cuenca, Ecuador
mamaya@ups.edu.ec

Abstract—The wide field of implementation of mobile robotics has increased the need for remote interconnection of a control and visualization station with the robot to monitor the data acquired by the robot's peripherals, with wireless networks based on the 802.11 protocol being the most optimal for its ease of implementation, bandwidth and operating frequencies. Within these networks, effective service transfer between the Access Points and the robot must be sought so as not to lose communication and for it to send the data in real time by it. For this reason, in this article the implementation of a ROS node that performs the Roaming process based on the intensity of the signal received RSSI from the Access Points of the network is presented. This node runs in parallel to the teleoperated and autonomous operation of the robot. With this node, it was possible to maintain an average link status of 81.15%, signal strength -53.18 dBm, and throughput for the download link of 163.45 Kbps and upload link of 13.45 Mbps, sufficient to maintain a stable interconnection between the control and visualization station and the robot.

Index Terms—Roaming, Sensor Networks, Telerobotics and Teleoperation, Wheeled Robots, Wi-Fi Network.

I. INTRODUCTION

Today mobile robots come to comprise a very broad field of research due to their applications in the industrial, social and medical field [1], [2]. Some of the most prominent research fields are: robots with legs, flying robots, wheeled robots, application of artificial intelligence (AI) for decision making, interconnectivity between the robot and visualization and control station, navigation, etc., which involve different technological areas such as computing, mechanics and electronics [2], [3].

Among the fields named above, the interconnectivity of the robot is crucial, due to a wide range of applications of mobile robotics, this class of robots can operate over long distances

*This work was supported by Universidad Politécnica Salesiana
[1] Luis Calle is a member of the Research Group on Simulation, Optimization and Decision Making (GID-STD), Universidad Politécnica Salesiana sede Cuenca. lcallea@ups.edu.ec
[2] Jordi Castel is a member of the Research Group on Simulation, Optimization and Decision Making (GID-STD), Universidad Politécnica Salesiana sede Cuenca. jcastel@ups.edu.ec
[3] Marco Amaya is a member of the Research Group on Simulation, Optimization and Decision Making (GID-STD), Universidad Politécnica Salesiana sede Cuenca. mamaya@ups.edu.ec

and remotely, both in a teleoperated and autonomous mode [4], [5], where there must be a stable and constant connection between the robot and the remote viewing and control station where the user is located, in such a way that the robot sends information about the environment so that the user can monitor the processes, trajectories, battery levels and data that the robots process in order to determine its constant operability [4].

Advances in wireless networks have made it possible to increase the distance range of operability of mobile robots and improve network infrastructure by increasing bandwidth and transmission rates to reduce latency and network saturation [3]. Among the wireless network protocols, one of the most used in mobile robotics is the IEEE 802.11 protocol, which has many benefits such as: easy to implement, a wide range of network topologies, affordable, and the hardware that represents low energy consumption [3], [5].

For the interconnectivity of mobile robots, the network infrastructure must be capable of performing a service transfer between the different nodes of the network in such a way that the robot does not lose connectivity with the monitoring and control station. For this reason, in other networks, such as mobile networks, there is a roaming process in which the transfer of service between base stations is sought without losing the connectivity of the mobile device [6]; however, in recent studies the application or implementation of roaming in mobile robots is not presented [7].

For this reason and given that the interconnection of the robot plays a crucial role to integrate the robot in an internal space, in this paper is presented the implementation of roaming through a ROS (Robot Operating System) node which allows transferring the service and coverage between the different AP (Access Points) of the fixed network established for the operation of the mobile robot, in such a way that the transmission rate reaches the required speed and the throughput maintains the necessary data load for the teleoperated and autonomous operation of the robot in order to reduce latency or loss of connection with the remote viewing and control station.

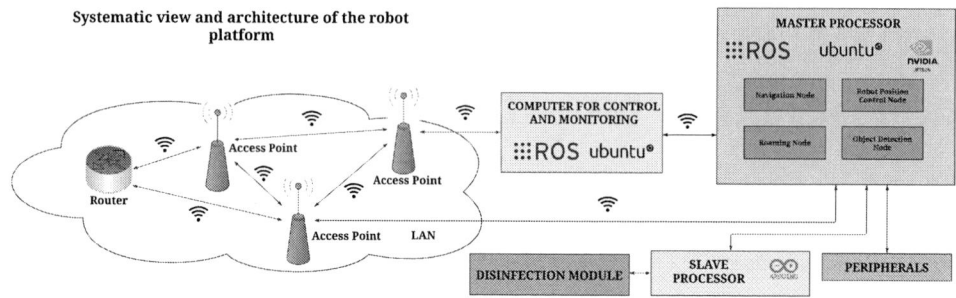

Systematic view and architecture of the robot platform

Fig. 1. Systematic view of the architecture of the disinfection robot.

II. SCENARIO DESCRIPTION

The differential robot is focused on the industrial field for the autonomous or teleoperated disinfection of different spaces such as classrooms, laboratories and corridors. The robot architecture consists of a central processor that executes the ROS nodes based on the data from the robot's peripherals necessary for its modes of operation, and a slave processor that is in charge of the operation of the disinfection modules. This robot is implemented in a LAN (Local Area Network) to have connectivity with the control and monitoring computer. The systematic view of the robot's architecture is presented in Figure 1, and in Figure 2 the distribution of the robot's elements is presented.

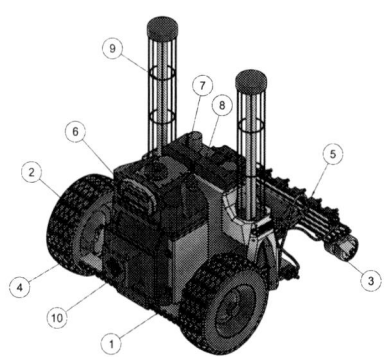

Fig. 2. CAD model of the disinfection robot. 1. Robot base, 2. Front wheels, 3. Beaver wheels, 4. Motors, 5. Liquid disinfection system, 6. Sensors (cameras and LIDAR), 7. UV disinfection system, 8. Tank, 9. UV-C lamps.

The robot has an Intel Dual Band Wireless-AC 8265 Wi-Fi card in its central processor, while the network nodes are made up of an ASUS-ROG-Rapture-GT-AX11000 router, and ASUS-RP-AC55 repeaters were used for the APs. The implemented network topology is meshed so the location of the router is close to the monitoring computer and the repeaters were deployed in the rest of the disinfection zone to cover the classrooms and corridors. The deployment of the network is presented in Figure 4, where the location of the router and the APs are indicated. The approximate distance between the router and the first AP is 11.5 m, the distance between AP1 and AP2 is 17 m and between AP2 and AP3 it is 12.21 m.

III. ROAMING ALGORITHM FOR ROS NODE

Commonly in 802.11 WLAN networks, wireless devices tend to connect to the AP that has the highest signal strength, described by the received signal strength indicator RSSI, and roam only when the RSSI value is below certain threshold. This scenario is presented in Figure 3, where the trajectory of the robot is presented and how, with the passage of time, the robot discovers possible links with other APs with which it must perform a service transfer in an instant of time to avoid disconnection and inactivity in the network.

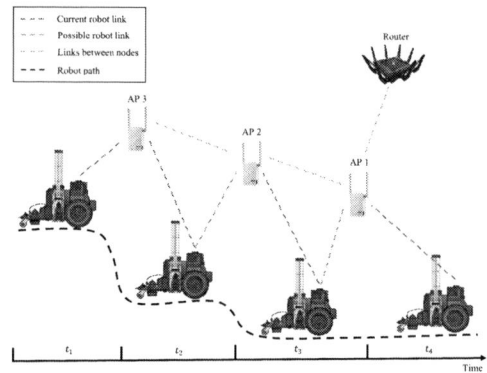

Fig. 3. Mobile robot roaming scenario.

In this scenario, the implemented ROS node first performs a scan among all the networks that share the same SSID to obtain the MAC address of the APs that provide coverage at that instant of time, so that in the roaming process the robot distinguishes the AP to which it should connect. Subsequently, the RSSI received signal strength is calculated between all the APs that provide coverage to the robot using Equation 1.

$$RSSI[dBm] = \phi(d) = r_0 + 10\alpha \log_{10}(d/d_0) \qquad (1)$$

Where d corresponds to the distance from the robot to the AP; d_0 is the reference distance with a value of one meter commonly; r_0 is the strength of the signal based on the distance d_0; and α is the path loss component which indicates the rate at which the path loss increases as a function of distance. Both r_0 and α can be obtained empirically, although

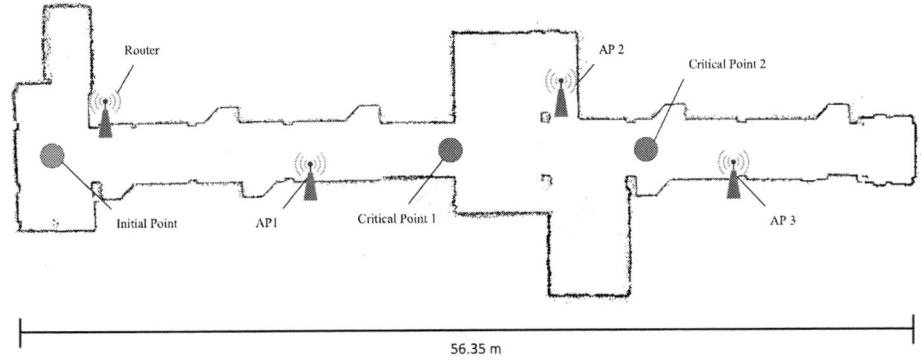

Fig. 4. Map of the deployment of the LAN network for the operation of the robot.

in the case of r_0 its value can be obtained from the hardware data indicated by the vendor.

Once the RSSI values of each AP have been determined, a comparison is made between these values to detect the highest, and then the transfer of service to that AP is carried out. This process is carried out constantly and parallel to the functions of the robot's autonomous or teleoperated modes. Figure 7 shows the relationship of each node used for the autonomous mode of the robot. This scheme indicates that the roaming node does not publish or subscribe to any topic since this process is independent of the perception and navigation of the robot.

IV. RESULTS

For the functional tests, a round trip linear path of the robot was carried out through the map presented in Figure 4 from the indicated starting point. In the first instance, the route was carried out in the teleoperated mode without the operation of the roaming node and later the same route was carried out but executing the roaming node. With the data obtained, it can be seen that the robot took longer to complete the trajectory without executing the roaming node, since there were disconnection events that prevented the robot from receiving speed data until it reconnected to an AP.

When analyzing the state of the link in both routes from Figure 5, the changes of the state can be visualized presenting a triangular pattern, this is due to the handover of the next AP when its signal intensity is greater. The state of the link when the roaming node is not executed presents two instances where the state of the link remains constant. This occurs at the moment where the robot is at the critical points presented in Figure 4, where the robot disconnected from the network. With this ROS node it was possible to maintain an average link state percentage of 81.15% while without executing this node an average of 65.29% was obtained. The behavior of the state of the signal is reflected in the intensity of the signal, since these parameters present a directly proportional relationship. Through the ROS node an average of -53.18 [dBm] was obtained, while in the route without the execution of the node an average of -64.34 [dBm]. Figure 6 presents

the signal strength data and the instants where roaming was performed between the APs in the network.

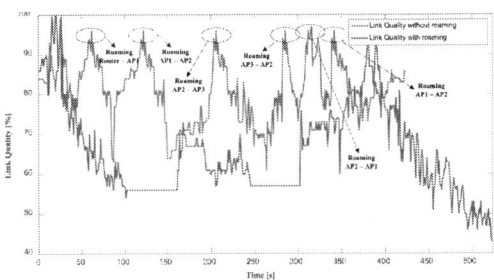

Fig. 5. Signal status during tours.

On the other hand, when analyzing the throughput we can notice that the traffic in both paths is similar since the same ROS nodes are executed for teleoperated operation. In the case of the download link during disconnection events at the critical points indicated in Figure 4, throughput reached a minimum speed of 0.95 Kbps, and an average of 133.32 Kbps, while when executing the node, an average of 163.45 Kbps is obtained. Figure 8a shows the throughput data for the download link.

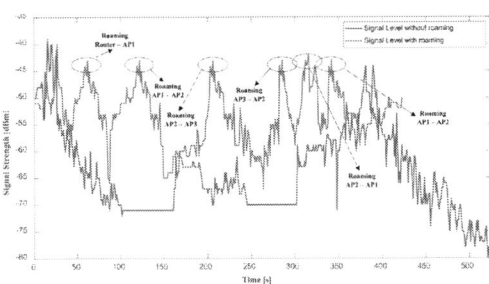

Fig. 6. Signal strength during trajectories.

Finally, Figure 8b shows the behavior of the throughput for the upload link, in which speeds greater than 10 Mbps were handled. As in the case of the download link, in disconnection events the minimum speed in this link reached 1.48 Kbps and

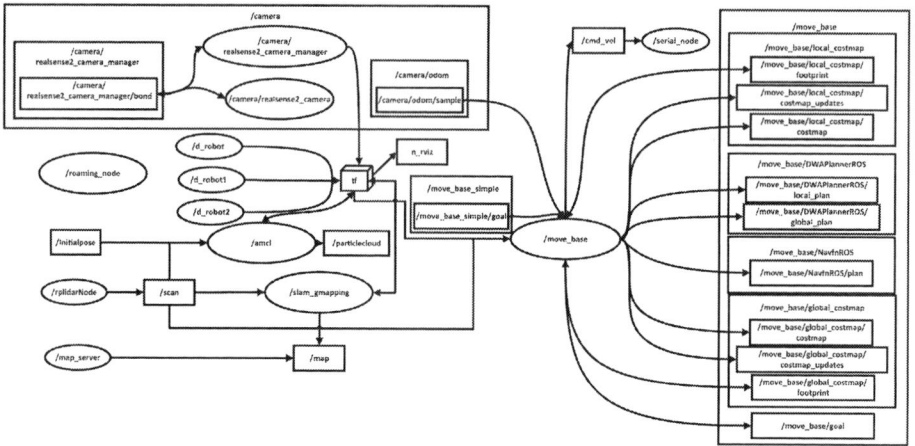

Fig. 7. Node tree for standalone mode.

an average of 8.8 Mbps, while when executing the roaming node an average of 13.45 Mbps was obtained.

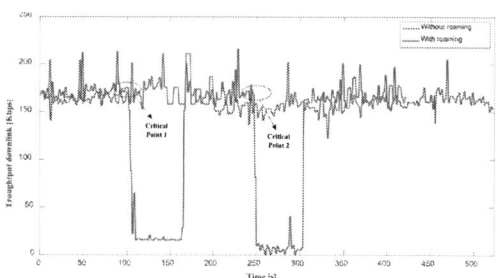

(a) Throughput downlink during trajectories.

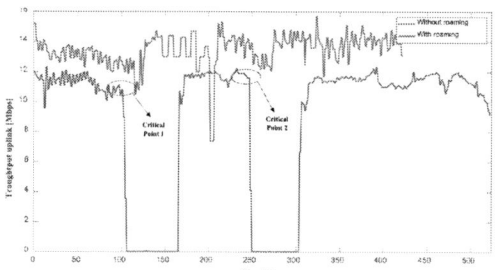

(b) Throughput uplink during trajectories.

Fig. 8. Throughput of upload and download links.

V. CONCLUSIONS

The roaming node implemented in the ROS ecosystem presents a solution for the transfer of services between the APs in the network, since it was possible to maintain an average of: link state 81.15%, signal strength - 53.18 dBm and throughput for the download link 163.45 Kbps and upload link 13.45 Mbps, sufficient to maintain a stable interconnection between the remote viewing and control station and the robot, in such a way that the user or operator of the robot has a good maneuverability of it and can go through the different disinfection areas. Although this node guarantees constant reconnection to the AP with the highest RSSI, reducing disconnection events, when the robot is implemented in a massive network with a large number of APs, the traffic would increase due to the number of APs that the robot must scan to identify the one with the highest RSSI, so the deployed network must consider the bandwidth required for the necessary requests during the scanning process. On the other hand, by implementing this node, the reconnection times are shorter since the roaming algorithm implemented in the ROS node is carried out at the moment that the next AP has a higher RSSI than the one the robot is connected to, while the roaming implemented by defect in Wi-Fi cards, they expect to exceed a threshold of approximately -70 dBm to perform roaming, at which time the link status is less than 60%, making it impossible to transmit data in real time.

REFERENCES

[1] S. G. Tzafestas, "Mobile Robot Control and Navigation: A Global Overview," J. Intell. Robot. Syst. Theory Appl., vol. 91, no. 1, pp. 35–58, 2018, doi: 10.1007/s10846-018-0805-9.

[2] F. Rubio, F. Valero, and C. Llopis-Albert, "A review of mobile robots: Concepts, methods, theoretical framework, and applications," Int. J. Adv. Robot. Syst., vol. 16, no. 2, pp. 1–22, 2019, doi: 10.1177/1729881419839596.

[3] S. Kahar, R. Sulaiman, A. S. Prabuwono, N. A. Ahmad, M. Ashri, and A. Hassan, "A Review of Wireless Technology Usage for Mobile Robot Controller," 2012 International Conference on System Engineering and Modeling (ICSEM 2012), no. April, pp. 7–12, 2012. [Online]. Available: https://www.researchgate.net/publication/232613949

[4] D. W. Gage, "Network Protocols for Mobile Robot Systems," vol. 3210, pp. 107–118.

[5] F. Zeiger, N. Kraemer, M. Sauer, and K. Schilling, "Challenges in realizing ad-hoc networks based on wireless LAN with mobile robots," Proc. 6th Int. Symp. Model. Optim. Mobile, Ad Hoc, Wirel. Networks, WiOpt 2008, pp. 632–639, 2008, doi: 10.1109/WIOPT.2008.4586151.

[6] C. Xue, W. Li, L. Yu, J. Shang, X. Chen, and S. Lu, "SERO: A model-driven seamless roaming framework for wireless mesh network with multipath TCP," IEEE Transactions on Communications, vol. 67, no. 2, pp. 1284–1296, 2019.

[7] N. Nguyen, M. Arifuzzaman, and T. Sato, "A novel WLAN roaming decision and selection scheme for mobile data offloading," J. Electr. Comput. Eng., vol. 2015, 2015, doi: 10.1155/2015/324936.

Thermal Effects in Fully-Depleted SOI Devices

Ziyi Wang
School of ECEE
Arizona State University
Tempe, AZ, USA
zwang581@asu.edu

Dragica Vasileska
School of ECEE
Arizona State University
Tempe, AZ, USA
vasileska@asu.edu

Caroline S. Soares
Programa de Pós-Graduação em
Microeletrônica
Universidade Federal do Rio
Grande do Sul
Porto Alegre, Brazil
santos.soares@ufrgs.br

Gilson Wirth
Programa de Pós-Graduação em
Microeletrônica
Universidade Federal do Rio
Grande do Sul
Porto Alegre, Brazil
gilson.wirth@ufrgs.br

Jairo Mendez Villanueva
Department of Electronics
National Institute of
Astrophysics, Optics and
Electronics
Tonantzintla-Puebla, Mexico
jairo_mendez@inaoep.mx

Marcelo A. Pavanello
Electrical Engineering
Department
Centro Universitario FEI
Sao Bernardo do Campo, Brazil
pavanello@fei.edu.br

Michael Povolotskyi
Jacobs Engineering Group Inc.
Naval Research Laboratory
Hanover, MD, USA
mpovolot@gmail.com

Abstract—Recently, excellent characteristics were reported for Fully-Depleted (FD) SOI devices operated under cryogenic temperatures. It was also observed that self-heating effects (SHE) play a crucial role to the FD SOI device operation. The goal of this work is to examine the role of the self-heating effects in 28 nm technology node FD SOI devices operated down to 78 K and compare our simulation results with available experimental data. Simulation results confirm experimental findings that the temperature increase in the active channel region of the device is more significant at low temperatures.

Keywords—Self-heating, FD SOI Devices, Monte Carlo Device Simulator, Cryogenic Temperatures

I. INTRODUCTION

The excellent electrical characteristics of Fully-Depleted (FD) SOI device technologies at temperatures down to 4.2 K opened a plethora of opportunities to use advanced CMOS in low-temperature environments such as space applications. Moreover, the growing interest in bringing down quantum computing into mainstream technology led to increased research and development of cryogenic control electronics for MOSFET operation at low temperatures. The low-temperature operation of MOSFETs is vital for the operation of quantum computers, as the qubits (quantum bits) used in these systems are extremely sensitive to temperature fluctuations.

In FDSOI devices, the heat generated primarily at the drain side of the channel during the device operation, represents a significant challenge due to the low thermal conductivities of the buried oxide (BOX) and the thin Si layer that constitutes the channel. The limited thermal conductivity of the thin silicon layers is attributed to the phonon boundary scattering. As a result, self-heating effects (SHE) lead to a significant temperature

increase in the channel region of the device when the device is in its on-state. The increase in channel temperature caused by SHE is known to have a severe impact on the device performance by reducing the carrier mobility. Numerous studies were conducted and published on the operation of FDSOI devices at room temperature, including both experimental [1] and theoretical [2] investigations. Also, experimental studies were carried out to investigate the SHE in FDSOI devices at cryogenic temperatures [3]. The goal of the above studies was to gain a better understanding of the thermal behavior of FDSOI devices under extreme conditions, which is crucial for their potential applications in low-temperature electronics and quantum computing. Therefore, there is a pressing need for simulation software that can accurately describe the existing experimental data and, at the same time, provide reliable predictions for FDSOI devices with different channel lengths, silicon film thicknesses and BOX thicknesses across a wide range of operating temperatures.

II. THEORETICAL MODEL

To accurately explain thermal effects without relying on any approximation, it is necessary to solve coupled Boltzmann Transport Equations (BTE) for both the electron and the phonon system. Specifically, the problem involves solving for the coupled electron-optical phonons-acoustic phonons-heat bath system, where each subprocess operates on a different time scale and requires individual treatment before being incorporated into the overall picture through a self-consistent loop. The BTEs for the distribution functions of electrons $f(\boldsymbol{k}, \boldsymbol{r}, t)$ and phonons $g(\boldsymbol{k}, \boldsymbol{r}, t)$ can be written as [2]:

979-8-3503-1191-4/23 $31.00 © 2023 IEEE

$$\left(\frac{\partial}{\partial t} + v_e(k) \cdot \nabla_r - \frac{e}{\hbar} E(r) \cdot \nabla_k\right) f =$$
$$= \sum_q \{W_{e/a,q}^{k\pm q \to k} - W_{e/a,q}^{k \to k \pm q}\} \tag{1}$$

$$\left(\frac{\partial}{\partial t} + v_p(q) \cdot \nabla_r\right) g =$$
$$= \sum_k \{W_{e,q}^{k+q \to k} - W_{a,q}^{k \to k+q}\} + \left(\frac{\partial g}{\partial t}\right)_{p-p} \tag{2}$$

Here, the probability of an electron transitioning from $k + q$ to k through the emission of phonon q is denoted as $W_{e,q}^{k+q \to k}$, while the process of absorption is represented by $W_{a,q}^{k-q \to k}$. These probabilities depend on the product of the electron and phonon distribution functions, which makes the system highly nonlinear. $\left(\frac{\partial g}{\partial t}\right)_{p-p}$ accounts for the phonon-phonon interaction.

As the system evolves, the electrons receive energy from the electric field E. The exchange of energy between electrons and phonons occurs through the terms W, which have a timescale on the order of a fraction of a picosecond. The transfer of energy from electrons to high-energy optical phonons is highly efficient when considering electron lattice coupling. However, optical phonons have minimal group velocity and therefore do not significantly contribute to heat diffusion. Hence, optical phonons must transfer their energy to acoustic phonons to diffuse heat. The transfer of energy between phonons is slower compared to the transport of energy between electrons and optical and acoustic phonons, thus leading to the possibility of thermal nonequilibrium between optical and acoustic phonons. Fig. 1 shows the thermal energy transport path and the associated timescales.

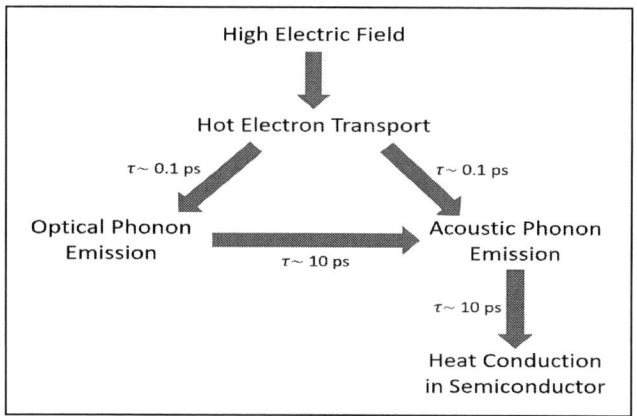

Fig. 1. Electron energy transfer to the acoustic and optical phonon baths, from the optical to acoustic phonons, and to the heat sink, in a semiconductor device.

To reiterate, Fig. 1 shows the primary path of energy transport that involves scattering between electrons and optical/acoustic phonons characterized with temperatures T_{LO} and T_A, respectively. Optical phonons near the zone center have almost zero group velocity. Hence, for the energy to be transported to the lattice, they must decay into acoustic phonons. Since this process is on the order of 10's of picoseconds, a hot spot forms [4].

The solution of the BTE for phonons is a formidable task. To overcome this difficulty, in this work we adopt a simplified

energy balance approach between the electron system and the optical and the acoustic phonon baths. In other words, we couple the BTE for the electrons with the energy balance equations for the optical and the acoustic phonons, respectively. These energy balance equations, in their most general form, are:

$$C_{LO} \frac{\partial T_{LO}}{\partial t} = \frac{3nk_B}{2}\left(\frac{T_e - T_{LO}}{\tau_{e-LO}}\right) + \frac{nm^* v_d^2}{2\tau_{e-LO}} - C_{LO}\left(\frac{T_{LO} - T_A}{\tau_{LO-A}}\right) \tag{3a}$$

$$C_A \frac{\partial T_A}{\partial t} = \nabla \cdot (k_A \nabla T_A) + C_{LO}\left(\frac{T_{LO} - T_A}{\tau_{LO-A}}\right) + \frac{3nk_B}{2}\left(\frac{T_e - T_L}{\tau_{e-L}}\right) \tag{3b}$$

In equation (3a), $\frac{3nk_B}{2}\left(\frac{T_e - T_{LO}}{\tau_{e-LO}}\right)$ and $\frac{nm^* v_d^2}{2\tau_{e-LO}}$ represent the energy gain from the electrons. $C_{LO}\left(\frac{T_{LO} - T_A}{\tau_{LO-A}}\right)$ is the energy loss to the acoustic phonons, which appears as a gain term in Eq. (3b). The first term on the RHS of equation (3b) describes the effect of heat diffusion, while the last term is a gain term from the electron bath to the acoustic phonons. The lattice temperature T_L in equation (3b) is assumed to be equal to the acoustic phonon temperature T_A. C_{LO} and C_A correspond to the heat capacity of optical and acoustic phonons, and k_A denotes the thermal conductivity of the silicon film. The electron temperature T_e is calculated through Ensemble Monte Carlo (EMC) time averages in our device simulator.

Due to the variable lattice temperature throughout the device and in the hot-spot region, a novel approach using temperature-dependent scattering tables was implemented into our in-house thermal device solver [2]. A unique scattering table is created for each combination of acoustic and optical phonon temperatures, which is energy dependent [2]. To randomly select a scattering mechanism for a given electron energy, it is imperative to identify the relevant scattering table, which requires additional steps in the MC phase (as shown in the right panel of Fig. 2).

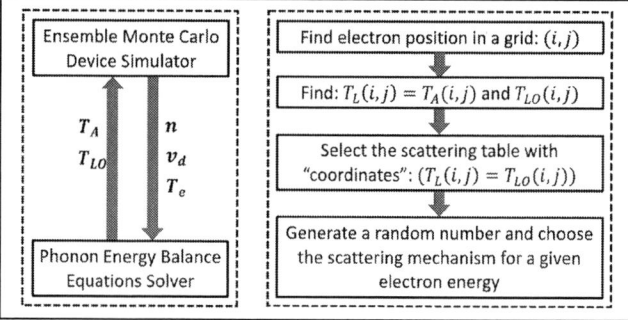

Fig. 2. Left panel: Exchange of variables between two solvers; Right panel: Choice of the proper scattering table.

In this methodology, the initial step is to locate the position of the electron on the grid, which allows for identification of the acoustic and optical phonon temperatures at that specific point. Once this information is obtained, the appropriate scattering table with "coordinates" (T_A, T_{LO}) is selected. A space-time averaging of electron density, drift velocity and electron energy is used to link the particle-based electron transport picture with the continuous phonon energy balance equations.

Thickness and temperature dependence of the thermal conductivity of the thin silicon layer is calculated using the methodology described in Ref. [5].

In the solution of the Poisson equation, partial ionization of the dopants is considered for temperatures down to 78 K.

III. SIMULATION RESULTS

The device structure considered in this study (28nm technology node with 30 nm physical channel length n-FDSOI device with 7 nm silicon film thickness, 1.2 nm effective gate oxide and 25 nm thickness of the BOX) is schematically shown in Fig. 3.

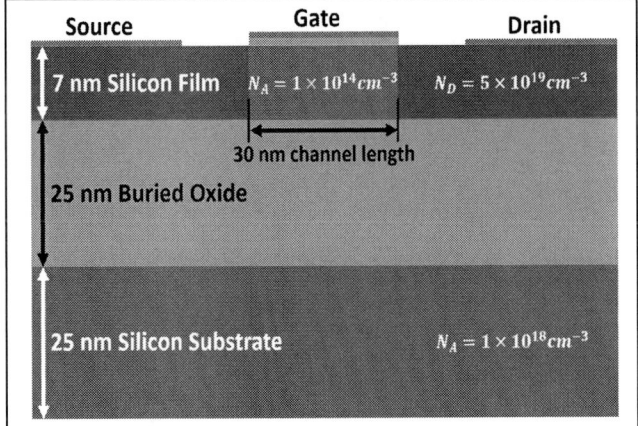

Fig. 3. FDSOI device structure being considered in this study.

Comparison between simulated transfer characteristics (shown in Fig. 4) and the available experimental data from Ref. [3], at temperature of 300 K, show excellent agreement in both the subthreshold and linear operation regions of the device.

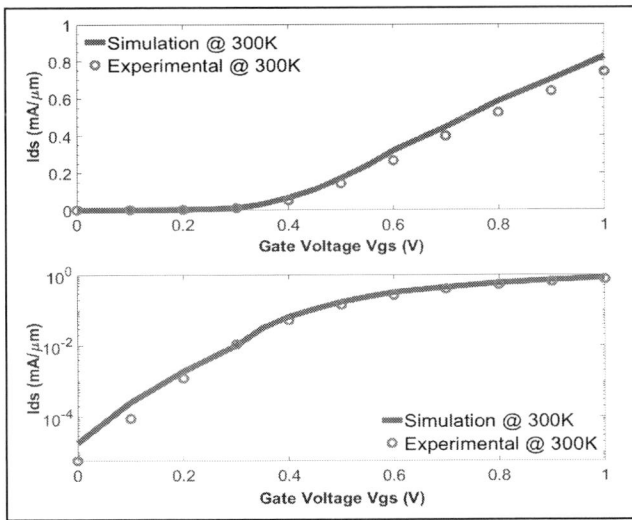

Fig. 4. Comparison of the simulated (lines) transfer characteristics and the available experimental data (open circles) from Ref. [3] at T=300K. The applied drain voltage is Vds=0.9V.

Fig. 5 shows the simulated transfer characteristics of the device at temperatures of 78 K, 150 K, and 300K. Our findings

indicate that the threshold voltage experiences a slight increase at lower temperatures due to the partial ionization of the dopants, which agrees with the experimental results [3].

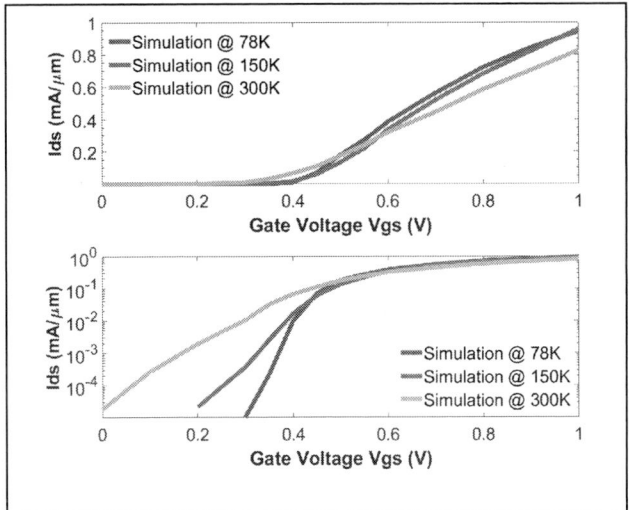

Fig. 5. Simulated transfer characteristics at 78K, 150K and 300K. The applied drain voltage is Vds=0.9V. Note that at low temperatures, due to the partial ionization of the dopants, the threshold voltage is shifted to a slightly higher value (top figure). This agrees with experimental trends [3]. Also, the subthreshold slope decreases (bottom figure) with decreasing temperature, as expected.

The lattice temperature profile at ambient temperature T=300K, and for Vgs=0.6V and Vds=0.9V, is shown in Fig. 6. The channel region experiences a significant amount of heat generation that is primarily concentrated on the drain side of the channel. In these simulations we assume that the source and drain contacts are perfect absorbers of heat.

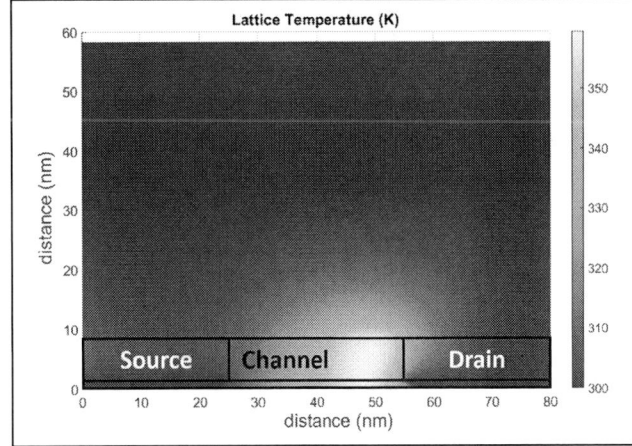

Fig. 6. Lattice temperature profile at ambient temperature T=300 K for Vgs=0.6V and Vds=0.9V. We assume T=300K at source and drain contacts, and zero heat flux at the gate contact.

A comparison between the simulated and the experimental excess temperatures in the channel under different input powers, and for ambient temperatures T=78 K, 150 K, and 300 K is shown in Fig. 7. To obtain self-heating measurements, the gate

979-8-3503-1191-4/23 $31.00 © 2023 IEEE

resistance thermometry technique was utilized in the experimental setup [1]. The thin gate dielectric layer in this method allows one to assume that the temperatures of the channel and gate electrode are almost identical. Although the simulated channel temperatures exceed the experimentally extracted values slightly at lower input powers, the thermal resistance slope remains quite similar. Discrepancies between the experimental and simulated data are believed to stem from uncertainties in the experimental device dimensions and the use of simplified zero heat flux boundary conditions for the lattice temperature on the gate electrode.

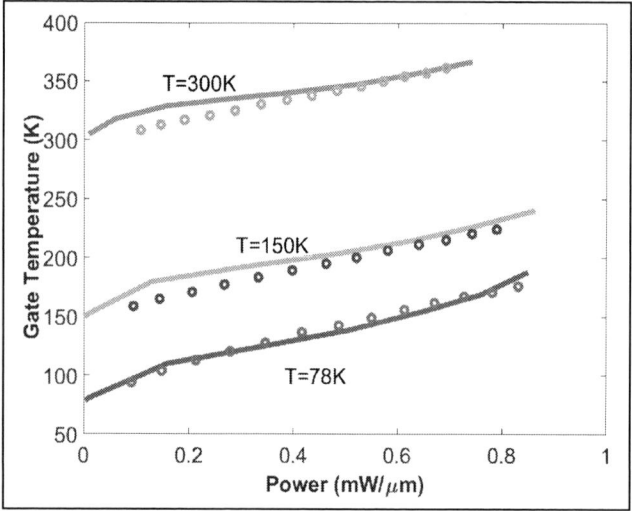

Fig. 7. Comparison of simulated (solid lines) and experimental (open circles) temperature under the gate for various input powers at T=78K, 150K and 300K.

IV. Conclusion

An in-house thermal device simulator was developed in which the electron Boltzmann Transport Equation is solved self-consistently with the energy balance equations for both the acoustic and optical phonons. At low temperatures, partial ionization of dopants is also considered. Thermal conductivity values are calculated according to the thickness of the thin silicon layer and the relevant temperature conditions.

Our simulation results for the device transfer characteristics are in excellent agreement with the experimental data. The gate temperature profile agrees with the experimental findings as well. In summary, this work has the potential to fill the gap in obtaining relevant simulation data concerning self-heating in advanced FDSOI transistors in a wide range of temperatures.

References

[1] K. Triantopoulos *et al.*, "Self-heating effect in FDSOI transistors down to cryogenic operation at 4.2K," IEEE Trans. Electron Devices, vol. 66, no. 8, pp. 3498-3505, August 2019.

[2] K. Raleva, D. Vasileska, S. M. Goodnick and M. Nedjalkov, "Modeling Thermal Effects in Nanodevices," IEEE Trans. Electron Devices, vol. 55, no. 6, June 2008.

[3] M. Casse *et al.*, "FDSOI for cyroCMOS electronics: device characterization towards compact model," IEDM 2022.

[4] C. L. Tien, A. Majumdar and F. M. Gerner, Eds., Microscale Energy Transport. New York: Taylor & Francis, 1998.

[5] K. Raleva and D. Vasileska, "The importance of thermal conductivity modeling for simulations of self-heating effects in FD SOI devices," J. Comput. Electron., vol. 12, no. 4, pp. 601-610, Dec. 2013.

2023 IEEE Latin American Electron Devices Conference (LAEDC)
Puebla, México, July 3-5, 2023

A general toolkit for advanced semiconductor transistors: from simulation to machine learning

Antonio J. García-Loureiro, Natalia Seoane, Julián G. Fernández and Enrique Comesaña

Abstract—**This work presents an overview of a set of in-house-built software intended for state-of-the-art semiconductor device modelling, ranging from simulators to post-processing tools and prediction codes based on statistics and machine learning techniques. First, VENDES is a 3D finite-element based quantum-corrected semi-classical/classical toolbox able to characterise the performance, scalability, and variability of transistors. MLFoMPy is a Python-based tool that post-processes IV characteristics, extracting the most relevant figures of merit and preparing the data for subsequent statistical or machine learning studies. FSM is a variability prediction tool that also pinpoints the most sensitive regions of a device to the source of fluctuation. Finally, we also describe machine learning-based prediction tools that were used to obtain full IV curves and specific figures of merit of devices suffering the influence of several sources of variability.**

Index Terms—**semiconductor devices, 3D modelling, variability, post-processing tools, machine learning.**

I. INTRODUCTION

Technology Computer-Aided Design (TCAD) based tools are essential for analysing the feasibility of new semiconductor device architectures, enabling cost reduction and accelerating device development time, thus, guiding the nanoelectronics industry and academy. Variability of transistor characteristics has become a common feature accompanying essential figures of merit for every technology solution aimed for future digital applications. The effect of variability worsens with device scaling and conditions the architecture choice for next-generation technology nodes [1]. To assess its impact, large ensembles of different device configurations need to be modelled to obtain statistical significance, which greatly increases simulation times and output data. Therefore, the development of new solutions to decrease computational costs and to automatize the handling of generated data is essential.

Work supported by the Spanish MICINN, Xunta de Galicia, and FEDER Funds under Grants RYC-2017-23312, PID2019-104834GB-I00, ED431F 2020/008 and ED431C 2022/16.

A. J. García-Loureiro, N. Seoane and J. G. Fernández are with the Centro Singular de Investigación en Tecnoloxías Intelixentes (CITIUS), Universidade de Santiago de Compostela, Spain; E. Comesaña is with the Escola Politécnica Superior de Enxeñaría, University of Santiago de Compostela, Spain; Corresponding author's e-mail: antonio.garcia.loureiro@usc.es.

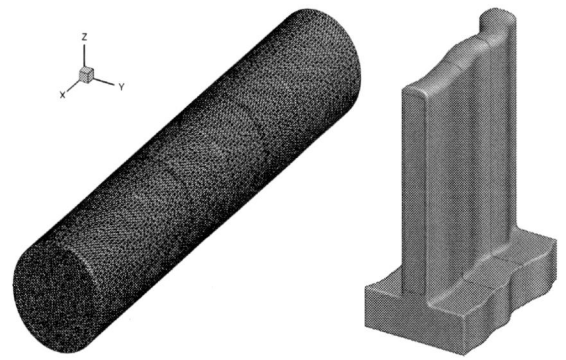

Fig. 1. Tetrahedral finite-element mesh used to model a 10 nm gate length gate-all-around nanowire FET (left). Example of a 12 nm gate length FinFET affected by line-edge roughness (LER) variability (right).

In this work, we present a set of tools aimed for semiconductor device modelling. First, we introduce a three-dimensional in-house-built simulator toolbox that allows modelling and characterizing new generation semiconductor devices and assessing the impact of variability effects. Next, we present a post-processing tool to efficiently manage the generated data, able to automatically extract the more relevant figures of merit that describe devices' performance. Finally, we conclude with some prediction tools, based on statistics and machine learning techniques, aimed to extract more information from the simulation data and to reduce the computational effort.

II. 3D SEMICONDUCTOR DEVICE MODELLING

VENDES (Variability Enabled Device Simulator) [2] is an in-house built three-dimensional semiconductor device simulation toolbox based on the finite-element method. The use of finite elements gives the software the ability to model complex structures with high accuracy (see examples in Fig. 1), being able to capture small deformations. VENDES combines Density-Gradient (DG) or 2D Schrödinger equation (SCH) quantum corrections with classical and semi-classical transport models, specifically, the drift-diffusion (DD) method and an ensemble Monte Carlo (MC). Previously, we have applied VENDES to MOSFETs [3], FinFETs [4], gate-all-around

979-8-3503-1191-4/23 $31.00 © 2023 IEEE

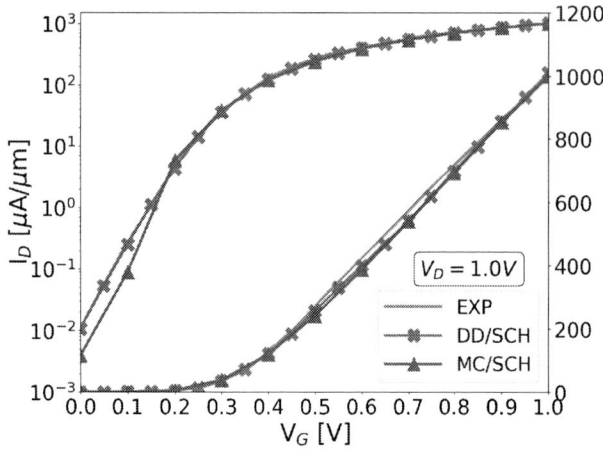

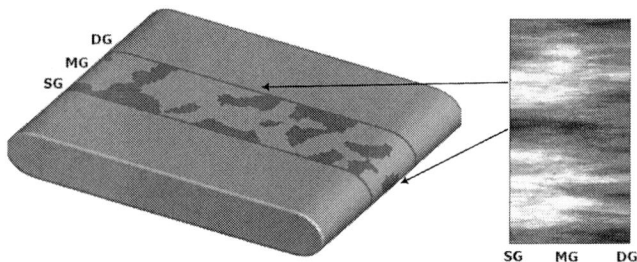

Fig. 3. Scheme of a 12 nm gate length gate-all-around nanosheet FET affected by TiN metal grain granularity (MGG) (right), and its associated threshold voltage fluctuation sensitivity map (FSM) (left). The drain bias is 1.0 V. The locations of the source/gate (SG) and the drain/gate (DG) borders are indicated together with the position of the middle of the gate (MG).

Fig. 2. I_D-V_G characteristics for a 22 nm gate length gate-all-around nanowire FET comparing experimental data against Schrödinger equation quantum corrected (SCH) drift-diffusion (DD) and Monte Carlo (MC) simulations, obtained with VENDES at a drain bias of 1.0 V.

nanowire FETs [5] and nanosheet FETs [6], among others. As an example, Fig. 2 shows the I_D-V_G characteristics for a 22 nm gate length gate-all-around nanowire FET comparing experimental data [7] against SCH quantum-corrected DD and MC simulations. Note that a very good agreement is achieved by both simulation methodologies against the experiment, with the exception of very low gate biases for the MC method (due to the reduced number of particles in the channel). The simulation toolbox incorporates several of the most relevant variability sources that affect nowadays semiconductor devices, namely metal grain granularity (MGG), line-edge roughness (LER), gate-edge roughness (GER), random discrete dopants (RDD) and oxide thickness fluctuations (OTF). An in-depth description of these sources of variability can be found at [2]. Fig. 1 (right) and Fig. 3 show examples of a FinFET device affected by LER and a gate-all-around nanosheet FET influenced by MGG, respectively.

III. POST-PROCESSING TOOLS

MLFoMPy [8] is an open-source Phyton-based post-processing tool that inputs I_D-V_G characteristics (accepting different data formats) and outputs several figures of merit (off-current, threshold voltage, sub-threshold slope, on-current, drain-induced barrier lowering) extracted from the IV curves. It includes different threshold voltage extraction methods (such as constant current, linear extrapolation and second derivative) [9] but also accepts user-defined criteria for parameter extraction. This tool is particularly useful for automatically processing large data sets, as is the case with variability studies where hundreds or thousands of simulations are considered; since it allows plotting the figures of merit statistical distributions and extracting different metrics (e.g. standard deviation, mean values) to characterize them. It also includes error-detection techniques to flag simulation results that may have failed to converge and otherwise would go unnoticed. MLFoMPy aims to standardise the figures of merit

extraction criteria, since, as previously demonstrated [10], it is a relevant factor when comparing the effect of variability on semiconductor devices performance. In addition, MLFoMPy processes the data in a format suitable to be used for predictive tools in later stages (see section IV), matching input variability parameters (e.g. deformation profiles, random dopant configurations, work-function distributions, etc.) with their correspondent figures of merit (outputs from the device simulation).

IV. PREDICTIVE TOOLS

TCAD-based variability studies can be very computationally demanding because large ensembles of devices need to be simulated in order to obtain statistical significance. For that reason, it is interesting to implement new tools able to predict the impact of variability while reducing the studies' computational cost, energy demand and carbon footprint.

The usual approach to analyse the effect of a variability source on the transistor performance is via the standard deviation of a figure of merit. However, to achieve a deeper understanding of the impact of a particular source of variability we have developed the Fluctuation Sensitivity Map (FSM) technique. The FSM is a multi-dimensional map that represents the spatial sensitivity of the different regions of the device to a variability source, which is advantageous when looking for the most variability-resistant device architecture. This technique has already been successfully applied to MGG [11] and LER [12]. Fig. 3 shows an example of a MGG-induced threshold voltage variability FSM. The regions of the map with lighter/darker colours represent the most/least sensitive areas of the gate to MGG. An important feature of the FSM is that since it represents the sensitivity of the device to a variability source (characterized by a specific profile) it should be possible to predict the value of a figure of merit, with great accuracy, without having to simulate the device by just multiplying the variability profile by the figure of merit FSM.

Currently, machine learning (ML) techniques are being heavily investigated since they can efficiently complement TCAD simulations and help to decrease the computational effort of variability studies [13]. We are currently developing

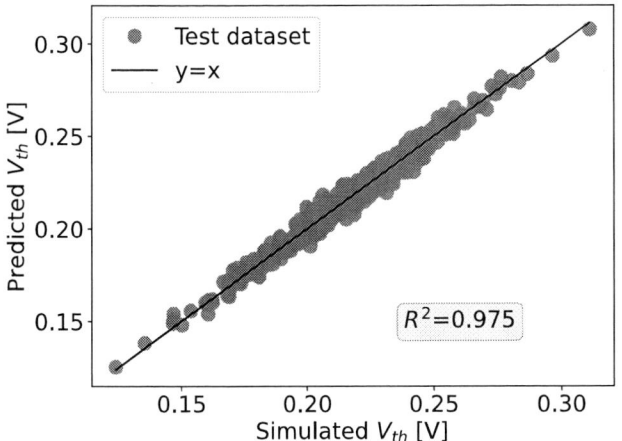

Fig. 4. Simulated threshold voltages versus predicted values using machine learning techniques for the 12 nm gate length gate-all-around nanosheet FET presented in Fig. 3 affected by MGG at a drain bias of 0.7 V. The coefficient of determination, R^2, is also indicated.

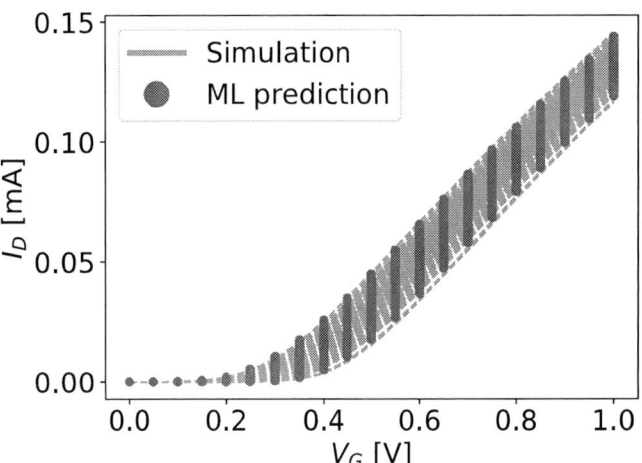

Fig. 5. Comparison of I_D-V_G characteristics simulated using VENDES versus predicted values using machine learning techniques for the 12 nm gate length gate-all-around nanosheet FET presented in Fig. 3 affected by MGG. The drain bias is 0.7 V.

ad-hoc multi-layer perceptron (MLP) neural networks (NN) to estimate the impact of different sources of variability (e.g. MGG and LER) on the performance of different multi-gate architectures (e.g. gate-all-around nanowire and nanosheet FETs). Some preliminary data and source codes that support the findings of this study are openly available in the Zenodo Repository [14] for gate-all-around nanowire FETs influenced by LER, considering two different gate lengths and a wide range of variability parameters (correlation lengths and root mean square heights). Fig. 4 shows an example, for a 12 nm gate length gate-all-around nanosheet FET affected by MGG, of the correlation between the threshold voltage values obtained via VENDES simulations and the ones obtained using machine learning techniques. Note that the coefficient of determination (R^2) is 0.975, proving its validity for this kind of studies. Another example of application is shown in Fig. 5, in which ML techniques are able to predict the complete I_D-V_G curves for the MGG-affected devices. The advantage of these methods lies in the relatively small number of device configurations that needed to be simulated in order to train the neural network. Once that process is done, the figures of merit prediction can be obtained without further simulations, greatly reducing computational times. When training data is expensive, as is in the case of the MC-based very time-consuming simulations, we develop transfer learning approaches that reuse the weights of one or more layers of pre-trained network models to reduce the number of simulations required for good-quality training.

V. CONCLUSION

In this work, we have presented a set of tools aimed for semiconductor device modelling. First, we have introduced VENDES, a three-dimensional in-house-built simulator toolbox that allows modelling and characterizing new generation semiconductor devices (e.g. FinFET, gate-all-around

nanowires, and nanosheet FETs) and assessing the impact of variability effects. Some of the most relevant variability sources that affect nanoscaled devices are incorporated in VENDES, i.e. metal grain granularity, line/gate edge roughness, random discrete dopants or oxide thickness fluctuations. Next, we have presented MLFoMPY, a post-processing tool to efficiently manage the generated data, able to automatically extract the main figures of merit (e.g. off- and on-currents, threshold voltage, sub-threshold slope, DIBL) that describe devices' performance. After that, we presented the FSM, a technique that allows us to determine the sensitivity of the different regions of the devices to a certain variability effect with high accuracy. Finally, we have described several machine learning (ML) prediction tools aimed to decrease the studies' computational cost. Using ML we were able to not only predict specific figures of merit related to variability-affected semiconductor devices, but also full IV curves, with great accuracy.

REFERENCES

[1] IEEE, *More than Moore. International Roadmap for Devices and Systems.* White paper, 2022.

[2] N. Seoane, D. Nagy, G. Indalecio, G. Espiñeira, K. Kalna, and A. J. García-Loureiro, "A Multi-Method Simulation Toolbox to Study Performance and Variability of Nanowire FETs," *Materials*, vol. 12, no. 15, pp. 2391–2406, 2019.

[3] A. Ruiz, C. Couso, N. Seoane, M. Porti, A. J. García-Loureiro, and M. Nafria, "Methodology for the Simulation of the Variability of MOSFETs With Polycrystalline High-k Dielectrics Using CAFM Input Data," *IEEE Access*, vol. 9, pp. 90 568–90 576, 2021.

[4] N. Seoane, G. Indalecio, M. Aldegunde, D. Nagy, M. A. Elmessary, A. J. García-Loureiro, and K. Kalna, "Comparison of Fin-Edge Roughness and Metal Grain Work Function Variability in InGaAs and Si FinFETs," *IEEE Trans. Electron Devices*, vol. 63, no. 3, pp. 1209–1216, March 2016.

[5] N. Seoane, K. Kalna, X. Cartoixà, and A. García-Loureiro, "Multilevel 3-D Device Simulation Approach Applied to Deeply Scaled Nanowire Field Effect Transistors," *IEEE Transactions on Electron Devices*, vol. 69, no. 9, pp. 5276–5282, 2022.

[6] N. Seoane, J. G. Fernandez, K. Kalna, E. Comesaña, and A. García-Loureiro, "Simulations of Statistical Variability in n-Type FinFET, Nanowire, and Nanosheet FETs," *IEEE Electron Device Letters*, vol. 42, no. 10, pp. 1416–1419, 2021.

[7] S. Bangsaruntip, K. Balakrishnan, S. L. Cheng, J. Chang, M. Brink, I. Lauer, R. L. Bruce, S. U. Engelmann, A. Pyzyna, G. M. Cohen, L. M. Gignac, C. M. Breslin, J. S. Newbury, D. P. Klaus, A. Majumdar, J. W. Sleight, and M. A. Guillorn, "Density scaling with gate-all-around silicon nanowire MOSFETs for the 10 nm node and beyond," in *Proc. IEEE Electron Devices Meeting (IEDM)*, Dec. 2013, pp. 20.2.1–20.2.4.

[8] (2023) The MLFoMPy website. [Online]. Available: https://gitlab.citius.usc.es/modev/mlfompy

[9] A. Ortiz-Conde, F. J. García-Sánchez, J. Muci, A. Terán Barrios, J. J. Liou, and C.-S. Ho, "Revisiting MOSFET threshold voltage extraction methods," *Microelectronics Reliability*, vol. 53, no. 1, pp. 90–104, 2013.

[10] G. Espiñeira, A. J. García-Loureiro, and N. Seoane, "Does the threshold voltage extraction method affect device variability?" *IEEE Journal of the Electron Devices Society*, vol. 9, pp. 469–475, 2021.

[11] G. Indalecio, N. Seoane, K. Kalna, and A. J. García-Loureiro, "Fluctuation sensitivity map: A novel technique to characterise and predict device behaviour under metal grain work-function variability effects," *IEEE Transactions on Electron Devices*, vol. 64, no. 4, pp. 1695–1701, 2017.

[12] G. Indalecio, A. J. García-Loureiro, M. A. Elmessary, K. Kalna, and N. Seoane, "Spatial Sensitivity of Silicon GAA Nanowire FETs Under Line Edge Roughness Variations," *IEEE Journal of the Electron Devices Society*, vol. 6, pp. 601–610, 2018.

[13] R. Butola, Y. Li, and S. R. Kola, "A Machine Learning Approach to Modeling Intrinsic Parameter Fluctuation of Gate-All-Around Si Nanosheet MOSFETs," *IEEE Access*, vol. 10, pp. 71 356–71 369, 2022.

[14] (2023) Line-edge roughness variability dataset for gate-all-around nanowire FETs. [Online]. Available: http://doi.org/10.5281/zenodo.7674909

2023 IEEE LATIN AMERICA ELECTRON DEVICES CONFERENCE (LAEDC) PUEBLA, MÉXICO, JULY 3-5, 2023

Modeling High Frequency Response of Nanometer SOI Devices Using Monte Carlo Transient Technique

J. Méndez-V, D. Vasileska *Fellow, IEEE*, E. A. Gutiérrez *Fellow Member, IEEE*,
and K. Raleva *Senior Member, IEEE*,

Abstract—The SOI technologies present several intrinsic advantages for analog and RF applications. Indeed, these technologies allow the reduction of the power consumption at a given operating frequency. Because of these properties, nanometer SOI devices are considered as one of the best options for the implementation of 5G communication technology. Hence, characterization of the high frequency performance of nanometer MOSFETs is needed. Ensemble Monte Carlo device simulator is implemented for this purpose to characterize the high frequency performance of 45 nm nanometer technology node partially-depleted SOI devices. Simulation results for the Y- and S-parameters agree with experimental data, which validates our theoretical model, and suggests that our simulator is able to predict high-frequency operation of a variety of SOI devices from different technology nodes.

Index Terms—High frequency response, Transient Ensemble Monte Carlo device simulations, S-parameters, SOI devices

I. INTRODUCTION

Silicon-on-Insulator (SOI) MOSFET technology has a potential for realization of high frequency commercial applications, reaching cutoff frequencies close to 500 GHz [1]. The 45-nm SOI technology node is, thus, considered for the implementation of 5G communication technology, which uses the bandwith corresponding to sub-6 GHz and mm-Wave range [2].

Besides the well known advantages of the ensemble Monte Carlo method for simulating static device characteristics, its transient response to a step excitation allows one to characterize the small signal response of nanometer SOI devices [3]. In this work we utilize the methodology described in [4] to study RF behavior of 45nm technology node SOI device structure.

The paper is organized as follows. In section II, we briefly discuss the theoretical model implemented in the in-house Ensemble Monte Carlo device simulator and the use of its transient response for characterization of rf behavior of 45 nm technology node PD SOI device. In section III, we present simulation results from our theoretical study and compare our simulation results with available experimental data. Conclusions regarding the work presented here are given in section IV.

This work was supported by CONACYT and in part (D.V.) by the US National Science Foundation under grant number NSF ECCS-2025490.

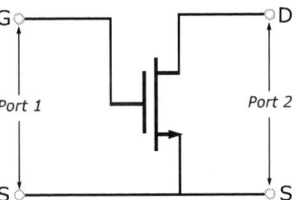

Fig. 1. MOSFET two port network

II. THEORETICAL MODEL

Solution of the Boltzmann Transport Equation (BTE), self-consistently coupled to a Maxwell solver, is the most fundamental semi-classical way of modeling charge transport in semiconductor devices. The Ensemble Monte Carlo (EMC) method in a stochastic way solves the BTE which is valid in the range from diffusive to ballistic transport regime [5]. For frequency range of interest of tens of GHz, for which the device length is smaller than the corresponding wavelength of the electromagnetic signal, it suffices to solve the Poisson's equation instead of the Maxwell equations. For example, for 28 GHz AC signal the corresponding wavelength is on the order of 10 mm, which is much larger than the channel length ($L = 45$ nm) of the device of interest in this work.

The transient response of the EMC device simulation procedure to a step excitation allows one to calculate the Y-parameters of the two port network (see Fig.(1)). For example, if one applies a step voltage on port q and monitors the current response on port p, the frequency dependent Y-parameters $Y_{p,q}$ are calculated using

$$Y_{p,q} = j\omega \frac{i_p(\omega)}{\Delta V_q} \qquad (1)$$

where, $i_p(\omega)$ is the Fourier transform of the current on port p, and ΔV_q is the step voltage applied to port q. The current through a particular contact is calculated via the net charge entering/exiting the contacts which including the displacement current is

$$Q(t) = q(N_{ent} - N_{ext}) + \epsilon\epsilon_0 \int E_y(x,y)dx, \qquad (2)$$

where, $N_{ent(ext)}$ is the net charge entering/exiting the source and drain electrodes, respectively. The last term on the

979-8-3503-1191-4/23 $31.00 © 2023 IEEE

RHS of Eq. (2) is the displacement charge, where $E_y(x, y)$ is the y component of the electric field (perpendicular component) integrated over the electrode surface. The current can be found as a derivative with respect to time of the total charge (cumulative plus displacement charge) entering/exiting the electrode in both transient and steady state conditions, $I(t) = dQ(t)/dt$. It is important to point out here that the displacement charge, and subsequently the displacement current ($\epsilon\epsilon_0 \frac{d}{dt} \int E_y(x,y)dx$), is relevant in the transient analysis because, for example, in the gate electrode only the displacement current exists. Note that in our present analysis we do not consider the leakage tunneling current at the gate electrode.

It is also important to point out that simulated Y-parameters correspond to the equivalent circuit enclosed by the dashed lines shown in Fig. (3). We denote those as simulated Y-parameters, Y^{sim}. If we convert the simulated Y^{sim} parameters into the simulated Z-impedance parameters Z^{sim}, we can easily substract the source R_S and drain R_D resistances. In this way we arrive at the intrinsic equivalent circuit model (see the circuit enclosed by the continuous red line shown in Fig. (3)). Hence, once R_S and R_D are extracted, we can easily find the values of the lumped elements forming the intrinsic circuit (C_{gs}, C_{ds}, C_{gd}, g_m and R_{ds}). In this way we will obtain more realistic comparison between our simulated parameters and the experimental data. In our simulations we do not take into account the gate and contact resistances (R_G and R_C), which are inherently present in a real device. To compensate this, we add these resistances to our simulation results (see Fig. (2)).

With regard to the Monte Carlo transport kernel implemented in our particle-based device simulator, we account for Coulomb, acoustic phonon and inter-valley scattering processes relevant for the silicon material system. Details about the EMC transport model can be found elsewhere [5].

III. SIMULATION RESULTS

Previously described methodology is applied to a 45nm SOI commercial device technology. A schematic description of the device is given in the inset of Fig. (4). The transfer and output characteristics at 300 K are shown in Fig. (4) and Fig. (5), respectively. As seen from the results presented in Fig.(5), there is a discrepancy between the experimental and the theoretical data at low values of the gate voltage V_G. This effect is attributed to the fact that, for this technology node, a high-k dielectric stack is used as a gate oxide [6] which is polar, so remote polar optical phonon scattering must be considered. As a result of this approximation, at low values of V_G, carriers in the channel region of the device exhibit less scattering and, therefore, more current. Nevertheless, a good agreement between the experimental and simulated data is observed.

The high frequency response of our structure is calculated for an operating point in the saturation region, $V_G = 1.0$ V and $V_D = 1.0$ V. To calculate the Y-parameters, we apply a step voltage $\Delta V = 0.1$ V on the port 1 (gate-source terminals) and, in order to get proper resolution of the transient in this

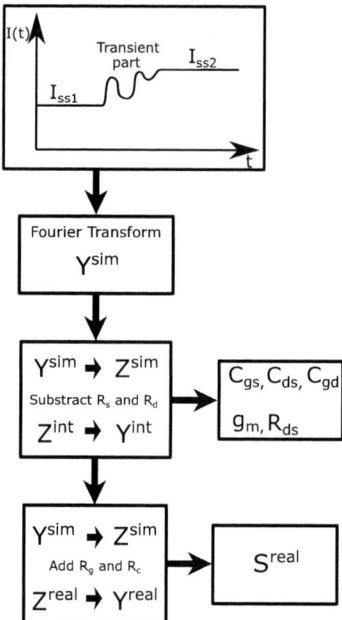

Fig. 2. Procedure of a MOSFET high frequency characterization. I_{SS1} (I_{SS2}) is the steady state current before (after) applying the step voltage ΔV on the corresponding port [4].

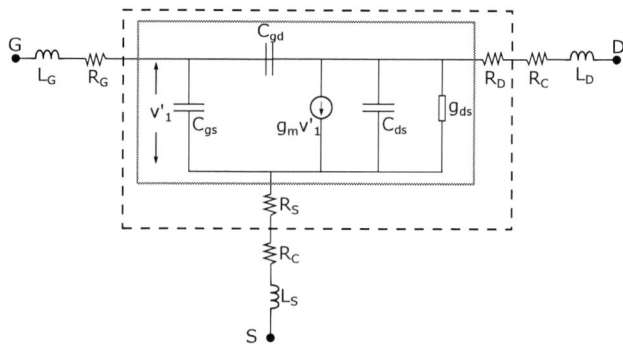

Fig. 3. Small-signal equivalent circuit model of the real FET. Dashed lines enclose the equivalent circuit model of the simulated FET. Red solid line encloses the intrinsic circuit [4].

saturation condition, a step voltage $\Delta V = 0.3$ on port 2 (drain-source terminals). The Y^{sim}-parameters are afterwards converted into Z^{sim}-parameters from which we subtract the source and drain resistances, with values $R_S = 5.178$, $R_D = 5.178$, respectively. Next, the values of the capacitances and transconductances, forming the intrinsic equivalent circuit model, are calculated (see Table I). Assuming $10\,\Omega$ gate resistance and $6\,\Omega$ contact resistance, we arrive at the S^{real}-parameters shown in Fig. (6). Under high frequency conditions, we observe good agreement between the experimental and simulated data. The small differences observed between the simulated and experimental data, and also in the capacitances C_{gd}, C_{gs}, C_{ds}, drain to source resistance R_{ds}, and transconductance g_m are attributed to the the lack of knowledge of the materials that constitute the gate oxide. For example, for this technology node the experimental devices have high-k dielectric as gate oxide [6]. High-k dielectric

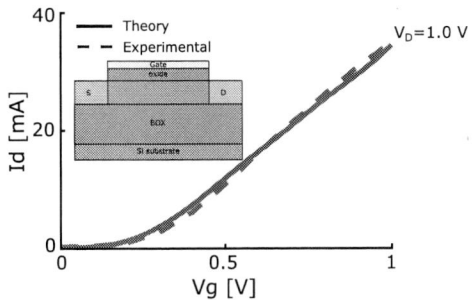

Fig. 4. I_D vs V_G at 300 K for a drain voltage $V_D = 1.0$ V.

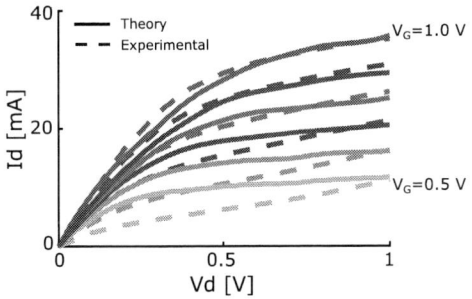

Fig. 5. I_D vs V_D at 300 K for different values of gate voltage, V_G. The increment of $\Delta V_G = 0.1$ V.

materials are highly polar [7]. This gives rise to remote polar optical phonon scattering [8]. This, when combined with the oversimplified geometry of the device and the simplistic doping profiles that we are using in our simulations, leads to the observed discrepancies.

TABLE I

LUMPED ELEMENTS VALUES OF THE INTRINSIC CIRCUIT MODEL, FOR THE BIAS CONDITIONS $V_G = 1$ V AND $V_D = 1$ V.

Data type	C_{gs} [fF]	C_{ds} [fF]	C_{gd} [fF]	g_m [Ω^{-1}]	R_{ds} [Ω]
Experimental	90	48	52	0.18	35
Theoretical	25	52	20	0.12	43

IV. CONCLUSIONS

We utilized the transient version of the Ensemble Monte Carlo device simulator for the characterization of the high frequency response of a 45nm SOI technology node device. Simulated transfer ($I_D - V_G$) and output characteristics ($I_D - V_D$) are in agreement with available experimental data measured in our Lab, which justifies the validaty of our simulation model. With the application of a step voltage of $\Delta V = 0.1$ V on port one and a step voltage of $\Delta V = 0.3$ V, on port two, after Fourier transforming the transient currents, we calculate the Y-parameters, and, therefore, find the high frequency response of our device structure of interest. In order to obtain more realistic results, in terms of comparison with measured data, we used a small signal equivalent circuit model, subtracted the source and drain resistances, R_S and R_D, and added gate and contact resistances, R_G and R_C, respectively. The values of the lumped elements forming the intrinsic circuit

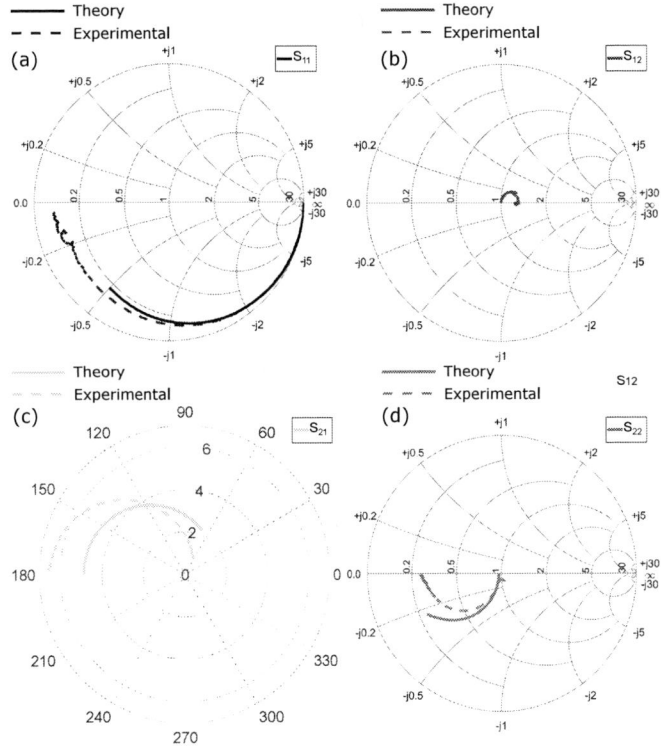

Fig. 6. Smith chart representation of the four theoretical and experimental *S-parameters*, for a bias condition of $V_G = 1.0$ V and $V_D = 1.0$ V. The inset figure (a) represents S_{11}, (b) S_{12}, (c) S_{21}, and (d) S_{22}, respectively. Note that S_{21} is plotted on a polar scale.

were found and compared with the experimental ones. The small differences between them are attributed to the lack of knowledge of the materials of which the gate oxide is made of and due to the oversimplified geometry were are using. Simulated S-parameters are in close agreement with experimental data, thus proving that our simulation tool is able to estimate the high-frequency response of a SOI-MOSFET.

REFERENCES

[1] J. -P. Raskin, "SOI technology: An opportunity for RF designers?," Proceedings of Technical Program - 2014 International Symposium on VLSI Technology, Systems and Application (VLSI-TSA), Hsinchu, Taiwan, 2014, pp. 1-2, doi: 10.1109/VLSI-TSA.2014.6839644.

[2] www.qualcomm.com

[3] C. C. Moglestue, "A Self-Consistent Monte Carlo Particle Model to Analyze Semiconductor Microcomponents of any Geometry," in IEEE Transactions on Computer-Aided Design of Integrated Circuits and Systems, vol. 5, no. 2, pp. 326-345, April 1986, doi: 10.1109/TCAD.1986.1270203.

[4] S. Babiker, A. Asenov, N. Cameron, S. P. Beaumont and J. R. Barker, "Complete Monte Carlo RF analysis of "real" short-channel compound FET's," in IEEE Transactions on Electron Devices, vol. 45, no. 8, pp. 1644-1652, Aug. 1998, doi: 10.1109/16.704358.

[5] Vasileska, D., Goodnick, S.M., Klimeck, G. (2010). Computational Electronics: Semiclassical and Quantum Device Modeling and Simulation (1st ed.). CRC Press. https://doi.org/10.1201/b13776

[6] Yim, K., Yong, Y., Lee, J. et al. Novel high- dielectrics for next-generation electronic devices screened by automated ab initio calculations. NPG Asia Mater 7, e190 (2015). https://doi.org/10.1038/am.2015.57

[7] Vladimir A. Gritsenko, Timofey V. Perevalov, Damir R. Islamov, Electronic properties of hafnium oxide: A contribution from defects and traps, Physics Reports, Volume 613, 2016, Pages 1-

979-8-3503-1191-4/23 $31.00 © 2023 IEEE

20, ISSN 0370-1573, https://doi.org/10.1016/j.physrep.2015.11.002. (https://www.sciencedirect.com/science/article/pii/S0370157315004986)

[8] Gamiz, Francisco Fischetti, Massimo. (2003). Remote Coulomb scattering in metal–oxide–semiconductor field effect transistors: Screening by electrons in the gate. Applied Physics Letters. 83. 4848-4850. 10.1063/1.1630169.

2023 IEEE Latin American Electron Devices Conference (LAEDC)
Puebla, México, July 3-5, 2023

Improving the performance of photovoltaic laser power converters using automatic global optimization techniques

Javier F. Lozano [1], Natalia Seoane [1], Enrique Comesaña [2], Florencia Almonacid [3], Eduardo F. Fernández [3] and Antonio García-Loureiro [1]

[1] Centro Singular de Investigación en Tecnoloxías de Información (CiTiUS), Departament of Electrónica e Computación,

Universidade de Santiago de Compostela, Santiago de Compostela, 15782, Spain

[2] Escola Politécnica Superior de Enxeñaría, Universidade de Santiago de Compostela, Campus Terra, Spain

[3] Advances in Photovoltaic Technology (AdPVTech), CEACTEMA, University of Jaén, Jaén, 23071, Spain
javier.fernandez.lozano@rai.usc.es

Abstract—To develop the potential of the High Power Laser Transmission technology, the efficiencies of the Laser Power Converters (LPCs) need to improve. New optimization methods are required to find the optimal configurations of a new generation of ultra-efficient LPCs. In this work we explore the suitability of the SHGO algorithm, a recently published black-box, global optimization algorithm, to optimize the TCAD modeling of a state-of-the-art GaAs-based Laser Power Converter. We found various configurations that improve the efficiency by a 7.3% of the original device structure in computational times under 17h.

Keywords—High Power Laser Transmission, Laser Power Converter, global optimization, SHGO.

I. INTRODUCTION

High power laser transmission (HPLT) is a key technology in the wireless power transfer field [1]. The main advantages are the absence of sparks and electromagnetic interferences when compared to traditional wiring [2]. Some applications are dual transmission of power and data [3] and optically power remote sensors, drones, or even satellites [4]. Even with the high efficiencies reached by the state-of-the-art GaAs-based laser power converters (LPCs), which are around ≈65% [5], the overall efficiency of the complete system is around ≈20% [1]. To overcome the main limitations of the technology, new architectures [6], [7] and materials [8], [9] have been proposed. In this scenario, TCAD modeling is as a powerful resource to aid the research and avoid lengthy trial and error processes in the developing of new generation ultra-efficient laser power converters.

However, the optimization of the device parameters may require a large number of simulations and some expertise, since the same parameter can affect various aspects of the LPC. For example, increasing a doping value may improves the light absorption and built-in voltage, but also reduce the carrier mobility and therefore the carrier diffusion length. This will require smaller layer thicknesses to prevent excessive carrier recombination, which will affect the beam absorption. These trade-offs can lead to a situation where various possible configurations are close to the maximum efficiency achievable, i.e., there might be various local maxima when optimizing an LPC. It is noteworthy that some of the potential optimal configurations may be more favorable to the manufacturing constraints.

In this regard, we require optimization algorithms that reduce the computational times of the study and obtain all the optimal configurations (Global Optimization (GO) algorithms) of the LPC. Also, as the derivatives of the objective function are unknown, we need the algorithm to solve a black-box optimization problem. Among this type of optimization algorithms, the simplicial homology global optimization (SHGO) algorithm [10] stands out. This recently published algorithm, that is fully available in open source, has proven to converge to the global optimum for a given function, even when this function is non-continuous, non-convex and non-smooth. This algorithm has proven great performance when compared with other state-of-the-art GO algorithms available. Therefore, in this work we present the SHGO algorithm as a practical and easy to use tool to automatically optimize modeled photovoltaic devices, applying it to a state-of-the-art GaAs-based LPC, showing noticeable efficiency improvements.

The structure of the paper is as follows. Section II presents the SHGO algorithm, the used TCAD tool and the GaAs LPC structure. Section III introduces the main results, where we perform a series of optimizations with the SHGO algorithm to find several optimum configurations of the device. Finally, the conclusions are summarized in Section IV.

II. METHODOLOGY

A. The optimization algorithm: SHGO

The SHGO algorithm is a highly competitive, open-source, general purpose global optimization (GO) algorithm [10]. SHGO is suitable for multidimensional, global, derivative free and black-box optimization problems. It uses combinatorial integral homology theory to find locally convex sub-domains and characterizes the objective function, finding both local and global minima. The algorithm can also be fed with information about the objective problem (such as bounds and symmetry) to speed up the optimization, which is desirable when solving high-dimensional problems. This method is very simple to use, since, to evaluate the objective function, it only requires the value range of the parameters to be optimized.

979-8-3503-1191-4/23 $31.00 © 2023 IEEE

We used the Sobol sequence [11] to generate the sampling points, since it produces better performance than the simplicial method [10] (both integrated in the library). The Sobol method requires two input parameters: 1) the initial number of sampling points in the search space (**n**) and 2) the number of iterations (**iters**), which are sequential refinements, internally performed by SHGO, of the initial **n** sampling points. To increase the sampling in the search space, the increment of the iterations is advised instead of increasing **n**, since the candidates to maximum found in an iteration will be re-evaluated in the next ones.

Fig. 1 shows the workflow of the optimization process. We consider as input parameters the thicknesses and doping values of the P and N layers (emitter and base of the LPC, respectively) and the wavelength of the input laser. The SHGO algorithm returns a list with all the maxima found and their corresponding device parameters, starting from the global maximum (GM) and followed by all the local maxima (LM) from the highest non-global (LM1) to the n-th one (LMN).

B. TCAD modeling

We carried out the TCAD modeling of this work with Silvaco Atlas Software [12], which has provided realistic and trustable results when modeling photovoltaic devices [13]–[15]. The characteristics of the device are obtained by solving Poisson and continuity equations. The concentration-dependent carrier mobility is extracted from Silvaco tabulated data. The doping concentration dependent Shockley-Read-Hall recombination parameters are obtained from available experimental data [16]. Optical and Auger recombinations are included in the simulation framework. The optical absorption coefficient depends both on wavelength and doping concentration have been obtained from experimental curves [17]. All simulations were performed at an input power density of 5 W/cm² and a temperature of 298 K.

To accurately test the optimization method in a realistic scenario, we modeled a state-of-the-art GaAs-based LPC, based on an experimental device [18]. The structure of the modeled LPC is shown in Fig. 2. The front contact is placed in spaced GaAs P++ type caps to ensure good ohmic contact. The device is composed, from top to bottom, of 1) an anti-reflection coating, 2) a P+ type GaInP window, 3) a P type GaAs emitter (thickness=0.5μm, doping=2·10¹⁸ cm⁻³), 4) an N type GaAs base (thickness=3.5μm, doping=1·10¹⁷ cm⁻³) and 5) an N+ type AlGaAs back surface field (BSF).

III. RESULTS

Fig. 3 shows the J-V curves for the experimental and modeled devices at an input power density of 5 W/cm². The modeled curve fits remarkably well the experimental one, and all the characteristic J-V points (J_{sc}, V_{oc}, V_m, FF, η) from the modeled device are within a 98% accuracy with respect to the experimental ones. These results support the reliability of the TCAD simulations.

Once we have established the validity of our simulation methodology, we perform three optimizations using the same set of device parameters and 60 initial sampling points, only varying the number of iterations between 1, 2 and 4. We carried out a preliminary test to determine the optimal initial number of sampling points to reduce the computational cost without losing accuracy in the results. The input parameter ranges are shown in Table I. For the doping concentration we explored a range of values wide enough to account for several optimum configurations. For the layer thicknesses and input laser wavelength we chose a range of physically meaningful values according to the GaAs absorption and bandgap properties. Note that this set of parameters is the only information required by the algorithm, which makes the SHGO a user-friendly and intuitive tool.

TABLE I. SET OF DEVICE AND SHGO PARAMETERS USED IN THE THREE OPTIMIZATIONS PERFORMED FOR THE GAAS LPC.

		Min	*Max*
Device	*Pthick [μm]*	0.1	2.0
	Nthick [μm]	1.0	4.0
	Pdop [cm⁻³]	$1\cdot10^{15}$	$7\cdot10^{18}$
	Ndop [cm⁻³]	$1\cdot10^{15}$	$7\cdot10^{18}$
	λ [μm]	0.800	0.860
SHGO	*n*	60	
	iters	1, 2, 4	

The J-V curve for the GM found during the SHGO optimizations is also shown in Fig. 3. Table II shows the design parameters and the figures of merit for the experimental device and the GM found. The GM configuration provides very similar layer thicknesses to those of the experimental device, and significantly decreases the doping values (from $2\cdot10^{18}$ to $4.0\cdot10^{15}$ cm⁻³ for the P layer, from $1\cdot10^{17}$ to $1.8\cdot10^{15}$ cm⁻³ for the N layer), which greatly reduces the Shockley-Read-Hall recombinations. The optimum wavelength found is slightly closer to the GaAs

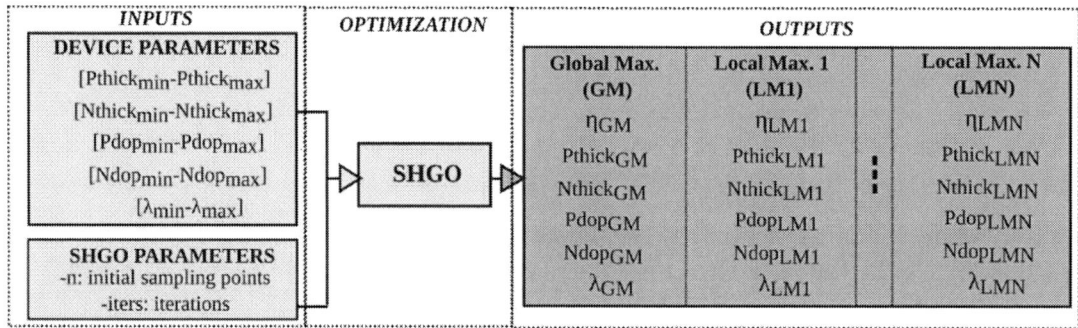

Fig 1. Workflow of the optimization algorithm. The device parameters Pthick, Nthick, Pdop and Ndop are the thickness and doping values of the P and N layers, respectively. λ is the input wavelength. The SHGO algorithm inputs are the ranges of the device parameters and the sampling parameters **n** (initial number of sampling points) and **iters** (iterations to refine the search space). The outputs are the global maximum (GM) and all the local maxima (LM) found for the efficiency (η), with the associated device parameters for each maxima.

979-8-3503-1191-4/23 $31.00 © 2023 IEEE

bandgap energy than the experimental one, which increases the efficiency. The optimized structure increases the J_{sc} value by ≈11% and the V_{oc} value by≈3%, which leads to an efficiency increase of 7.25% with respect to the experimental device.

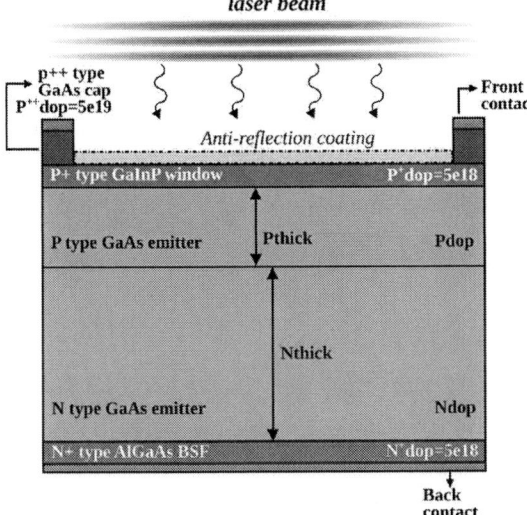

Fig 2. 2D scheme of the modelled structure. Pthick, Nthick, Pdop and Ndop are the targeted optimization parameters. The third dimension does not affect transport and is set to 1 μm.

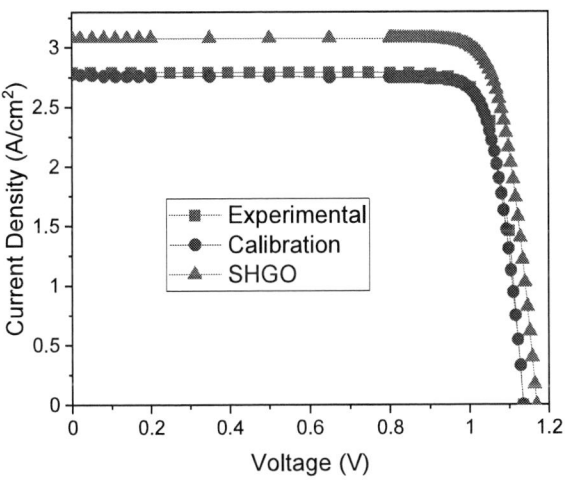

Fig 3. J-V curves for the experimental, calibrated and optimized devices.

TABLE II.   DESIGN PARAMETERS AND FIGURES OF MERIT FOR THE EXPERIMENTAL (EXP) AND OPTIMIZED (SHGO) GAAS DEVICE.

	Design parameters				
	Pthick [μm]	*Nthick [μm]*	*Pdop [cm⁻³]*	*Ndop [cm⁻³]*	*λ [μm]*
Exp	0.50	3.50	$2.0 \cdot 10^{18}$	$1.0 \cdot 10^{17}$	0.808
SHGO	0.36	3.87	$4.1 \cdot 10^{15}$	$1.8 \cdot 10^{15}$	0.850

	Figures of merit				
	J_{sc} [A/cm²]	*V_{oc} [V]*	*V_m [V]*	*FF [%]*	*Eff [%]*
Exp	2.77	1.137	0.996	84.56	53.3
SHGO	3.08	1.168	1.014	84.10	60.5

All the efficiency maxima found, the corresponding device parameters and the required time for the three optimizations are shown in Table III. The optimizations with 1 and 2 iterations of the Sobol sampling method provided the same two maxima and required around 6 and 10 hours respectively to complete. The optimization with 4 iterations took over 16 hours and found two more maxima, being one of them the absolute global maximum. These computational times are much lower than those required for a 5-dimensional brute method optimization (also known as a grid search). Note that, for the optimization performed with 4 iterations, there are three different configurations with very similar performances, all in the range of ±0.5% of efficiency. The lesser maxima, found by all optimizations, shows a doping value for the N layer two orders of magnitude higher than the rest of the maxima found.

To visualize the behavior of the SHGO algorithm, Fig. 4 shows the sampled points and the corresponding efficiency for the optimization performed with 1 iteration. From samplings 1 to 52, the algorithm sweeps the search space using the Sobol sampling method to select potential maxima candidates. After that, from samplings 53 to 96, it verifies that the candidates are actual maxima. Firstly, from samplings 53 to 74, the algorithm probes the candidate to GM by performing slight displacements around the initial point (increasing/decreasing the parameters under study). A larger efficiency was found by increasing the N layer thickness, so the algorithm swept this parameter to achieve the best configuration (see Fig. 4). Secondly, from samplings 75 to 96, the algorithm probes the only LM of this optimization (LM1) and simultaneously swept the P layer thickness and the wavelength.

TABLE III.   DESIGN PARAMETERS AND EFFICIENCY FOR ALL THE MAXIMA FOUND IN THE THREE OPTIMIZATIONS MADE WITH THE SHGO ALGORITHM. THE OPTIMIZATIONS MADE WITH 1 AND 2 ITERATIONS FOUND THE SAME MAXIMA.

Initial sampling points	60					
Iterations	*1 / 2*		*4*			
Maximum	*GM*	*LM1*	*GM*	*LM1*	*LM2*	*LM3*
Eff [%]	60.2	57.7	60.5	60.2	60.1	57.7
Pthick [μm]	1.64	1.76	0.36	1.64	0.29	1.76
Nthick [μm]	3.06	3.63	3.87	3.06	3.46	3.63
Pdop [cm⁻³]	$4.8 \cdot 10^{16}$	$3.0 \cdot 10^{15}$	$4.1 \cdot 10^{15}$	$4.8 \cdot 10^{16}$	$2.6 \cdot 10^{16}$	$3.0 \cdot 10^{15}$
Ndop [cm⁻³]	$1.7 \cdot 10^{15}$	$2.5 \cdot 10^{17}$	$1.8 \cdot 10^{15}$	$1.7 \cdot 10^{15}$	$1.4 \cdot 10^{15}$	$2.5 \cdot 10^{17}$
λ [μm]	0.856	0.823	0.850	0.856	0.842	0.823
Time [h]	5.82 / 9.95		16.26			

IV. CONCLUSIONS

In this work we studied the feasibility of the SHGO, which is a highly competitive, easy to use, open-source, general purpose global optimization algorithm, to optimize a Laser Power Converter (LPC). We calibrated a TCAD-modeled state-of-the-art GaAs LPC. Next, we applied the SHGO algorithm to optimize this LPC. We made three optimizations, varying the number of sampling points through refining the initial search space. The two optimizations with fewer sampling points found the same two maxima, with efficiency increments of ≈6.9% and ≈4.5% with respect to the calibrated TCAD device. The computational times for these

979-8-3503-1191-4/23 $31.00 © 2023 IEEE   19

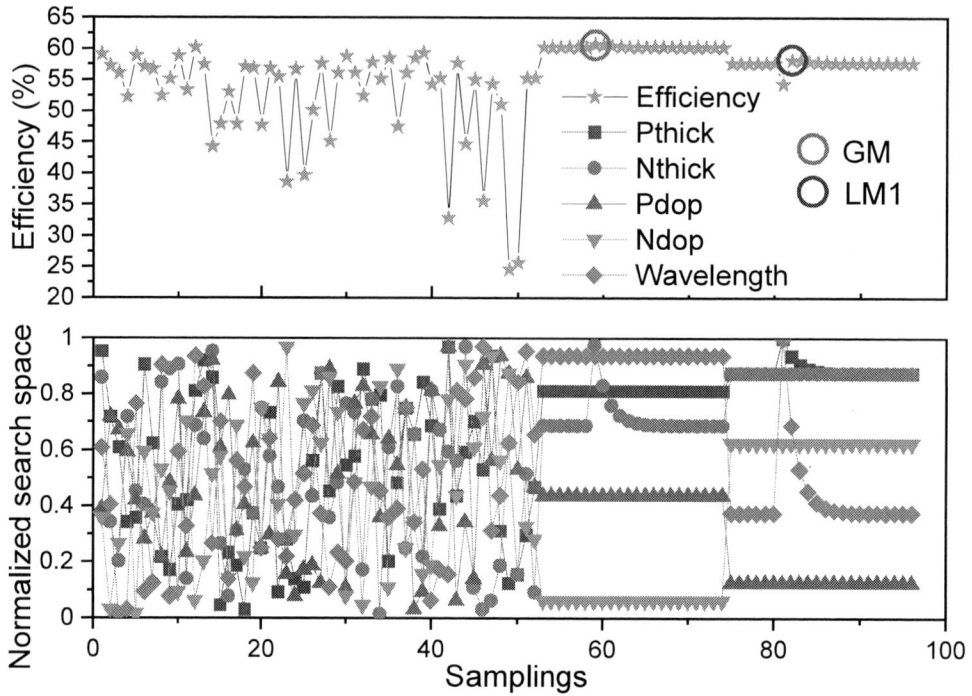

Fig 4. Evolution of the sampled points. From samplings 1) 1-52: search space with Sobol sampling method, 2) 53-74: probe the Global Maximum (GM), found on sampling n° 59, and 3) 75-96: probe the only Local Maximum (LM1), found on sampling n°82.

optimizations were 5.82 h and 9.95 h. The optimization with more sampling points found two more maxima, one of them the global one, achieving an efficiency gain of ≈7.3% with respect to the TCAD calibration. This optimization took around 16 hours to complete. In summary, the algorithm found four maxima with similar efficiency values and different device configurations. These results, together with the easy implementation of the algorithm and the wide range of its foreseeable applications make SHGO an interesting tool to optimize photovoltaic devices, from solar cells to power converters.

REFERENCES

[1] K. Jin and W. Zhou, "Wireless Laser Power Transmission: A Review of Recent Progress," *IEEE Trans. Power Electron.*, vol. 34, no. 4, pp. 3842–3859, 2019, doi: 10.1109/TPEL.2018.2853156.

[2] D. Krut, R. Sudharsanan, W. Nishikawa, T. Isshiki, J. Ermer, and N. H. Karam, "Monolithic multi-cell GaAs laser power converter with very high current density," *Conf. Rec. IEEE Photovolt. Spec. Conf.*, pp. 908–911, 2002, doi: 10.1109/pvsc.2002.1190727.

[3] M. Matsuura, H. Nomoto, H. Mamiya, T. Higuchi, D. Masson, and S. Fafard, "Over 40-W Electric Power and Optical Data Transmission Using an Optical Fiber," *IEEE Trans. Power Electron.*, vol. 36, no. 4, pp. 4532–4539, 2021, doi: 10.1109/TPEL.2020.3027551.

[4] D. Shi, L. Zhang, H. Ma, Z. Wang, Y. Wang, and Z. Cui, "Research on Wireless Power transmission system between satellites," in *2016 IEEE Wireless Power Transfer Conference, WPTC 2016*, May 2016, vol. 3, pp. 1–4, doi: 10.1109/WPT.2016.7498851.

[5] H. Helmers *et al.*, "68.9% Efficient GaAs-Based Photonic Power Conversion Enabled by Photon Recycling and Optical Resonance," *Phys. Status Solidi - Rapid Res. Lett.*, vol. 15, no. 7, pp. 1–7, 2021, doi: 10.1002/pssr.202100113.

[6] S. Fafard *et al.*, "High-photovoltage GaAs vertical epitaxial monolithic heterostructures with 20 thin p/n junctions and a conversion efficiency of 60%," *Appl. Phys. Lett.*, vol. 109, no. 13, p. 131107, Sep. 2016, doi: 10.1063/1.4964120.

[7] N. Seoane, E. F. Fernández, F. Almonacid, and A. García-Loureiro, "Ultra-efficient intrinsic-vertical-tunnel-junction structures for next-generation concentrator solar cells," *Prog. Photovoltaics Res. Appl.*, vol. 29, no. 2, pp. 231–237, 2021, doi: 10.1002/pip.3369.

[8] J. F. Lozano, N. Seoane, E. Comesaña, F. Almonacid, E. F. Fernández, and A. García-Loureiro, "Laser Power Converter Architectures Based on 3C-SiC with Efficiencies >80%," *Sol. RRL*, vol. 6, no. 8, p. 2101077, Aug. 2022, doi: 10.1002/solr.202101077.

[9] E. F. Fernández, A. García-Loureiro, N. Seoane, and F. Almonacid, "Band-gap material selection for remote high-power laser transmission," *Sol. Energy Mater. Sol. Cells*, vol. 235, no. September 2021, p. 111483, 2022, doi: 10.1016/j.solmat.2021.111483.

[10] S. C. Endres, C. Sandrock, and W. W. Focke, "A simplicial homology algorithm for Lipschitz optimisation," *J. Glob. Optim.*, vol. 72, no. 2, pp. 181–217, Oct. 2018, doi: 10.1007/s10898-018-0645-y.

[11] I. . Sobol, "On the distribution of points in a cube and the approximate evaluation of integrals," *USSR Comput. Math. Math. Phys.*, vol. 7, no. 4, pp. 86–112, Jan. 1967, doi: 10.1016/0041-5553(67)90144-9.

[12] Silvaco, "Silvaco software (version 5.30.0.R)," *https://www.silvaco.com*, 2020.

[13] C. Outes, E. F. Fernandez, N. Seoane, F. Almonacid, and A. J. Garcia-Loureiro, "GaAs Vertical-Tunnel-Junction Converter for Ultra-High Laser Power Transfer," *IEEE Electron Device Lett.*, pp. 1–1, 2021, doi: 10.1109/led.2021.3121501.

[14] M. Ochoa, E. Barrigón, L. Barrutia, I. García, I. Rey-Stolle, and C. Algora, "Limiting factors on the semiconductor structure of III-V multijunction solar cells for ultra-high concentration (1000-5000 suns)," *Prog. Photovoltaics Res. Appl.*, vol. 24, no. 10, pp. 1332–1345, Oct. 2016, doi: 10.1002/pip.2791.

[15] M. C. A. York and S. Fafard, "High efficiency phototransducers based on a novel vertical epitaxial heterostructure architecture (VEHSA) with thin p/n junctions," *J. Phys. D. Appl. Phys.*, vol. 50, no. 17, p. 173003, May 2017, doi: 10.1088/1361-6463/aa60a6.

[16] G. B. Lush *et al.*, "A study of minority carrier lifetime versus doping concentration in n-type GaAs grown by metalorganic chemical vapor deposition," *J. Appl. Phys.*, vol. 72, no. 4, pp. 1436–1442, 1992, doi: 10.1063/1.351704.

[17] H. C. Casey, D. D. Sell, and K. W. Wecht, "Concentration dependence of the absorption coefficient for n − and p −type GaAs between 1.3 and 1.6 eV," *J. Appl. Phys.*, vol. 46, no. 1, pp. 250–257, Jan. 1975, doi: 10.1063/1.321330.

[18] T. Shan and X. Qi, "Design and optimization of GaAs photovoltaic converter for laser power beaming," *Infrared Phys. Technol.*, vol. 71, pp. 144–150, 2015, doi: 10.1016/j.infrared.2015.03.010.

2023 IEEE Latin American Electron Devices Conference (LAEDC)
Puebla, México, July 3-5, 2023

Device parameters estimation of GFETs at temperatures below 300 K

Leslie M. Valdez-Sandoval, Eloy Ramírez-García, Nikolaos Mavredakis, David Jiménez, Anibal Pacheco-Sanchez

Abstract—Low-temperature performance of four different graphene field-effect transistor (GFET) technologies previously reported in the literature is studied here by means of device transport parameters such as the intrinsic and extrinsic mobility degradation coefficients and the contact resistance. Model-based extraction methodologies are used for obtaining the parameter values. A mobility degradation-based transport model describes accurately the experimental ambipolar *I-V* data of devices with different gate lengths and at temperatures below 300 K with the extracted parameters. The temperature dependence of both the low-field and effective mobility, calculated based on the validated parameters, enables to obtain an insight on relevant scattering mechanisms at different device conditions, e.g., temperature and bias. From the extracted data it is suggested that, below a certain threshold temperature, the extrinsic mobility degradation improves, i.e., the gate control over the channel, for devices with low scattering mechanisms at room temperature than the ones with significant values of mobility degradation at 300 K.

Index Terms—graphene FET, mobility degradation, temperature, contact resistance

I. INTRODUCTION

Graphene field-effect transistors (GFETs) are considered promising candidates for analog/high-frequency (HF) applications due to their promising inherent properties such as their saturation velocity and room-temperature (RT$\sim$300 K) high-mobility [1], [2]. Mobility enhancing scenarios such as low-temperature conditions become an attractive option to exploit further the GFETs properties, e.g., at high-frequency where higher mobility generally improves the dynamic performance of incumbent technologies [3]. However, contraintuitive results have been found in the scarce related literature: a weak temperature-dependence of the dynamic performance of GFETs have been reported in [4] and [5] for temperatures T below 300 K. This apparent dynamic thermal stability of graphene transistors below RT lacks of a sound explanation

This work has received funding from Consejo Nacional de Ciencia y Tecnología, Mexico, under grant agreement 006405, from Instituto Politécnico Nacional under project SIP/20230362, from the European Union's Horizon 2020 research and innovation programme under grant agreement No GrapheneCore3 881603, from Ministerio de Ciencia, Innovación y Universidades under grant agreements PID2021-127840NB-I00 (MCIN/AEI/FEDER, UE) and FJC2020-046213-I. This article has been partially funded by the European Regional Development Funds (ERDF) allocated to the Programa Operatiu FEDER de Catalunya 2014-2020, with the support of the Secretaria d'Universitats i Recerca of the Departament d'Empresa i Coneixement of the Generalitat de Catalunya for emerging technology clusters to carry out valorization and transfer of research results. Reference of the GraphCAT project: 001-P-001702.

L. Valdez-Sandoval and E. Ramírez-García are with Instituto Politécnico Nacional, UPALM, Edif. Z-4 3er Piso, Mexico, email: lvaldezs1001@alumno.ipn.mx

N. Mavredakis, D. Jiménez and A. Pacheco are with the Departament d'Enginyeria Electrònica, Escola d'Enginyeria, Universitat Autònoma de Barcelona, Bellaterra 08193, Spain, e-mail: AnibalUriel.Pacheco@uab.cat.

and hence, a comprehensive investigation on the temperature-dependent transport phenomena, e.g., mobility and contact interfaces [6], [7], influencing the dynamic GFET performance is required.

Phenomena at interfaces between graphene and metal contacts captured by a contact resistance R_c parameter have been extensively studied as well as their response at low-temperatures [7]-[12]. Values of R_c in such studies are mainly reported by using the transfer length method (TLM). The latter is a useful approach to reveal, in most of the cases, the tendency of R_c with respect to bias, or as in these cases, with T. However, TLM fails to provide information regarding other mobility degradation parameters. Results have shown that R_c decreases with T as long as the injection mechanisms remain diffusive, otherwise, there is a weak T-dependence due to quasi-ballistic junctions [8], [9], [11].

Bulk or low-field mobilities of graphene transistors, characterized below RT and in general obtained via four-probe measurements[1], have been reported in the literature [13]-[17] showing practically no dependence on T. In contrast, above RT, the bulk mobility has been reported to be inversely proportional to T [14]-[20]. This onset of the mobility decay above RT in GFETs has been phenomenologically described by scattering mechanisms dominated by phonons -activated at $T \gtrsim 300$ K-, rather than by Coulomb interaction [17], [18], [20]. All of these studies, however, fail to report an effective device mobility μ_{eff} [21], i.e., carrier concentration dependent, in which degradation effects are present [22].

In this work, effective mobility-related parameters and contact resistance values are extracted from experimental temperature-dependent transfer characteristics of different graphene transistor technologies. The latter is obtained by considering an underlying transport model including mobility degradation effects. Velocity saturation effects are not considered in this work since the focus is to evaluate the GFETs performance at low-temperature and low-bias where such phenomena are not significant.

II. EXTRACTION METHODOLOGY

The unipolar drain current of GFETs at room temperature after considering mobility degradation effects ($\mu_{\text{eff}} = \mu_0/(1 + \theta_{\text{ch}}\sqrt{V_0^2 + V_{\text{GSO}}^2})$) reads as [22]

$$I_{\text{D}} = \beta \frac{\sqrt{V_0^2 + V_{\text{GSO}}^2}}{1 + \theta\sqrt{V_0^2 + V_{\text{GSO}}^2}} V_{\text{DS}}, \quad (1)$$

where $\beta = \mu_0 C_{\text{ox}} w_{\text{g}}/L_{\text{g}}$ is the transconductance parameter [23] with μ_0 as a low-field mobility, C_{ox} the oxide capacitance per unit area, w_{g} and L_{g} as the gate width and

[1]With the exception of [7] where the mobility has been obtained by a resistance model fitting.

979-8-3503-1191-4/23 $31.00 © 2023 IEEE

length, respectively, V_0 is a parameter related to the residual charge density at the Dirac voltage $V_{\mathrm{Dirac}}(= V_{\mathrm{GS}}|_{\min I_{\mathrm{D}}})$ [22], $V_{\mathrm{GSO}} = V_{\mathrm{GS}} - V_{\mathrm{Dirac}}$ is the extrinsic overdrive voltage with $V_{\mathrm{GS/DS}}$ the extrinsic gate-to-source/drain-to-source voltage, and θ is the extrinsic mobility degradation coefficient given by [23], [24] $\theta = \theta_{\mathrm{ch}} + \beta R_{\mathrm{c}}$, where θ_{ch} is the intrinsic mobility degradation due to vertical fields.

The parameters defining the transport in GFETs under the above conditions and given by Eq. (1), i.e., θ, θ_{ch}, β, R_{c} and V_0, have been extracted by methodologies presented elsewhere [25], [26]. For a GFET technology of devices with different L_{g} and considering the experimental I_{D} obtained at a low V_{DS}, the parameters can be extracted as (methodology A) [25]:

i) obtain $g_{\mathrm{m}}(= \partial I_{\mathrm{D}}/\partial V_{\mathrm{GS}})$, $R_{\mathrm{tot}}(= V_{\mathrm{DS}}/I_{\mathrm{D}})$ and the Y-function [23], [27] $(Y = I_{\mathrm{D}}/\sqrt{g_{\mathrm{m}}})$,
ii) obtain β from slope of Y^2 vs. V_{GSO}^2 (cf. [10, Eq. (4)] for each device,
iii) obtain θ/β from slope of $R_{\mathrm{tot}}Y$ vs. Y (cf. [10, Eq. (5)] for each device, and calculate θ
iv) obtain θ_{ch} and R_{c} from the plot of θ vs. β
v) obtain V_0 from $R_{\mathrm{tot}}|_{V_{\mathrm{Dirac}}}$, and,
vi) compare Eq. (1), using the extracted parameters, with experimental data for validation.

On the other hand, for an individual device, the parameters θ, β and V_0 can be obtained from the transfer characteristics at a given V_{DS} by following the steps above described for a single curve, with the exception that the effects described by θ_{ch} and R_{c} can not be provided [26] (methodology B).

From a practical point of view, Eq. (1) is able to describe the transfer characteristics of a GFET under any conditions, e.g., doping, temperature, by adjusting accordingly its parameters. In the case of the latter effect, the temperature dependence is embraced by the model parameters as inferred also in [22] (see Fig. 3c). Hence, in this work, the change of values of the device model parameters, at temperatures equal and below 300 K, is quantified in order to shed light on the reason behind an increased I_{D} at low temperatures while unraveling the temperature dependence of the mobility. The development of a device I_{D} model at low temperatures is out of the scope of this work. Notice that a comparison between methodologies A and B is not intended in this work, they are just used here in order to extract different kind of information regarding the transport in GFETs according to the available data found in the literature.

III. RESULTS AND DISCUSSION

A. With methodology A

Extraction methodology A has been applied to two different global-back-gate (GBG) GFET technologies containing devices with different lengths, which have been characterized elsewhere in the literature at different temperatures [12], [28]. Devices of technology T1 [12] (T2 [28]) with gate lengths of 5 μm, 10 μm, 15 μm and 20 μm (1 μm, 2 μm, 3 μm and 4 μm) have been measured at temperatures of 35 K, 150 K and 300 K (77 K and 300 K). Device schematic and information on their different materials are included in the Appendix. Further details on the fabrication process and measurement procedure for each technology can be found in [12] and [28].

Transfer characteristics of devices from T1 at room temperature are shown in Fig. 1 along with the required curves

for the parameters extraction in order to show the step-by-step procedure of methodology A (steps i) to iv) mentioned above). A good agreement between experimental data and Eq. (1) using the corresponding extracted parameters is shown in Fig. 1(a) for the T1 devices[2] and in Fig. 2 for the devices from T2 at different T. Such an accurate description of the experimental data, validates the extracted parameters. Values of θ_{ch} and R_{c} of these technologies, for both ambipolar regions, are listed in Table I. Contact resistance values are expected to differ between unipolar regions in GFETs due to the inherent strong p-doping of graphene and to metal-channel phenomena [8], [29].

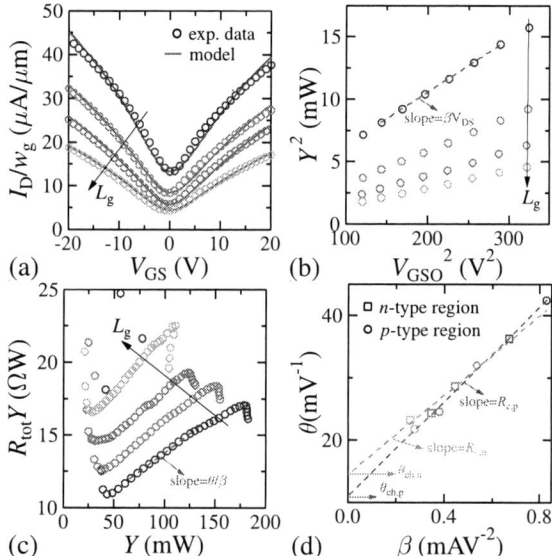

Fig. 1. Parameter extraction process for the GFET technology reported in [12]. (a) Transfer characteristics at room temperature. Markers are experimental data and lines are modeling results. L_{g}=5 μm, 10 μm, 15 μm and 20 μm. (b) Y^2 versus V_{GSO}^2 plots (only p-type region shown) required for the extraction of β. (c) $R_{\mathrm{tot}}Y$ versus Y plots (only p-type region shown) required for the extraction of θ/β. (d) θ versus β plot showing the extraction of R_{c} and θ_{ch} at different operation regimes. $V_{\mathrm{DS}} = 0.05$ V for all data

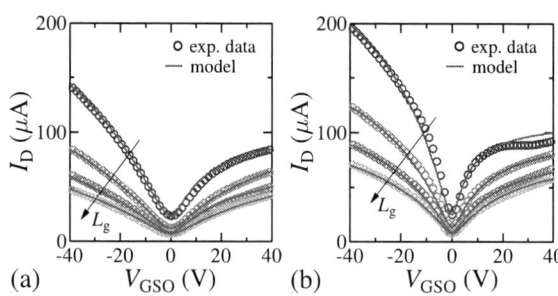

Fig. 2. Transfer characteristics of the devices reported in [28] at (a) 300 K and (b) 77 K. L_{g}=1 μm, 2 μm, 3 μm and 4 μm. $V_{\mathrm{DS}} = 0.05$ V for all data

Regarding the extracted values for the contact resistances, the ones obtained for the T1 devices at 35 K are similar to those obtained with the transfer length method (TLM), i.e., $R_{\mathrm{c,p(n)}}w_{\mathrm{g}}|_{@35\,\mathrm{K}}$ extracted here is 150 Ω μm (190 Ω μm) which is similar to the ~180 Ω μm (~200 Ω μm) reported in [12] at

---

[2]Similar results have been obtained for the transfer characteristics at other temperatures (not shown here).

TABLE I
EXTRACTED VALUES OF GFET TECHNOLOGIES WITH METHODOLOGY A

tech. [ref.]	T (K)	$\theta_{\text{ch,p}}$ (mV^{-1})	$R_{\text{c,p}}$ (Ω)	$\theta_{\text{ch,n}}$ (mV^{-1})	$R_{\text{c,n}}$ (Ω)
T1 [12]	35	38	15	42	19
	150	37	17	40	20
	300	12	30	14	32
T2 [28]	77	26	36	12	418
	300	9	70	3	449

the same V_{GSO} range used here -between ±15 V and 25 V- for the p-type (n-type) region. A similar minimum difference is obtained between the extracted results obtained here and the ones reported[3] in [28] via TLM. Notice that TLM is not able to provide information regarding mobility degradation in contrast to the method used here. From the results of the T1 devices, it can be observed that R_c at both unipolar regimes drops drastically below 300 K and remains almost unchanged at lower temperatures, explained by a quasi-ballistic injection, as observed in previous works [8], [11]. The higher impact of the scattering processes within the devices channels on the wider T1 devices ($w_g = 10\,\mu$m) than on T2 ($w_g = 2.2\,\mu$m) is captured by higher values of θ_{ch} for the former technology, specially at lower temperatures.

The p-type low-field mobility of the devices can be obtained from the corresponding β parameter extracted for each device (not shown here). Fig. 3(a) shows μ_0 of T1 devices at different T obtained by considering[4] a C_{ox} of 1.06 F/m^2. The weak T-dependence is clearly observed for the available points below 300 K as reported widely in the literature [13]-[17]. The more visible difference between μ_0 at RT and lower temperature for the longer devices might be explained by higher Coulomb scattering due to their larger area.

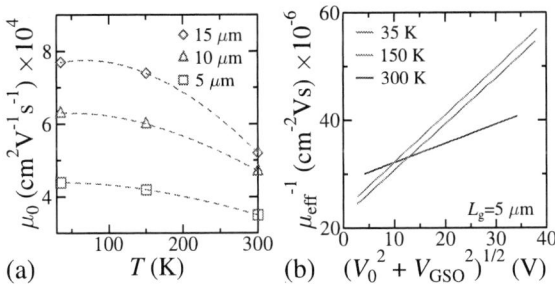

Fig. 3. (a) Low-field mobility obtained for T1 devices [12] with different L_g over T. Dashed lines are added as a guide for the eyes. (b) Inverse of μ_{eff} over V_{GS} for the 5 μm-long T1 device at different T. All results obtained for the p-type region.

Two scattering mechanisms can be identified from a plot of the inverse of μ_{eff} (by considering the Mathiessen's rule), one dominated by $1/\mu_0$ and the other one by the carrier density [22]. The inverse of μ_{eff}, calculated with μ_0 (cf. Fig. 3(a)) and the corresponding θ_{ch} extracted value, is shown in Fig.3(b) for the shorter T1 device at different T. It can be observed that scattering dominated by μ_0 is stronger at RT and at the p-type region away from V_{Dirac}; whereas the larger θ_{ch} obtained at lower T (cf. Table I) is reflected by

[3]Notice that Fig. 1c in [28] reports half of the value of the contact resistance.
[4]The devices have a w_g of 10 μm and a 325 nm-thick back-gate SiO2 dielectric.

the sharper slopes of the corresponding curves at the same bias regions, hence suggesting that the carrier density-induced scattering dominates at such scenario. At bias closer to V_{Dirac}, the model fails to capture the mobility degradation due to residual charges as discussed elsewhere [22].

B. With methodology B

Individual devices, with a unique gate length, can be characterized with methodology B yielding an estimated value for the extrinsic mobility degradation coefficient. Hence, such methodology has been applied to GBG devices from technologies 3 (T3) and 4 (T4) reported separately in [8] and [11], respectively. Both GFETs are 1 μm-long and 2 μm-wide (see Appendix for a schematic of the device and material details). T3 and T4 have been measured at 6 K, 40 K, 75 K, 125 K, 200 K, 250 K and 300 K and at 77.5 K, 110 K, 150 K, 200 K, 250 K and 300 K, respectively. The experimental ambipolar total resistance of each device at different temperatures is successfully described by considering Eq. (1) and the extracted parameters, i.e., θ and β for the case of this methodology B, as shown in Fig. 4.

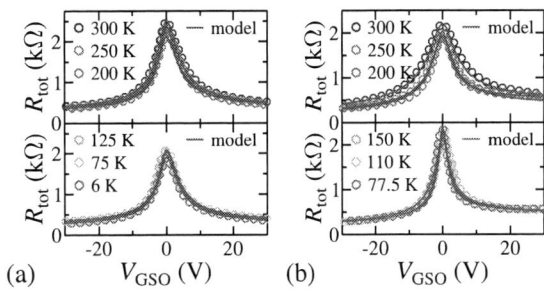

Fig. 4. Total resistance over V_{GSO} at different T of devices from (a) T3 [8] ($V_{\text{DS}} = 0.001$ V) and (b) T4 [11] ($V_{\text{DS}} = 0.05$ V).

Contact resistance values have been provided for each device in the corresponding publication and used in the calculations here: a direct temperature dependent R_c is reported for the T3 device [8] whereas constant R_c values at each unipolar regions for the T4 device [11] at all T. The difference has been explained elsewhere [11] by the quasi-ballistic injection at low-T in the first case and diffusive interfaces for the latter regardless T. Hence, θ and its two contributions, θ_{ch} and βR_c, have been obtained here as shown in Fig. 5, for both T3 (Fig. 5(a)) and T4 (Fig. 5(b)) devices over T. Despite the almost identical device structure and fabrication process, notice that, in addition to having a different oxide thickness (cf. Table II in Appendix), V_{DS} for the experimental data from T3 and T4 devices is equal to 1 mV and 50 mV, respectively, and hence, the impact of bias-dependent phenomena might differ between each of the results as discussed below.

The non-monotonical response of θ over T for the T3 device at both unipolar regimes, observed in Fig. 5(a), can be explained by the strong T-dependence of θ_{ch} rather than by βR_c. The latter term contributes with an important quasi-constant value since the T-dependence of R_c is somehow compensated by the variations in T of μ_0 (c.f. Fig. 3(a)) captured by β, however, the combined effect is weaker in temperature than that of θ_{ch}. The drop of θ below a threshold $T(\sim 125$ K$)$ for both cases indicates an improved gate control, since θ_{ch}

979-8-3503-1191-4/23 $31.00 © 2023 IEEE

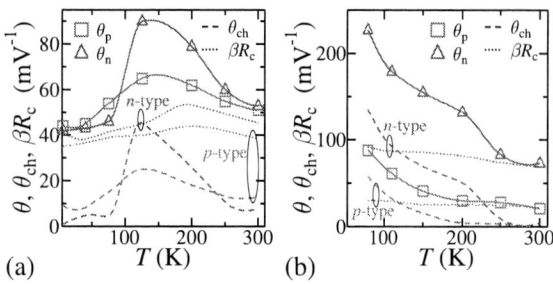

Fig. 5. Extrinsic mobility degradation and its individual contributions ($\theta = \theta_{ch} + \beta R_c$) over different temperature for the devices (a) T3 [8] ($V_{DS} = 0.001\,\text{V}$) and (b) T4 [11] ($V_{DS} = 0.05\,\text{V}$). Symbols are extracted values, dashed and dotted lines are modeling results and solid line is added as a guide for the eye.

follows this decrease as well, and transport dominated by quasi-ballistic injection via R_c [8].

For the T4 device it can be observed in Fig. 5(b) that the mobility within both unipolar regimes is degraded mainly by the contacts at RT and by vertical fields (via θ_{ch}) at lower T. Strong scattering (i.e., higher θ) is observed for T4 than for T3, which can be related to the difference of applied lateral fields in each characterization and to the thickest oxide of T4 which is almost three times the thickness of the dielecetric used for T3 (cf. Table II in Appendix). These observations suggest that the temperature required for the gate control improvement for T4 at such bias is lower than that of the lowest T reported for the measurements in [11], i.e., no drop of θ is observed for T4.

IV. CONCLUSION

The temperature-dependent transport performance of four different GFET technologies at and below RT and at both unipolar bias regimes has been evaluated in this work by means of their device parameters, e.g., the intrinsic and extrinsic mobility degradation coefficients and the contact resistance. For this purpose, two model-based extraction methodologies have been used which consider mobility degradation effects. The extracted parameters have been validated by the accurate description of the experimental data by the underlying transport model using them. The decrease with temperature of the contact resistance found here for the characterized technologies agrees with findings in the literature for diffusive contacts. Based on the extracted parameters, low-field mobility and effective mobility due to vertical fields effect have been calculated for one of the technologies. The saturation of μ_0, reported with other methods in the literature, is observed here as well. The sharper slopes of μ_{eff}^{-1}, i.e., θ_{ch} in the model used here, at low T reveals a higher impact of scattering effects related to the carrier density rather than to the material as in the case of room temperature data. The extrinsic mobility degradation has a non-monothonical behaviour at low fields whereas data at higher electric field reveal an inverse dependence of temperature of this parameter. For the former case, a drop of θ and θ_{ch} below a threshold suggests an improvement of the transport properties at low temperature.

APPENDIX

GFET technologies studied in this work, i.e., [12] and [28] (for applying methodology A) and [8] and [11] (for applying methodology B), have been fabricated with a common global-back-gate as shown in Fig. 6. Notice that while TLM has been used in the original works where the devices are reported, the methodologies used here are different.

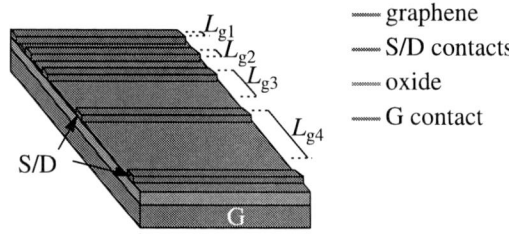

Fig. 6. Schematic device representation of global-back-gate GFETs with different gate lengths.

Gate lengths and widths of these technologies have been defined in the main text. Material characteristics of all technologies are listed in Table II.

TABLE II
CHARACTERISTICS OF GFET TECHNOLOGIES STUDIED HERE
(– REPRESENTS NOT REPORTED INFORMATION).

tech. [ref.]	gate oxide (thickness)	gate contact (height)	S/D contacts (height)
T1 [12]	SiO$_2$ (85 nm)	Si (–)	Au (–)
T2 [28]	SiO$_2$ (285 nm)	Si (–)	Pd/Au (30/50 nm)
T3 [8]	SiO$_2$ (90 nm)	Si (–)	Pd/Au (25/25 nm)
T4 [11]	SiO$_2$ (285 nm)	Si (–)	Pd/Au (30/50 nm)

REFERENCES

[1] G. C. Ghivela, J. Sengupta, "The Promise of Graphene: A Survey of Microwave Devices Based on Graphene". *IEEE Microwave Magazine*, vol. 21, no. 2, pp. 48-65, Feb. 2020. DOI: 10.1109/MMM.2019.2951967

[2] M. Saeed, P. Palacios, M.-D. Wei, E. Baskent, C.-Y. Fan, B. Uzlu, K.-T. Wang, A. Hemmetter, Z. Wang, D. Neumaier, M. C. Lemme, R. Negra, "Graphene-Based Microwave Circuits: A Review", *Advanced Materials*, vol. 34, no. 48, 2108473, Dec. 2022. DOI: 10.1002/adma.202108473

[3] E. Cha, N. Wadefalk, G. Moschetti, A. Pourkabirian, J. Stenarson, J. Li, D.-H. Kim, J. Grahn, "Optimization of Channel Structures in InP HEMT Technology for Cryogenic Low-Noise and Low-Power Operation", *IEEE Transactions on Electron Devices*, Mar. 2023. DOI: 10.1109/TED.2023.3255160

[4] N. Meng, J. F. Fernandez, D. Vignaud, G. Dambrine, H. Happy. "Influence of Temperature on High Frequency Performance of Graphene Nano Ribbon Field Effect Transistor", in Proc. *IEEE Device Research Conference*, Notre Dame, IN, USA, Jun. 2010. DOI: 10.1109/DRC.2010.5551934

[5] Y. Wu, Y. Lin, A. A. Bol, K. A. Jenkins, F. Xia, D. B. Farmer, Y. Zhu, P. Avouris, "High-frequency, scaled graphene transistors on diamond-like carbon", *Nature*, vol. 472, pp. 74-78, Apr. 2011. DOI: 10.1038/nature09979

[6] M. Bonmann, A. Vorobiev, M. A. Andersson, J. Stake, "Charge carrier velocity in graphene field-effect transistors", *Applied Physics Letters*, vol. 111, 233505, Dec. 2017. DOI: 10.1063/1.5003684

[7] M. E. Ramón, H. C. P. Movva, Sk. F Chowdhury, K. N. Parrish, A. Rai, C. W. Magnuson, R. S. Ruoff, D. Akinwande, S. K. Banerjee, "Impact of contact and access resistances in graphene field-effect transistors on quartz substrates for radio frequency applications", *Applied Physics Letters*, vol. 104, 073115, Feb. 2014. DOI: 10.1063/1.4866332

[8] F. Xia, V. Perebeinos, Y. Lin, Y. Wu, P. Avouris, "The origins and limits of metal–graphene junction resistance", *Nature Nanotechnology*, vol. 6, pp. 179-184, Mar. 2011. DOI: 10.1038/NNANO.2011.6

[9] L. Wang, I. Meric, P. Y. Huang, Q. Gao, Y. Gao, H. Tran, T. Taniguchi, K. Watanabe, L. M. Campos, D. A. Muller, J. Guo, P. Kim, J. Hone, K. L. Shepard, C. R. Dean, "One-Dimensional Electrical Contact to a Two-Dimensional Material", *Science*, vol. 342, pp. 614-617, Nov. 2013. DOI: 10.1126/science.1244358

[10] M. Zhu, J. Wu, Z. Du, S. Tsang, E. H. T. Teo, "Gate voltage and temperature dependent Ti-graphene junction resistance toward straightforward p-n junction formation", *Journal of Applied Physics*, vol. 124, 215302, Dec. 2018. DOI: 10.1063/1.5052589

[11] H. Zhong, Z. Zhang, B. Chen, H. Xu, D. Yu, L. Huang, L. Peng, "Realization of low contact resistance close to theoretical limit in graphene transistors", *Nano Research*, vol. 8, pp. 1669-1679, May 2015. DOI: 10.1007/s12274-014-0656-z

[12] A. Gahoi, S. Kataria, M. C. Lemme, "Temperature dependence of contact resistance for gold-graphene contacts", in Proc. *IEEE European Solid-State Device Research Conference (ESSDERC)*, Leuven, Belguim, Sep. 2017. DOI: 10.1109/ESSDERC.2017.8066604

[13] S. V. Morozov, K. S. Novoselov, M. I. Katsnelson, F. Schedin, D. C. Elias, J. A. Jaszczak, A. K. Geim, "Giant Intrinsic Carrier Mobilities in Graphene and Its Bilayer", *Physical Review Letters*, vol. 100, 016602, Jan. 2008. DOI: 10.1103/PhysRevLett.100.016602

[14] J.-H. Chen, C. Jang, S. Xiao, M. Ishigami, M. S. Fuhrer, "Intrinsic and extrinsic performance limits of graphene devices on SiO2", *Nature Nanotechnology*, vol. 3, pp. 206–209, Apr. 2008. DOI: 10.1038/nnano.2008.58

[15] W. Zhu, V. Perebeinos, M. Freitag, P. Avouris, "Carrier scattering, mobilities, and electrostatic potential in monolayer, bilayer, and trilayer graphene", *Physical Review B*, vol. 80, 235402, Dec. 2009. DOI: 10.1103/PhysRevB.80.235402

[16] Sk. F. Chowdhury, L. Tao, S. Banerjee, D. Akinwande, "Enhancement of graphene field-effect transistor by surface treatment", in Proc. *IEEE International Conference on Nanotechnology*, Toronto, Canada, Aug. 2014. DOI: 10.1109/NANO.2014.6968125

[17] J. H. Gosling, S. V. Morozov, E. E. Vdovin, M. T Greenaway, Y. N. Khanin, Z. Kudrynskyi, A. Patanè, L. Eaves, L. Turyanska, T. M. Fromhold, O. Makarovsky "Graphene FETs with high and low mobilities have universal temperature-dependent properties", *Nanotechnology*, vol. 34, no. 12, 125702, Jan. 2023. DOI: 10.1088/1361-6528/aca981

[18] V. E. Dorgan, M.-H. Bae, E. Pop, "Mobility and saturation velocity in graphene on SiO2", *Applied Physics Letters*, vol. 97, 082112, Aug. 2010. DOI: 10.1063/1.3483130

[19] S. L. Rumyantsev, G. Liu, M. S. Shur, A. A. Balandin, "Observation of the memory steps in graphene at elevated temperatures", *Applied Physics Letters*, vol. 98, 222107, Jun. 2011. DOI: 10.1063/1.3596441

[20] M. Bonmann, M. Krivic, X. Yang, A. Vorobiev, L. Banszerus, C. Stampfer, M. Otto, D. Neumaier, J. Stake, "Effects of Self-Heating on fT and fmax Performance of Graphene Field-Effect Transistors", *IEEE Transactions on Electron Devices*, vol. 67, no. 3, pp. 1277-1284, Mar. 2020. DOI: 10.1109/TED.2020.2965004

[21] D. K. Schroder, "Semiconductor Material and Device Characterization", 3rd ed., Jown Wiley & Sons, NJ, USA, 2006.

[22] K. Jeppson, M. Asad, J. Stake, "Mobility Degradation and Series Resistance in Graphene Field-Effect Transistors", *IEEE Transactions on Electron Devices*, vol. 68, no. 6, pp. 3091-3095, Jun. 2021. DOI:10.1109/TED.2021.3074479

[23] G. Ghibaudo, "New method for the extraction of MOSFET parameters", *Electronics Letters*, vol. 24, no. 9, pp. 543-545, 1988. DOI: 10.1049/el:19880369

[24] L. Risch, "Electron mobility in short-channel MOSFET's with series resistances", *IEEE Transactions on Electron Devices*, vol. 30, no. 8, pp. 959-961, Aug. 1983. DOI: 10.1109/T-ED.1983.21246.

[25] A. Pacheco-Sanchez, N. Mavredakis, P. C. Feijoo, D. Jiménez, "An extraction method for mobility degradation and contact resistance of graphene transistors," *IEEE Transactions on Electron Devices*, vol. 69, no. 7, pp. 4037-4041, Jul. 2022. DOI: 10.1109/TED.2022.3176830

[26] A. Pacheco-Sanchez, N. Mavredakis, D. Jiménez, "Extrinsic mobility degradation extraction in GFETs for technologies benchmarking", *IEEE Spanish Conference on Electron Devices* (accepted), 2023.

[27] S. Jain, "Measurement of threshold voltage and channel length of submicron MOSFETs", *IEE Proceedings I (Solid-State and Electron Devices)*, vol. 135, no. 6, pp. 162–164, 1988. DOI: 10.1049/ip-i-1.1988.0029

[28] H. Zhong, Z. Zhang, H. Xu, C. Qiu, L.-M. Peng, "Comparison of mobility extraction methods based on field-effect measurements for graphene", *AIP Advances*, vol. 5, 057136, May 2015. DOI: 10.1063/1.4921400

[29] B. Huard, N. Stander, J. A. Sulpizio, D. Goldhaber-Gordon, "Evidence of the role of contacts on the observed electron-hole asymmetry in graphene", *Physical Review B*, vol. 78, 121402, Sep. 2008. DOI: 10.1103/PhysRevB.78.121402

979-8-3503-1191-4/23 $31.00 © 2023 IEEE

Experimental Characterization of a Thermoelectric Generator System

1st Daniel Sanin-Villa
Dept. of Mechatronics-Electromechanics
Instituto Tecnológico Metropolitano
Medellín, Colombia
ORCID: 0000-0001-6853-340X

2nd Elkin Henao-Bravo
Dept. de Ingeniería Eléctrica, Electrónica y Computación
Universidad Nacional de Colombia
Manizales, Colombia
ORCID: 0000-0001-9663-1082.

3rd Juan Pablo Villegas-Ceballos
Dept. of Electronics-Telecommunications
Instituto Tecnológico Metropolitano
Medellín, Colombia
ORCID: 0000-0002-2598-4486

4th Farid Chejne
Processes and Energy Department
Universidad Nacional de Colombia
Medellín, Colombia
ORCID: 0000-0003-0445-7609

Abstract—This research paper presents a detailed investigation of the performance of a thermoelectric generator (TEG) system. The TEG system is composed of multiple thermoelectric legs connected in series, a heat source, and an electric load. The experimental setup is designed to simulate real-world operating conditions and allow for the measurement of the TEG's power output as a function of the temperature difference between the hot and cold sides, and the load conditions. The performance evaluation of the TEG system is conducted using a commercial thermoelectric generator module (TEM) and several measurement instruments such as power analyzers, thermocouples, and electronic devices. The results of the study are used to plot the TEG's operating curves, which show the power output as a function of the temperature difference and the TEM current including the identification of the maximum power point (MPP). The operating curves allow for the evaluation of the TEG's behavior and thermal characteristics under different conditions. Additionally, this result provides valuable insights into the performance of TEGs and contribute to the advancement of this technology.

Index Terms—TEG, experimental setup, TEM, operating curves

I. INTRODUCTION

Thermoelectric generators (TEGs) have gained attention in recent years as a promising technology for converting waste heat into electricity). TEGs are based on the Seebeck effect, which states that a voltage is generated when two dissimilar materials are maintained at different temperatures [1]. This effect can be harnessed by connecting multiple thermoelectric modules (TEMs) in series to form a TEG. TEGs have the potential to be used in a wide range of applications, including automobiles, industrial processes, and even health applications [2]. However, the performance of TEGs is highly dependent on the TEMs used, the temperature difference between the hot and cold sides, and the load conditions [3]. To fully understand the capabilities and limitations of TEGs, it is necessary to conduct performance evaluations in a controlled laboratory environment.

In this research paper, the performance of a TEG system is evaluated in a laboratory environment using a commercially available TEM [4]. The TEG system is composed of multiple TEMs connected in series, a heat source, and a load. The heat source and load conditions are varied to investigate the effects on the TEG's performance.

Characterizing a thermoelectric module (TEM) is an essential step in understanding its performance and potential applications. One of the main parameters used to characterize a TEM is its electrical resistivity, which is a measure of the resistance of the material to the electric current flow [5]. A higher electrical resistivity results in a lower power output from the TEM. Another important parameter is thermal conductivity, which is a measure of the ability of the material to transfer heat. A TEM with high thermal conductivity will have a lower temperature difference between the hot and cold sides, which in turn will result in a lower power output.

The Seebeck coefficient, also known as the thermoelectric power, is a measure of the voltage generated per unit temperature difference between the two sides of the TEM. A higher Seebeck coefficient results in a higher power output from the TEM. Characterizing a TEM can help in understanding the potential performance of the module and the optimal operating conditions for a specific application. It can also be helpful in the selection of TEMs for a particular application, for example, higher electrical conductivity and lower thermal conductivity TEMs may be preferred for high-temperature applications.

One way to evaluate the behavior of TEMs is using a testing setup known as the "two-thermocouple method" [6]. This method involves attaching thermocouples to both the hot and cold sides of the TEM and measuring the voltage and temperature difference between them. By measuring the voltage and temperature difference, it is possible to calculate the Seebeck coefficient, electrical resistivity, and thermal conductivity of the TEM. Another way to evaluate TEMs is by using a testing setup known as the "three-point method" [7]. This method involves attaching thermocouples to the hot side, cold side, and the center of the TEM, measuring the voltage and temperature difference between them. By measuring the voltage and temperature difference, it is possible to calculate the thermal conductivity and Seebeck coefficient of the TEM. In addition to these testing methods,

979-8-3503-1191-4/23 $31.00 © 2023 IEEE

2023 IEEE Latin American Electron Devices Conference (LAEDC)
Puebla, México, July 3-5, 2023

it is also important to measure the power output of the TEM under different loading conditions [8].

The experimental setup must be designed to minimize errors and uncertainties in the measurements, to obtain accurate and reliable results. Once the operating curves have been obtained, the TEM's performance can be evaluated under different conditions. Additionally, by comparing the operating curves at different temperatures, it is possible to investigate the effects of temperature on the TEM's performance.

The experimental assessment of TEG systems can help in the design and optimization of TEGs for various applications. This work presents a comprehensive experimental setup for evaluating the power output of a TEM under controlled conditions. The establishment of this methodology is important for characterizing TEMs and understanding the performance of thermoelectric generators.The characterization of TEMs and the knowledge of their operative curves are crucial for the design and optimization of TEGs, reaching the maximum power point (MPP) and increasing their efficiency. This paper will provide valuable insights into the performance of TEGs and will contribute to the advancement of this technology.

This research paper presents a detailed experimental study on the performance evaluation of a commercial TEM: Section II provides a description of the experimental setup, including the heat source, load, and temperature measurement devices and the measurement protocol for obtaining the operating characteristics of the TEM. Section III presents the characteristic power curves of the TEM, which show the power output as a function of temperature difference and load, and a detailed analysis of the curves. Finally, Section IV summarizes the main findings and highlights the importance of the work and suggests possible future work.

II. EXPERIMENTAL SETUP FOR TEM CHARACTERIZATION

The experimental setup for evaluating the performance of a thermoelectric module is presented in Figure 1. The setup is composed of two main parts: the power generation system and the instrumentation.

The power generation system is composed of the TEM manufactured by Thermal Electronics Corp. [4], a heat source, a heat sink, and a load. The TEM is the main component of the power generation system, and it converts heat energy into electrical energy. The heat source, which is a hot plate or a heater, is used to provide the heat energy to the TEM; it includes an electrical resistance with a value of 100 used to heat the hot side of the TEM. The load is a programmable electronic load used to consume the electrical energy generated by the TEM. The hot side temperature is controlled by an on-off temperature controller that regulates the power supplied to the electrical resistor. The cold side of the TEM is maintained at a constant temperature by using a large reservoir of water (tank) that is continuously pumped by a pump. The tank of water provides a large thermal mass that allows the cold side of the TEM to maintain a constant temperature, which is important for accurate measurement of the TEM's performance.

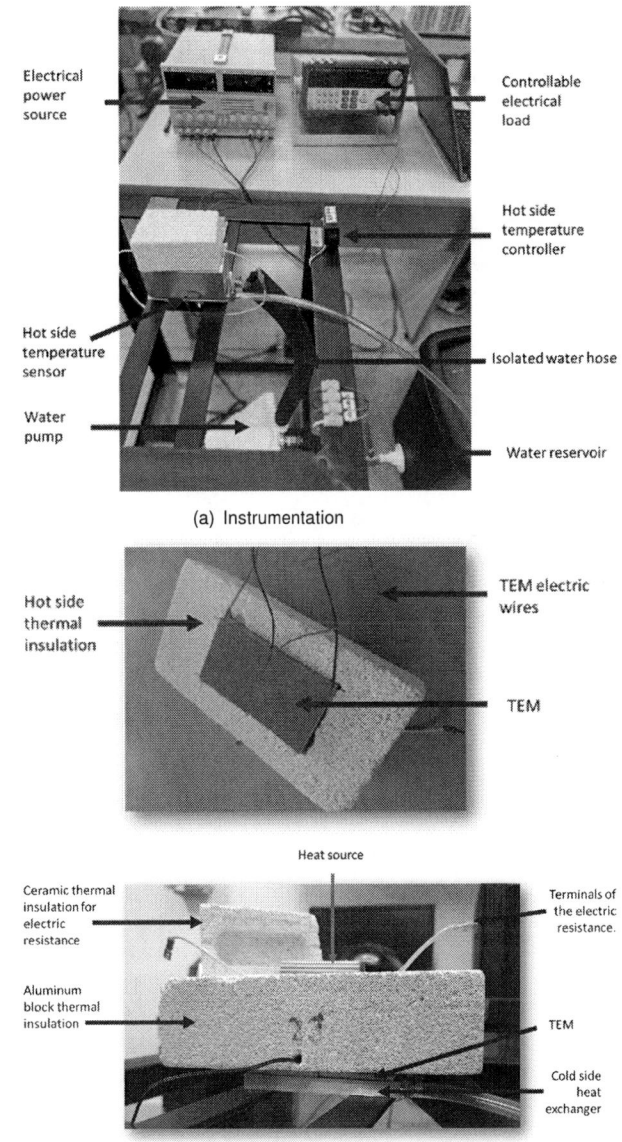

(a) Instrumentation

(b) Power generation system

Fig. 1: Components of the experimental setup to evaluate TEG power generation at different temperature boundaries.

One of the key aspects of the experimental setup is the use of a controlled heat source and a load to simulate real-world operating conditions. This allows for the measurement of the TEM's power output as a function of the temperature difference between the hot and cold sides and the load conditions. To obtain the operating characteristics of the TEM, a series of measurements are taken at different temperature differences and load conditions. These measurements are then used to plot the TEM's operating curves, which show the power output as a function of the temperature difference and the load. The operating curves are important because they allow for the identification of the maximum power point (MPP), which is the point on the curve where the TEM produces the

979-8-3503-1191-4/23 $31.00 © 2023 IEEE

2023 IEEE Latin American Electron Devices Conference (LAEDC)
Puebla, México, July 3-5, 2023

maximum power output. The on-off temperature controller allows for precise control of the hot side temperature, while the large reservoir of water ensures that the cold side temperature remains constant. This combination of temperature control and measurement allows for accurate characterization of the TEM's performance and the determination of its operating characteristics.

The instrumentation part is composed of devices used to measure the performance of the TEM. These include an oscilloscope, a multimeter, and thermocouples. Auxiliary control electronics are also used in the experimental setup to control the temperature of the heat source and to automate the measurement process.

The experimental procedure for evaluating the performance of a thermoelectric module using an oscilloscope involved the following steps:

- The temperature gradient across the TEM was set to a specific value by adjusting the temperature of the heat source using the on-off temperature controller and maintaining the cold side temperature constant using the large reservoir of water.
- The load was programmed with a current profile consisting of step changes in current, with each step being $50mA$.
- The oscilloscope was used to record the voltage and current generated by the TEM for a period.
- The instant power consumption was computed by multiplying the voltage and current values and recorded and saved for further analysis.
- The above steps were repeated for different temperature gradients and load conditions.
- The data obtained from the oscilloscope was analyzed to determine the TEM's operating characteristics, such as the maximum power point (MPP) for each temperature difference.
- The oscilloscope was used to dynamically observe the behavior of the voltage, current, and power consumption, allowing the detection of any unwanted behavior that can occur in the TEM, such as ripple, voltage sag, etc.

Figure 2 depicts the oscilloscope screen during a single experimental test, where the green trace corresponds to the current supplied by the thermoelectric module (TEM). In contrast, the yellow and red traces represent the measured voltage and calculated power output, respectively. The overall power output profile exhibits a parabolic shape, suggesting that the maximum power point is attained at a specific current and voltage value. Notably, a noticeable inverse relationship is observed between the current and voltage, as an increase in the module's current leads to a corresponding decrease in the voltage.

III. EXPERIMENTAL RESULTS FOR TEM CHARACTERIZATION

In this research, a digital filter process is applied to a current signal to remove unwanted noise and improve signal quality. The filter used is a Butterworth filter, a type of low-pass filter that is characterized by a flat frequency response in the passband and a roll-off that is -3dB per octave in the stopband. The filter order is specified as 3, and the cutoff frequency is specified as 0.05 Hz. These values are used to calculate the filter coefficients 'b' and 'a'.

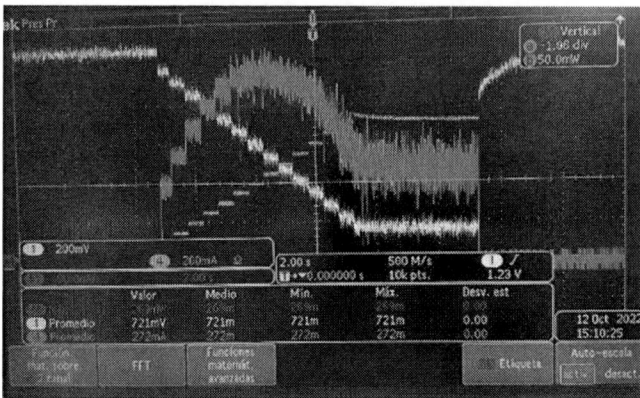

Fig. 2: Experimental measurement of Voltage, Current, and Power Output of a TEM in the oscilloscope.

Next, the filter is applied to the current signal.

Figure 3 shows the filtered current signal and its corresponding power output over time. The red and black lines in indicate the filtered signals, while the blue lines indicate the raw signals. The filtered signals are smoother and less noisy than the raw signals. The y-axis on the left side of the plot shows the current in amperes, while the y-axis on the right side of the plot shows the power output in watts. The x-axis represents time in seconds. Overall, the figure demonstrates the effectiveness of the filtering process in reducing noise and improving the accuracy of the data.

After applying the digital filter process to the current signal, the filtered signals for each temperature gradient can be further analyzed by fitting them to a polynomial curve. The method of fitting a polynomial curve to the filtered signal is carried out by utilizing the least-squares polynomial fit algorithm. In order to select the most appropriate polynomial model for the filtered signal, various polynomial models, such as quadratic or cubic equations, were compared based on their goodness of fit, which can be evaluated using the coefficient of determination (R^2). Figure 3 shows the polynomial curve that best represents the filtered signal and can be used as a model of the actual thermoelectric module (TEM) behavior under different temperature gradients, with a constant cold side of $24.3\,°C$. Table 1 shows the coefficients of the polynomial $y = ax^2 + bx + c$ and the corresponding R^2 values for each different temperature curves.

TABLE I: Polynomial coefficients fitted from experimental and filtered data.

	a	b	c	R^2
$T_H=60\,°C$	-1.2459	1.0118	-0.0004	0.997
$T_H=70\,°C$	-1.1596	0.8698	0.0003	0.989
$T_H=80\,°C$	-1.1861	0.7737	-0.0001	0.978
$T_H=90\,°C$	-1.2159	0.6819	-0.0002	0.996
$T_H=100\,°C$	-1.0748	0.5259	3.00E-06	0.984
$T_H=110\,°C$	-1.1398	0.4204	7.00E-05	0.992

2023 IEEE Latin American Electron Devices Conference (LAEDC)
Puebla, México, July 3-5, 2023

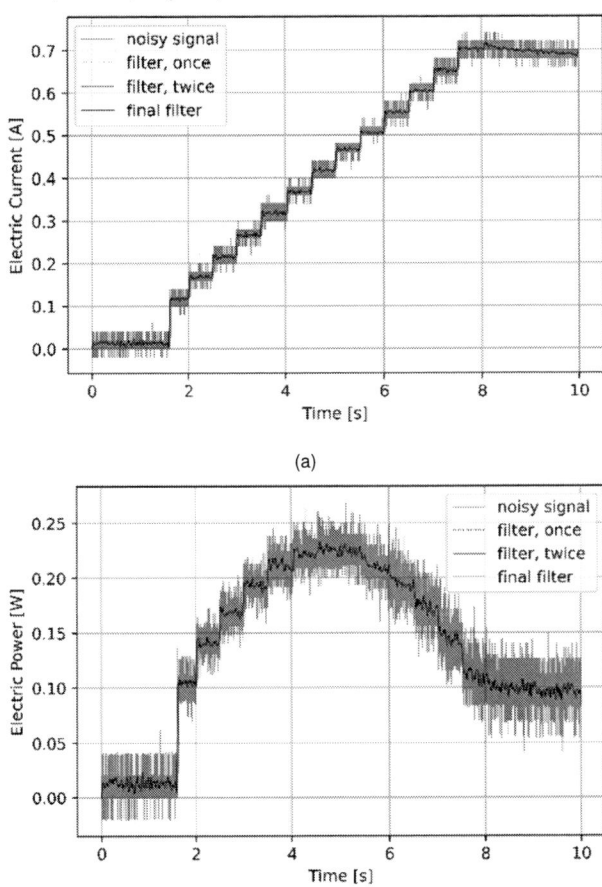

(a)

(b)

Fig. 3: Improved Accuracy of Current and Power Output Measurements with Signal Filtering: a) current profile, b) power output.

The R^2 values range from 0.978 to 0.997, indicating a high degree of correlation between the temperature and the data being measured. In general, the data suggests that as the temperature increases, the R^2 values also tend to increase, indicating that the correlation between the temperature predicted by the polynomial approximation and the measured data becomes stronger. However, there is some variability in the data, with the R^2 value for $90\,°C$ being significantly higher than the surrounding values.

These models can be used to predict the TEM's performance under different conditions and can also be used to optimize the TEM's performance. The polynomial curve can be used to determine MPP of the TEM, which is the temperature gradient at which the TEM generates the maximum power. The polynomial curve can also be used to calculate the TEM's performance, which is the relationship between the power generated by the TEM and the electric current delivered at a fixed temperature and specific load.

IV. CONCLUSIONS

The results of this research paper provide valuable insights into the behavior and thermal characteristics of a thermoelectric generator (TEG) system. The experimental setup designed and utilized in

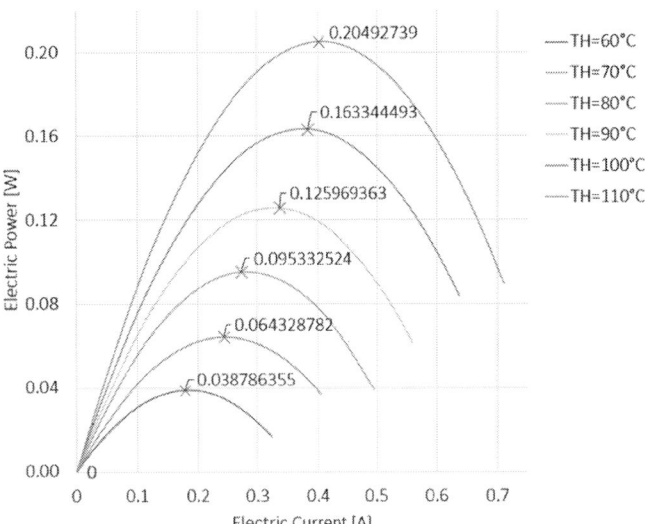

Fig. 4: Polynomial curves for filtered current vs. power at different temperature gradients for different temperatures between the hot and cold sides. Cold side temperature 24.3 °C.

this study provides an accurate simulation of real-world operating conditions and allows for the measurement of the TEG's power output under different load conditions and temperature differences. The operating curves obtained through the study show the TEG's behavior and thermal characteristics, including the maximum power point (MPP), under varying conditions. This study contributes to the advancement of TEG technology by providing a comprehensive analysis of the TEG system's behavior and thermal characteristics. Furthermore, the experimental setup designed and utilized in this study provides a valuable tool for evaluating the power generation of TEGs under known boundary conditions. As future work, the evaluation of the pressure over the hot and cold side of the TEG and the effect of transient thermal conditions over the power generation could be investigated. These additional studies would provide a more comprehensive understanding of the behavior and performance of TEG systems and contribute to the advancement and optimization of this technology.

ACKNOWLEDGMENT

This work was supported by Minciencias, Universidad Nacional de Colombia, Universidad del Valle, and Instituto Tecnológico Metropolitano under the research project "Estrategias de dimensionamiento, planeación y gestión inteligente de energía a partir de la integración y la optimización de las fuentes no convencionales, los sistemas de almacenamiento y cargas eléctricas, que permitan la generación de soluciones energéticas confiables para los territorios urbanos y rurales de Colombia", (Minciencias code 71148), which belongs to the research program "Estrategias para el desarrollo de sistemas energéticos sostenibles, confiables, eficientes y accesibles para el futuro de Colombia", (Minciencias code 1150-852-70378, ITM code RC 80740-178-2021-1.). The authors also thank the Alliance for Biomass and Sustainability Research–ABISURE-Universidad Nacional

979-8-3503-1191-4/23 $31.00 © 2023 IEEE

2023 IEEE Latin American Electron Devices Conference (LAEDC)
Puebla, México, July 3-5, 2023

de Colombia, Hermes code 53024, for its support in the realization of this study.

REFERENCES

[1] D. Sanin-Villa, "Recent developments in thermoelectric generation: A review," *Sustainability*, vol. 14, p. 16821, 12 2022.

[2] D. Sanin-Villa, O. D. Monsalve-Cifuentes, and J. S. D. Rio, "Early fever detection on covid-19 infection using thermoelectric module generators," *International Journal of Electrical and Computer Engineering (IJECE)*, vol. 11, p. 3828, 10 2021.

[3] D. Sanin-Villa, O. D. Monsalve-Cifuentes, and E. E. Henao-Bravo, "Evaluation of thermoelectric generators under mismatching conditions," *Energies*, vol. 14, p. 8016, 12 2021.

[4] T. solistate power generation, "Specifications teg module teg1-12611-6.0," 2021.

[5] S. Zhang and X. Liao, "The test structures to measure resistivity and contact resistance of poly-si for thermoelectric-photoelectric integrated generator." IEEE, 4 2019, pp. 443–446.

[6] S. Dalola, M. Ferrari, V. Ferrari, M. Guizzetti, D. Marioli, and A. Taroni, "Characterization of thermoelectric modules for powering autonomous sensors," *IEEE Transactions on Instrumentation and Measurement*, vol. 58, pp. 99–107, 1 2009.

[7] H.-J. Kim, J. R. Skuza, P. Yeonjoon, G. C. King, H. C. Sang, and A. Nagavalli, "System to measure thermal conductivity and seebeck coefficient for thermoelectrics," pp. 1–39, 2012.

[8] E. I. Ortiz-Rivera, A. Salazar-Llinas, and J. Gonzalez-Llorente, "A mathematical model for online electrical characterization of thermoelectric generators using the p-i curves at different temperatures." IEEE, 2 2010, pp. 2226–2230.

979-8-3503-1191-4/23 $31.00 © 2023 IEEE

2023 IEEE Latin American Electron Devices Conference (LAEDC)
Puebla, México, July 3-5, 2023

Real-Time Wheelchair Controller Based on POF-Based Pressure Sensors

González-Cely, A.X.[1,2], Blanco-Díaz, C.F.[1], Camilo A.R. Díaz[2], and Bastos-Filho T.[1]

Telecommunications Laboratory [1], Robotics and Assistive Technology Laboratory [2],
Graduate Program in Electrical Engineering, Federal University of Espírito Santo, Av. Fernando Ferrari, 514-Goiabeiras,
Vitoria, Brazil
aura.cely@edu.ufes.br

Abstract—The design of electrical wheelchair controllers remains a challenge for the scientific community. In recent years, Polymer Optical Fiber (POF)-based sensors have been used for biomedical applications, however, the application of these types of sensors for the control of assistive devices has been little explored. For this reason, this work proposes the design and validation of a fuzzy logic controller that uses information from an array of POF-based sensors located in a pillow, which is used for wheelchair control in four directions (right, left, forward and stop) by using neck movements. The results allow concluding that the system is fast, accurate, and reliable. This work presents a contribution related to the design of assistive devices based on optical fiber sensors and focuses on people with upper- and lower-limb mobility restrictions. Future studies will address the evaluation of the system for people with disabilities.

Index Terms—Polymer Optical Fiber (POF), Fuzzy Logic Controller, Wheelchair, Real-time, Pressure sensors, Neck motion.

I. INTRODUCTION

Real-time implementation of wheelchair controllers has allowed users to generate more independence. In addition, several researchers have been developing innovative assistive devices to help people with physical restrictions in their upper and/or lower limbs [1]. The literature review presents different types of instrumentation for users to command a wheelchair through head [2]–[4], finger [5], eye [6], and hand movements [4] as well as alternative ways as the use of electroencephalography (EEG) [7], [8], electromyography (EMG) [9] signals, and voice sounds [10] among others. Researchers also have used obstacle avoidance systems [11], virtual reality controllers [12], and cameras [13] integrated in wheelchairs. Considering some wheelchair applications that use Artificial Intelligence (AI), Ambarwati *et al.* addressed a classification methodology using a processed dataset and a feature matrix to classify wheelchair commands [14]. Several studies have been reported in the literature related to the control of wheelchairs using strategies to improve the quality of life of individuals with reduced mobility. Among these alternatives, it is possible to highlight the use of inertial sensors, such as the study by Binte *et al.*, who implemented the prototype of a wheelchair controlled by head movement through this kind of sensors located on a cap [2]. Other types of sensors, such as joysticks, gesture recognition, and voice recognition have been reported for wheelchair controllers. For

instance, Sutikno *et al.* achieved accuracy of approximately 100% using artificial neural networks [10], and Rafiul *et al.* achieved the same performance by using computer vision techniques [1]. This high accuracy has been obtained when the user is not sitting in the wheelchair, however, these prototypes were not evaluated in a real electric-powered wheelchairs. Other techniques related to biosignals have been reported, such as Xu *et al.*, who implemented a controller based on EEG, and Tsung *et al.* , who implemented controllers based on EMG with accuracy close to 80% [7], [9]. Nevertheless, this type of sensor has a low signal-to-noise ratio and inter-subject variability, providing low accuracy, which is not desirable for real-time implementation.

The literature review also presents the use of instrumentation based on Optical Fiber Sensors (OFS) for different biomedical applications [15], [16]. In assistive devices, OFS have been used for posture monitoring [17], physiological monitoring [18], and biosensing [19], among others. Polymer Optical Fiber (POF) made of Polymethyl Methacrylate (PMMA) is a type of optical fiber that is commonly used in biomedical applications owing to its advantages such as electromagnetic immunity, maneuverability, and high sensitivity. POF-based sensors has been cited in the literature for posture monitoring of wheelchair users [20].

Although the literature mentions several implementations that use electronic sensors for wheelchair control or OFS for physiological monitoring, there is still a lack of a combination of OFS with wheelchair controllers to operate electric-powered wheelchairs in real time focused on people with disabilities in their upper and/or lower limbs.

Thus, this research aims to develop a real-time wheelchair controller based on OFS to improve the quality of life of wheelchair users and generate independence for them. To deal with this issue, this work uses a multiplexed line of POF-based pressure sensors located in a neck pillow to detect pressure signals of the user's neck in four directions: normal position or stopped, right, left, and forward. The user's intentions are then processed using a fuzzy logic controller to command the wheelchair in real-time.

979-8-3503-1191-4/23 $31.00 © 2023 IEEE

II. Materials and Method

A. Materials

1) POF-based Sensors: The multiplexed line of the sensors was built using five sensors based on POF. Two photodetectors IF-D92 [21] are connected at the end of the optical fiber, which transform optical signals into electrical signals. The POF has a transversal cut working as a sensitive zone, in which a high brightness Light Emitting Diode (LED) is located in front of it. Through a 3D malleable mold that joins the sensitive area of the fiber with the LED, pressure variations are realized to generate light variations. Figure 1 shows the pressure sensor developed here. The sensor has an additional mold on the top to concentrate the force in the middle of the sensitive zone, making the sensor more precise in measuring pressure variations [22]–[24].

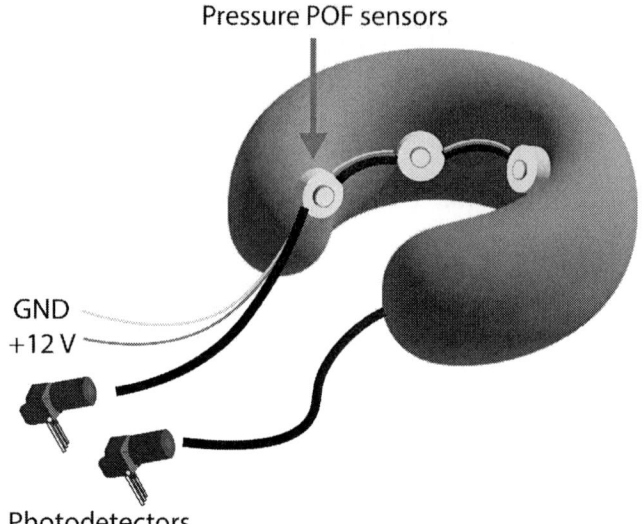

Figure 2: Neck pillow system.

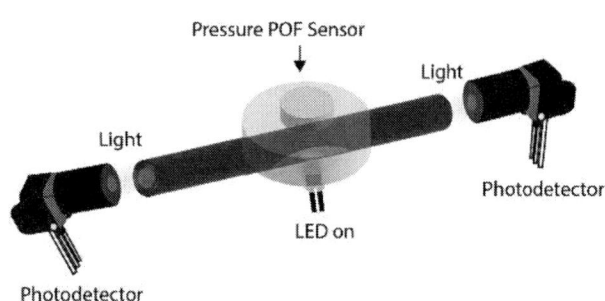

Figure 1: POF-based pressure sensor.

The multiplexing sensors located in the neck pillow, as shown in Figure 2, have a circuit that sends signals to the computer through a ESP32. A user datagram communication protocol is implemented for the communication. The computer processes the information and sends commands to the wheelchair using Wi-Fi.

2) Electronic instrumentation: The electronic instrumentation of the system is composed by five LEDs, POF-based pressure sensors, communication devices, and control unit, which are presented in Figure 3. Figure 3a exhibits the LEDs control circuit, which employs an isolation circuit to prevent eddy currents using an optoisolator MOC3021 and a Metal Oxide Semiconductor Field-Effect Transistor (MOSFET) IRFZ44N. This circuit amplifies the control signal for the LEDs. In Figure 3b, the pressure sensors are presented, consisting of photodetectors IFD-92 with a pull-up connection for the neck sensors and the wheelchair. Lastly, Figure 3c depicts the complete PCB of the neck pillow where the microcontroller is a Teensy 3.6 [25]. The Freedom Series Gold XL D10 wheelchair used in previous studies [23] is commercially controlled by a 5V joystick, which was modified to connect the developed system based on information obtained from the neck pillow system.

B. Methods

The control technique implemented in this research uses a fuzzy logic controller based on the evaluation of motion ranges

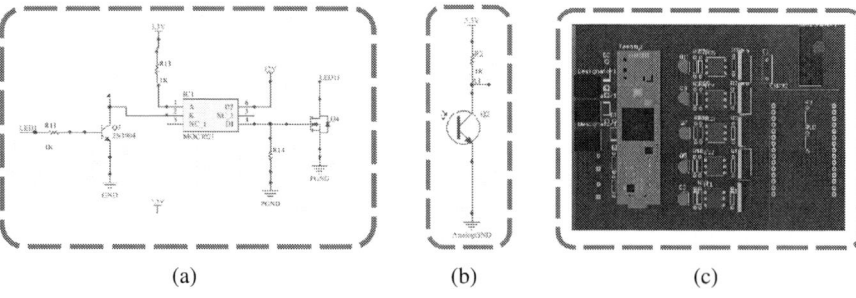

(a)   (b)   (c)

Figure 3: Electronic instrumentation for electrical wheelchair control. (a) LEDs power circuit. (b) PDs circuit. (c) Complete PCB of the neck pillow.

per direction. This type of controller has demonstrated feasibility, speed and accuracy in real-time implementations, using electronic sensors in comparison to other type of controllers [26]. The controller was composed by a pre-processing stage to avoid noise and a membership functions construction which allows the controller tuning.

1) Pre-processing stage: A Direct Current (DC) filter was implemented before the controller implementation. Equation 1 presents the DC filter.

$$V_i^{DC} = V_i - (1/M \sum_{j=1}^{M} V_j), \qquad (1)$$

where V_j represents the vector of M samples, which is used to obtain the mean, and V_i represents the actual value subtracted from the mean value of each M sample.

2) Membership functions construction: The membership functions are based on the selection of values per direction. Therefore, a simulation was conducted using Google Colaboratory to obtain the controller response and analyze the range of values, and with this information, the controller was tuned considering previous studies reported in the literature for wheelchair control design and fuzzy logic [26]. Membership functions for the stop or normal position were created allowing the user to move the neck in case of performing other movements that do not involve the wheelchair controller. In addition, the membership functions also involve wrong-measured values, in whose case a stop command is sent to the wheelchair to avoid wrong executions of directions. A Mandani technique is used for the fuzzy controller using a defuzzified output to command the wheelchair.

3) Performance metrics: For system evaluation, the accuracy metric was used where an experimental design of movements was executed and validated in offline phase. Also, an experimental design for wheelchair control in 4 directions by using neck movements was implemented in real-time where the accuracy was computed to verify the correct intention of the user and the direction executed by the wheelchair. The accuracy was calculated by using the equation 2.

$$Acc = \frac{Number - of - correct - executed - tasks}{Total - tasks} \quad (2)$$

III. RESULTS AND DISCUSSION

The users execute neck movements as shown in Figure 4a) to control the system in different directions, whose signals are adjusted for the membership functions of the fuzzy controller. Initial experiments were conducted with the user sitting outside the wheelchair to analyze the real-time implementation, such as illustrated in Figure 4b), which was mentioned in previous studies [1], [2], [10].

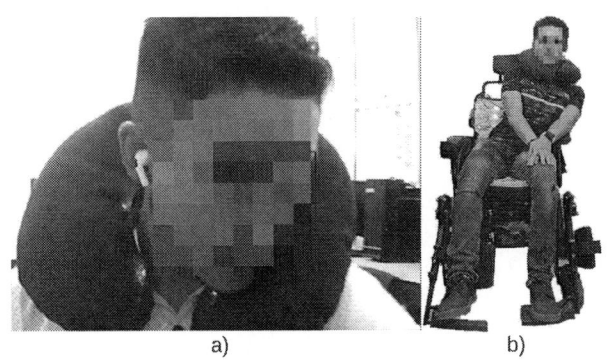

Figure 4: Neck controller implemented. a) User executing a movement; b) User sitting in a wheelchair.

Figure 5 represents the behavior results of the digital signals converted by both photodetectors. This figure shows the difference between the two photodetectors, whose behavior is different owing to their position with respect to the POF-based pressure sensors. From these outcomes, it was possible to find patterns in the signals, which are discriminant to identify the different neck movements. In each figure, the stop position is graphed before the trigger, which was represented at 1000 Kilosamples per second (Ksps), and after the trigger, the neck movement behavior is represented. Figure 5a shows the signal behavior when the wheelchair user is executing the forward neck movement, where a decreasing value measured by the Analog-to-Digital Converter (ADC) of the microcontroller is detected with the Photodetector 2 (PD2). Figure 5b shows the signal behavior when the user is executing the right neck movement, and the ADC value increment was detected with the Photodetector 1 (PD1). Finally, figure 5b represents the signal behavior for left neck movement, which generates an ADC value increase for both Photodetectors (PD1 and PD2). According to these values obtained from the signal analysis, the membership functions were adjusted.

Table I indicates the range of values used per direction in terms of both photodetectors of the line of sensors located in the neck pillow for the fuzzy controller.

Table I: Membership functions values per direction.

DirectionPD values	PD1 [counts]	PD2 [counts]
Forward	[-1000,-600]	[-250,250]
Right	[-2500,-1000]	[-500,0]
Left	[-400,-50]	[1500,3000]
Stop	[-250,300]	[-250,250]
Stop	[-30000,-2500]	[-30000,-800]
Stop	[2000,30000]	[-800,250]
Stop	[300,2000]	[3000,30000]

The controller was tested in real-time and the user was video-recorded to analyze the response of the controller in terms of accuracy and response time. The execution time of the controller was 6 ms, and the execution time per command was 20 ms. Therefore, the total response time of the system was 26 ms, as the signals were read until the wheelchair received

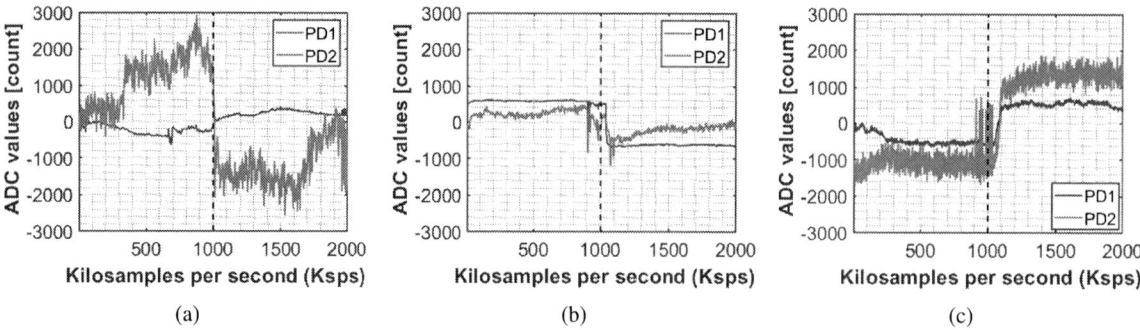

Figure 5: Processed data from neck sensors located in the neck pillow. a) Forward position signals; b) Right position signals; c) Left position signals.

a command. Here, it was possible to highlight an accuracy of 100% in detection, according to the experimental design.

The accuracy achieved in this work is similar to those reported in the literature, considering that several controllers have achieved results of approximately 100% with algorithms of high computational cost or with sensors based on electronic instrumentation [1], [2], [10]. In addition, it is possible to highlight that the response time allows a fast response with adequate accuracy. However, it is worth mentioning that it was possible to observe a response of the POF-based sensors that was affected by the mechanical movement of the wheelchair.

Although the use of OFS has been reported in the literature for both posture monitoring of wheelchair users [20] and for measuring physiological parameters [27], the use of sensors based on optical fibers in wheelchair controllers has not been extensively explored. For instance, some works have conducted experiments with the user seated in the wheelchair, however, in these studies was not possible to evaluate conditions related to vibrations, user physical conditions, and dynamic compensations, among others, which limit the electrical wheelchair usability [1], [2], [10]. In this case, a real-time control strategy was evaluated with the user seated in the wheelchair, but findings generated for the POF-based sensors due to the wheelchair movements were also observed, which to the best of the authors' understanding, has not been reported before in literature. For this reason, future works will focus on the implementation of robust pre-processing techniques using adaptive filters such as Kalman Filters or the use of deep learning-based controllers.

IV. CONCLUSION

The objective of this work was to design a controller based on fuzzy logic for the control of an electric-powered wheelchair by using neck movements and POF-based pressure sensors located in a pillow. The results allow concluding that the methodology is adequate for user intention recognition with an accuracy of 100% and response time of ≈ 26 ms, which is useful for real-time control of devices without involving limb movements.

The use of POF-based pressure sensors provides physical advantages such as electromagnetic immunity, high sensitiv-

ity, and high maneuverability in comparison to conventional sensors reported in the literature. Additionally, the use of fuzzy logic controllers is computationally low-cost compared to AI algorithms, which generates a faster real-time response and improves the interaction between the users and the system. Future studies will focus on the implementation of this strategy with an extension of the controller for speed control and the validation with people with mobility restrictions seated in a wheelchair.

REFERENCES

[1] M. R. Huda, M. L. Ali, and M. S. Sadi, "Real-time hand-gesture recognition for the control of wheelchair," in *2022 12th International Conference on Electrical and Computer Engineering (ICECE)*, 2022, pp. 384–387.

[2] F. B. Haque, T. H. Shuvo, and R. Khan, "Head motion controlled wheelchair for physically disabled people," in *2021 Second International Conference on Smart Technologies in Computing, Electrical and Electronics (ICSTCEE)*, 2021, pp. 1–6.

[3] M. A. M. Azraai, S. Yahaya, I. A. Chong, Z. C. Soh, Z. Hussain, and R. Boudville, "Head gestures based movement control of electric wheelchair for people with tetraplegia," in *2022 IEEE 12th International Conference on Control System, Computing and Engineering (ICCSCE)*, 2022, pp. 163–167.

[4] R. Kumawat, A. Dayal, and S. Srinivasan, "Implementation of self-controlled wheelchairs based on joystick, gesture motion and voice recognition," in *2021 IEEE International Symposium on Smart Electronic Systems (iSES)*, 2021, pp. 240–243.

[5] S. Pawar, S. Pandey, S. Kedar, M. Jadhav, and S. Sahu, "Finger motion controlled biomedical wheelchair," in *2022 5th International Conference on Advances in Science and Technology (ICAST)*, 2022, pp. 637–641.

[6] K. Rajkumar, S. Mohan, A. N. Begum, T. K, and V. U, "Electronic wheelchair with eye motion control for physically hectic people," in *2022 International Conference on Power, Energy, Control and Transmission Systems (ICPECTS)*, 2022, pp. 1–6.

[7] Y. Xu, X. Shi, and Z. Li, "Research on intelligent wheelchair control based on eeg," in *2020 IEEE 9th Joint International Information Technology and Artificial Intelligence Conference (ITAIC)*, vol. 9, 2020, pp. 1620–1627.

[8] T. Saichoo, P. Boonbrahm, and Y. Punsawad, "Facial-machine interface-based virtual reality wheelchair control using eeg artifacts of emotiv neuroheadset," in *2021 18th International Conference on Electrical Engineering/Electronics, Computer, Telecommunications and Information Technology (ECTI-CON)*, 2021, pp. 781–784.

[9] C.-T. Chang, L.-C. Hung, and F.-H. Xu, "Controlling the electric wheelchair using the occlusal myoelectric signal," in *2021 9th International Conference on Orange Technology (ICOT)*, 2021, pp. 1–4.

[10] Sutikno, K. Anam, and A. Saleh, "Voice controlled wheelchair for disabled patients based on cnn and lstm," in *2020 4th International*

979-8-3503-1191-4/23 $31.00 © 2023 IEEE

Conference on Informatics and Computational Sciences (ICICoS), 2020, pp. 1–5.

[11] N. Farheen, G. G. Jaman, and M. P. Schoen, "Object detection and navigation strategy for obstacle avoidance applied to autonomous wheel chair driving," in *2022 Intermountain Engineering, Technology and Computing (IETC)*, 2022, pp. 1–5.

[12] K. Naito, R. Katsube, J. Aoki, K. Cheng, S. Masuko, and T. Uchiyama, "Otomo: Electric wheelchair-type remote communication mobility," in *2022 IEEE 11th Global Conference on Consumer Electronics (GCCE)*, 2022, pp. 513–516.

[13] I. Yamamoto, K. Nakamura, N. Matsunaga, and H. Okajima, "Crowd tracking of electric wheelchair using rgb-d camera with median of candidate vectors observer," in *2022 61st Annual Conference of the Society of Instrument and Control Engineers (SICE)*, 2022, pp. 174–178.

[14] E. Ambarwati, A. Arifin, M. H. Fatoni, I. W. N. B. Pradivta, T. A. Sardjono, F. Arrofiqi, and A. Risciawan, "Subject intention speed control of electric wheelchair for person with disabilities using myoelectric signals," in *2020 International Conference on Computer Engineering, Network, and Intelligent Multimedia (CENIM)*, 2020, pp. 12–17.

[15] P. Roriz, S. Silva, O. Frazão, and S. Novais, "Optical fiber temperature sensors and their biomedical applications," *Sensors*, vol. 20, no. 7, 2020. [Online]. Available: https://www.mdpi.com/1424-8220/20/7/2113

[16] E. Vavrinsky, N. E. Esfahani, M. Hausner, A. Kuzma, V. Rezo, M. Donoval, and H. Kosnacova, "The current state of optical sensors in medical wearables," *Biosensors*, vol. 12, no. 4, 2022.

[17] M. Rocha, C. Tavares, C. Nepomuceno, P. F. d. C. Antunes, M. de Fátima Domingues, and N. J. Alberto, "Fbgs based system for muscle effort monitoring in wheelchair users," *IEEE Sensors Journal*, vol. 22, no. 13, pp. 12886–12893, 2022.

[18] R. Kuang, Y. Ye, Z. Chen, R. He, I. Savović, A. Djordjevich, S. Savović, B. Ortega, C. Marques, X. Li, and R. Min, "Low-cost plastic optical fiber integrated with smartphone for human physiological monitoring," *Optical Fiber Technology*, vol. 71, 2022.

[19] S. Akgönüllü and A. Denizli, "Recent advances in optical biosensing approaches for biomarkers detection," *Biosensors and Bioelectronics: X*, vol. 12, 2022.

[20] A. González-Cely, T. Bastos-Filho, and C. A. Díaz, "Wheelchair posture classification based on pof pressure sensors and machine learning algorithms," in *2022 IEEE Latin American Electron Devices Conference (LAEDC)*, 2022, pp. 1–4.

[21] "IF-D92 plastic fiber optic phototransistor. available online:," https://www.digchip.com/datasheets/parts/datasheet/1777/IF-D92-pdf.php , (accessed on April 10, 2023).

[22] A. X. González-Cely, C. A. R. Díaz, T. F. Bastos-Filho, M. E. V. Segatto, and M. J. Pontes, "Design, manufacturing and characterization of a polymer optical fiber pressure sensor for to posture monitoring applied in a wheelchair," in *Latin America Optics and Photonics (LAOP) Conference 2022*. Optica Publishing Group, 2022, p. M3B.5. [Online]. Available: https://opg.optica.org/abstract.cfm?URI=LAOP-2022-M3B.5

[23] A. González-Cely, A. Natali, C. Díaz, E. Salles, and T. Bastos-Filho, "Design and manufacturing of polymer optical fiber (pof) pressure sensors for user posture monitoring in a wheelchair," in *27th International Conference on Optical Fiber Sensors*. Optica Publishing Group, 2022, p. W4.75. [Online]. Available: https://opg.optica.org/abstract.cfm?URI=OFS-2022-W4.75

[24] L. Avellar, A. Leal-Junior, C. Marques, E. Rocon, and A. Frizera, "Proof-of-concept of pof-based pressure sensors embedded in a smart garment for impact detection in perturbation assessment," in *Converging Clinical and Engineering Research on Neurorehabilitation IV*, D. Torricelli, M. Akay, and J. L. Pons, Eds. Springer International Publishing, 2022.

[25] "Teensy3.6 teensy® 3.6 development board. available online:," https://www.pjrc.com/store/teensy36.html, (accessed on April 10, 2023).

[26] M. Callejas-Cuervo, A. X. González-Cely, and T. Bastos-Filho, "Design and implementation of a position, speed and orientation fuzzy controller using a motion capture system to operate a wheelchair prototype," *Sensors*, vol. 21, no. 13, p. 4344, 2021.

[27] A. Leal-Junior, L. Avellar, J. Jaimes, C. Díaz, W. dos Santos, A. A. G. Siqueira, M. J. Pontes, C. Marques, and A. Frizera, "Polymer optical fiber-based integrated instrumentation in a robot-assisted rehabilitation smart environment: A proof of concept," *Sensors*, vol. 20, no. 11, 2020. [Online]. Available: https://www.mdpi.com/1424-8220/20/11/3199

ACKNOWLEDGMENT

González-Cely A.X. acknowledges the financial support from CAPES (001). C.F.Blanco-Díaz would like to thank the FAPES/I2CA (Resolution N° 285/2021). Camilo A. R. Diaz acknowledges the financial support of FAPES (459/2021), CNPq (310668/2021-2), and MCTI/FNDCT/FINEP (2784/20). Bastos-Filho T. acknowledges the financial support from CNPq (301233/2018-7).

AUTHOR CONTRIBUTIONS

Conceptualization, A.X.G.-C.; methodology, A.X.G.-C., C.A.R.-D. and C.F.B.-D.; software, C.F.B.-D. and A.X.G.-C.; validation, A.X.G.-C., C.A.R.-D., C.F.B.-D. and T.B.-F.; formal analysis, C.A.R.-D., C.F.B.-D. and A.X.G.-C.; investigation, A.X.G.-C., C.A.R.-D., C.F.B.-D. and T.B.-F.; resources, T.B.-F. and C.A.R.-D.; data curation, C.A.R.-D., C.F.B.-D., A.X.G.-C. and T.B.-F.; writing— original draft preparation, A.X.G.-C.; writing—review and editing: C.F.B.-D., C.A.R.-D., A.X.G.-C. and T.B.-F.; visualization,A.X.G.-C.; supervision, A.X.G.-C., C.A.R.-D., C.F.B.-D. and T.B.-F.; project administration, T.B.-F. and C.A.R.-D.; funding acquisition, C.A.R.-D. and T.B.-F..All authors have read and agreed to the published version of the manuscript.

Transconductance-to-Current Ratio Based Threshold Voltage Extraction in MOSFETs from Linear to Saturation Modes

Matthias Bucher[1], Nikolaos Makris[1,2], and Loukas Chevas[1]

[1]School of Electrical & Computer Engineering, Technical University of Crete, GR-73100 Chania, Greece
European University on Responsible Consumption and Production (EURECA-PRO) (Joint affiliation)
[2]Foundation for Research and Technology-Hellas, GR-71110 Heraklion, Greece {mbucher, nmakris, lchevas}@tuc.gr

Abstract—The present work describes a technique for extracting MOSFET threshold voltage from linear to saturation modes. Based on the charge-based model, transconductance-to-current ratio provides a direct way to determine threshold voltage analytically. Experimental data from advanced bulk CMOS processes corroborate the advantages of the present technique, such as consistent estimation of drain induced barrier lowering and reverse short-channel effects.

Index Terms—Charge-Based Model, Compact Model, MOSFET, Parameter Extraction, Threshold Voltage.

I. INTRODUCTION

Transconductance-to-current ratio is a key figure-of-merit (FoM) for MOSFET operation and has therefore been the basis in particular for low-power, analog/RF integrated circuit design for bulk CMOS [1] - [4], as well as SOI CMOS [5], [6]. The moderate inversion region in MOSFETs is the result of the gradual transition from diffusion- to drift-carrier transport, contains the threshold region, and is characterized by reduced impact of high-field effects.

The transconductance-to-current ratio G_m/I_D is furthermore the basis for threshold voltage extraction techniques [7] - [15]. The latter techniques use G_m/I_D, or its derivatives, as a criterion for determining threshold conditions. Some of the above techniques are potentially applicable at low and high drain voltage, however, none of them has been demonstrated in detail over the full range of drain voltage.

Many other threshold voltage extraction techniques exist, most of which are applicable only in a certain region of operation (linear mode or saturation mode) [14]. One example is the well-known Y-function method [16] operating in linear mode. Certain aspects of this method and of some G_m/I_D-related methods have been discussed in [17].

In the present work, an analytical transconductance-to-current ratio based criterion to obtain MOSFET threshold voltage for the whole range of drain voltage, from linear to saturation modes [18] will be presented. The technique

This work was co-funded by the ERASMUS+ Programme of the European Union (Contract number: 101004049 — EURECA-PRO — EAC-A02-2019 / EAC-A02-2019-1).

will be investigated for several different CMOS processes. In particular, the technique is interesting in providing insight into the behavior of halo-doped devices. Furthermore, the technique is used to investigate drain-induced barrier lowering (DIBL) and reverse short-channel effects (RSCE).

II. MOSFET VT EXTRACTION TECHNIQUE BASED ON GM/ID

In the charge-based MOSFET model, the transconductance-to-current ratio is expressed in all regions of operation as [19],

$$\frac{G_m \cdot U_T}{I_D} = \frac{1}{n \cdot (1 + q_s + q_d)} \quad (1)$$

where n is the slope factor and U_T is the thermal voltage. The inversion charge density q_x, at any point x in the channel, is related to pinch-off voltage $V_P = (V_G - V_T)/n$ (where V_T is the threshold voltage) and channel potential V_x via the charge-voltage relationship [20] as,

$$V_P - V_x = U_T \cdot [2q_x + ln(q_x)] \quad (2)$$

which may be solved,

$$q_x = 0.5 \cdot W_0 \left(2e^{(V_P - V_x)/U_T}\right) \quad (3)$$

where W_0 is the Lambert W function. Considering the threshold condition $V_P = V_S$ at both source and drain sides in (3), using $V_{DS} = V_D - V_S$ and inserting the result in (1) we obtain the analytical transconductance-to-current ratio at threshold [18],

$$\left.\frac{G_m \cdot U_T}{I_D}\right|_{V_T} = \frac{1}{n \cdot \left(1.4263 + 0.5 \cdot W_0 \left(2e^{-V_{DS}/U_T}\right)\right)} \quad (4)$$

In (4), the slope factor n may be estimated from the weak inversion slope, as the inverse of the maximum of $G_m U_T/I_D$. Therefore, threshold voltage V_T is easily determined as the gate voltage V_G at which $G_m U_T/I_D$ equals (4), which is valid for the full range of V_{DS}. For every $I_D - V_G$ characteristic at each value of V_{DS}, one value of V_T is determined via the transconductance-to-current ratio only. Note that this procedure is fully analytic.

979-8-3503-1191-4/23 $31.00 © 2023 IEEE

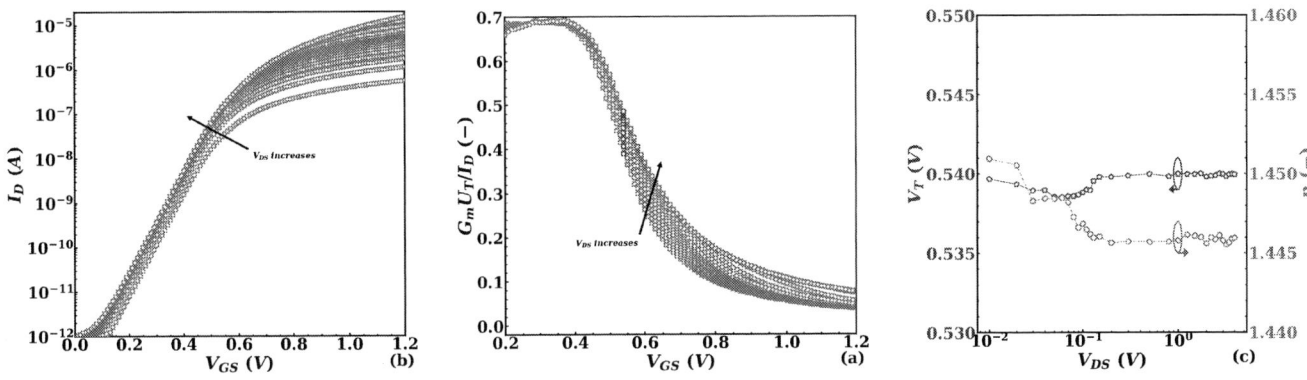

Fig. 1. Extraction of threshold voltage, covering linear to saturation modes for an n-MOSFET ($T_{ox} = 20nm$) with W=20um and L=20um, at room temperature. (a) measured $I_D - V_G$ characteristics at different V_{DS}, (b) corresponding transconductance-to-current ratio $G_m U_T / I_D$ vs. V_G with criterion for the extraction, (c) extracted threshold voltage V_T and slope factor n vs. V_{DS} ranging from $10mV$ to $4V$. Either show a very limited variation over the entire V_{DS} range, which may be attributed mostly to measurement noise.

III. EXPERIMENTAL RESULTS

In this Section, the technique will be demonstrated on different silicon bulk CMOS technologies. DC measurements are performed on-wafer at room temperature, using an HP4142B semiconductor parameter analyzer. All $I_D - V_G$ data (with variable V_{DS} and zero back-bias) are taken with a step of $10mV$ and long integration time. Note the importance of using the derivative of the natural logarithm of I_D to obtain G_m/I_D,

$$\frac{\partial ln(I_D)}{\partial V_G} = \frac{1}{I_D} \cdot \frac{\partial(I_D)}{\partial V_G} = \frac{G_m}{I_D} \quad (5)$$

to reduce noise in weak-moderate inversion regions.

The $I_D - V_G$ characteristics (Fig. 1(a)) for different V_{DS} values ($10mV$ to $4V$) of a long-channel n-MOSFET using thick oxide with uniform doping profile and without source/drain pocket/halo implants, are utilized to validate our approach. From these transfer characteristics we obtain $G_m U_T / I_D$ (Fig. 1(b)) and using (4), we determine the normalized transconductance that corresponds to threshold voltage for each V_{DS} (blue markers). Finally the extracted V_T and n corresponding to each V_{DS} are demonstrated in Fig. 1(c). The extracted quantities deviate marginally from the mean values of $V_T = 539.5mV$ (within less than $\pm 1mV$) and $n = 1.447$ (within $\pm 0.2\%$), respectively, over the range of $V_{DS} = 10mV$ to $4V$. The observed variations may therefore be attributed mostly to measurement noise, which remains in reasonable limits. The constancy of extracted V_T and n over the whole range of V_{DS} is quite remarkable, and this result constitutes an important indication for the consistency of the method. Indeed, in an ideal, uniformly doped, long-channel device without halo/pocket implant, a single value of threshold voltage is expected to hold, irrespectively of applied drain voltage.

The extracted V_T from linear to saturation mode for devices of a 110nm CMOS technology with different channel length (from L=110nm to 8um) are presented using logarithmic and linear x-axis in Fig. 2(a) and (b). V_T in long-channel devices is approximately constant throughout saturation. However, V_T increases as V_{DS} is lowered from saturation towards linear

mode, with the transition occurring at about $V_{DS} \approx 6 \cdot U_T$, which corresponds approximately to saturation voltage in moderate/weak inversion. This effect is similarly present in the linear mode from long- to short-channel devices. The increase of V_T in linear mode with decreasing V_{DS} is attributed to the halo implants of the devices and it is interpreted as a long-channel DIBL or DITS effect [21]. TCAD simulations have confirmed this hypothesis [18]. The extent of this V_T shift at longer channel lengths is about $15mV$, and has a slightly more pronounced amplitude for short-channel devices of about $30mV$. In saturation, DIBL effect dominates in short-channel devices causing a significant drop of V_T, which is almost linearly related to V_{DS}, as seen in Fig. 2(b).

The behavior of weak inversion slope factor determined as $n = [max(G_m U_T / I_D)]^{-1}$ is shown in Fig. 2(d), (e). A general trend is the gradual increase of weak inversion slope factor towards shorter devices. The slope factor n shows a slight increase towards low V_{DS} that is of similar magnitude at all channel lengths. A slight increase in n is also observed at high V_{DS} in short-channel devices. The V_{DS}-dependent variation of the slope factor n is nevertheless quite limited, within typically less than about 5%. This change in slope factor n has a slight impact on the extracted V_T (on the order of a few mV), and hence cannot be held responsible for the overall observed behavior of V_T vs. V_{DS}. The rather low V_{DS}-dependence of the slope factor n could therefore be neglected in the extraction procedure; the channel-length dependence of n is comparatively more significant, on the order of $15...25\%$, as observed in Fig. 2(d), (e).

The scaling of extracted threshold voltage versus channel length is illustrated in Fig. 2(c) for linear ($V_{DS} = 10mV$) and saturation ($V_{DS} = 1.5V$) regions. Reverse short-channel effect, attributed to pocket/halo doping, is clearly apparent. On the other hand, the drain-induced barrier lowering (DIBL) phenomenon, defined as $\eta_{DIBL} = \Delta V_T / \Delta V_{DS}$, is non-linear, as is shown in Fig. 2(a), (b). The DIBL effect evaluated in the saturation part of the device (for V_{DS} ranging from 0.5 to 1.5V) in Fig. 2(f) shows an exponential channel-length

979-8-3503-1191-4/23 $31.00 © 2023 IEEE

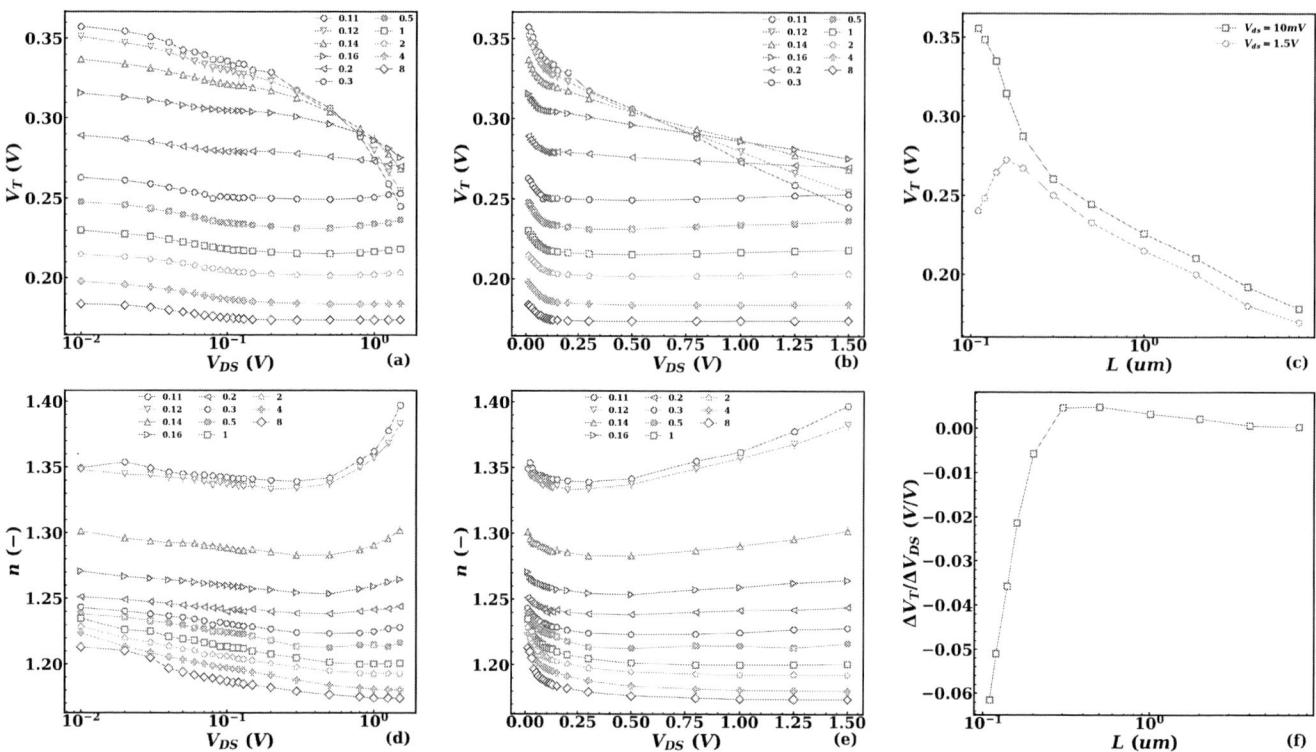

Fig. 2. Extraction of threshold voltage from transconductance-to-current ratio covering linear to saturation modes, for n-MOSFETs of a 110nm CMOS process, with fixed width ($W = 10um$) and different channel lengths. (a),(b) Extracted threshold voltage V_T vs. V_{DS}. (d),(e) Extracted weak inversion slope n vs. V_{DS}. (c) Length scaling of V_T at $V_{DS} = 10mV$ and 1.5V. (f) Length scaling of drain induced barrier lowering $\eta_{DIBL} = \Delta V_T / \Delta V_{DS}$ in saturation.

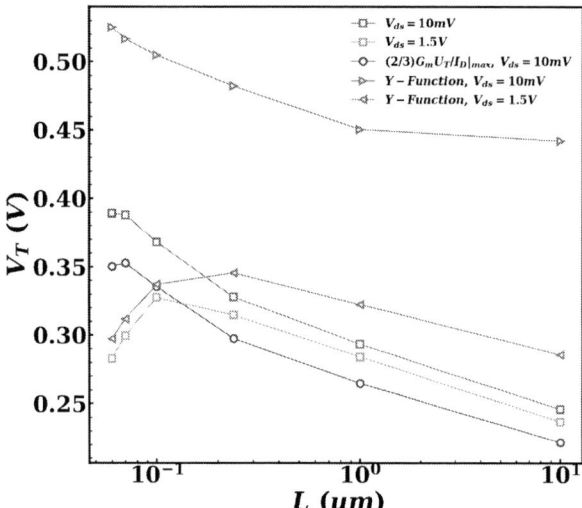

Fig. 3. Extraction of threshold voltage for n-MOSFETs of a 55nm CMOS technology using the present method in linear and saturation modes, [16] and [11] in linear and [14] in saturation modes.

A comparison of different V_T extraction methods applied to n-MOSFETs of a 55nm CMOS technology is demonstrated in Fig. 3. Our transconductance-to-current ratio method shows similar behavior for V_T vs. V_{DS} (not shown here) as the 110nm technology. V_T as extracted in both linear and saturation modes is shown. Threshold voltage is also extracted in linear mode utilizing a criterion of $(2/3)G_m U_T / I_D|_{max}$ as in [11], as well as via the Y-function method [16]. V_T from [11] is behaviorally similar to this work but provides a threshold voltage lower by almost $50mV$ compared to the present method. The Y-function method in linear mode seems to overestimate threshold voltage compared to both other methods. Furthermore, a modified Y-function method as in [14] is evaluated in saturation. V_T from [14] and this work happen to concur for shorter channel lengths, in the range of $L = 60nm$ to $100nm$, but deviate significantly as channel length increases. Clearly, the Y-function method could not provide a consistent information on the DIBL effect, in contrast to the present method.

IV. DISCUSSION

The present technique relies only on measured $I_D - V_G$ characteristics, measured at different values of V_{DS} as required for a specific application. The method derives a threshold voltage criterion directly from transconductance-to-current ratio given in (4), uniquely as a function of drain-source voltage. This

dependence at short-channel, as is expected. The observed behavior in both V_T and n versus V_{DS}, as well as the related scaling graphs for V_T and DIBL factor versus channel length, are expected to be useful in more detailed compact modeling.

979-8-3503-1191-4/23 $31.00 © 2023 IEEE

procedure may be considered as fully analytic by using the Lambert W function, available in mathematics software, to solve the basic voltage-charge relationship. The slope factor n contained in (4) is itself obtained from G_m/I_D evaluated in weak inversion.

As a result, the detailed experiment on a long-channel, practically ideal MOSFET in Fig. 1, shows an extracted V_T within less than $\pm 1mV$ for small V_{DS} steps from $4V$ to $10mV$. Hence, deviations from the nominal value of V_T w.r.t. V_{DS}, may be correctly interpreted in terms of DIBL and halo doping effects, as has been shown in Fig. 2.

The present method provides a framework for consistent modeling and parameter extraction for advanced MOSFETs. Last but not least, due to the inherent relationship of transconductance-to-current ratio with fundamental device physics, namely diffusion and drift transport in FETs, the present method is expected to be applicable also in the context of other types of field-effect transistors.

V. CONCLUSION

In summary, a novel extraction technique for MOSFET threshold voltage has been demonstrated analytically and has been confirmed on different CMOS technologies. Threshold voltage is obtained from measured $I_D - V_G$ characteristics over the whole range of drain voltages. The criterion for threshold voltage corresponds to the gate voltage at which the transconductance-to-current ratio G_m/I_D reaches a certain fraction with respect to its maximum in weak inversion. The criterion for G_m/I_D is obtained from the general charge-based model. The technique provides a detailed analysis of threshold voltage, from V_{DS} as low as a few mV to V_{DD}. As the extraction takes place in moderate inversion, the method has the advantage of limited impact of high-field effects, in particular vertical field mobility dependence, as well as series resistance. Due to its aptitude to cover the full range of V_{DS}, a detailed analysis of DIBL effect has been possible. The method reveals a distinct increase of V_T, on the order of $20mV$, present at all channel lengths, when V_{DS} is lowered below $6 \cdot U_T$ (non-saturated condition). This effect is attributed to halo doping. The technique is expected to be useful in detailed device compact modeling and related parameter extraction.

REFERENCES

[1] F. Silveira, D. Flandre and P. G. A. Jespers, "A g_m/I_D based methodology for the design of CMOS analog circuits and its application to the synthesis of a silicon-on-insulator micropower OTA," in IEEE Journal of Solid-State Circuits, vol. 31, no. 9, pp. 1314-1319, Sept. 1996. doi: 10.1109/4.535416.

[2] M. Bucher, D. Kazazis, F. Krummenacher, D. Binkley, D. Foty and Y. Papananos, "Analysis of transconductance at all levels of inversion in deep submicron CMOS", in Proc. IEEE Int. Conf. on Electronics, Circuits and Systems (ICECS), vol. 3, pp. 1183-1186, Sept. 2002. doi: 10.1109/ICECS.2002.1046464.

[3] C. C. Enz and E. A. Vittoz, "CMOS low-power analog circuit design," Emerging Technologies: Designing Low Power Digital Systems, 1996, pp. 79-133. doi: 10.1109/ETLPDS.1996.508872.

[4] C. Enz, F. Chicco and A. Pezzotta, "Nanoscale MOSFET modeling: Part 2: Using the inversion coefficient as the primary design parameter," IEEE Solid-State Circuits Mag., vol. 9, no. 4, pp. 73-81, 2017. doi: 10.1109/MSSC.2017.2745838.

[5] S. El Ghouli, D. Rideau, F. Monsieur, P. Scheer, G. Gouget, A. Juge, T. Poiroux, J.-M. Sallese and C. Lallement, "Experimental gm/ID invariance assessment for asymmetric double-gate FDSOI MOSFET," IEEE Trans. Electron Devices, vol. 65, no. 1, pp. 11-18, Jan. 2018. doi: 10.1109/TED.2017.2772804.

[6] J. R. R. de Oliveira Martins, A. Mostafa, J Juillard, R. Hamani, F. de Oliveira Alves and P. M. Ferreira, "A Temperature-Aware Framework on gm/ID-Based Methodology Using 180 nm SOI from 40 C to 200 C," IEEE Open Journal of Circuits and Systems, vol. 2, pp. 311-322, 2021. doi: 10.1109/OJCAS.2021.3067377.

[7] K. Aoyama, "A method for extracting the threshold voltage of MOSFETs based on current components," in H. Ryssel, P. Pichler (Eds.) Simulation of Semiconductor Devices and Processes, pp. 118–121, Springer, Vienna, 1995. doi: 10.1007/978-3-7091-6619-2_28.

[8] M. C. Schneider, C. Galup-Montoro, M. B. Machado and A. I. A. Cunha, "Interrelations between threshold voltage definitions and extraction methods", Workshop on Compact Models - NSTI Nanotech 2006, vol. 3, pp. 868-871, 2006. [Online Available:] briefs.techconnect.org.

[9] D. Flandre, V. Kilchytska, and T. Rudenko, "gm/Id method for threshold voltage extraction applicable in advanced MOSFETs with nonlinear behavior above threshold", IEEE Electron Device Lett., vol. 31, no. 9, pp. 930-932, Sept. 2010. doi: 10.1109/LED.2010.2055829.

[10] A. Bazigos, M. Bucher, J. Assenmacher, S. Decker, W. Grabinski and Y. Papananos, "An adjusted constant-current method to determine saturated and linear mode threshold voltage of MOSFETs", IEEE Trans. Electron Devices, vol. 58, no. 11, pp. 3751–3758, Nov. 2011. doi: 10.1109/TED.2011.2164080.

[11] T. Rudenko, V. Kilchytska, M. K. Md Arshad, J.-P. Raskin, A. Nazarov and D. Flandre, "On the MOSFET threshold voltage extraction by transconductance and transconductance-to-current ratio change methods: Part II—Effect of drain voltage", IEEE Trans. Electron Devices, vol. 58, no. 12, pp. 4180-4188, Dec. 2011. doi: 10.1109/TED.2011.2168227.

[12] A. I. A. Cunha, M.-A. Pavanello, R. D. Trevisoli, C. Galup-Montoro, M. C. Schneider, "Direct determination of threshold condition in DG-MOSFETs from the gm/ID curve", Solid-State Electron., vol. 56, pp. 89–94, 2011. doi: 10.1016/j.sse.2010.10.011.

[13] O. Franca Siebel, M. C. Schneider and C. Galup-Montoro, "MOSFET threshold voltage: definition, extraction, and some applications", Microelectron. J., vol. 43, no. 5, pp. 329–336, May 2012. doi: 10.1016/j.mejo.2012.01.004.

[14] A. Ortiz-Conde, F. J. García-Sánchez, J. Muci, A. Terán Barrios, J. J. Liou and C.-S. Ho, "Revisiting MOSFET threshold voltage extraction methods", Microelectron. Reliab., vol. 53, no. 1, pp. 90–104, 2013. doi: 10.1016/j.microrel.2012.09.015.

[15] M. Bucher, N. Makris and L. Chevas, "Generalized constant current method to determine MOSFET threshold voltage", IEEE Trans. Electron Devices, vol. 67, no. 11, pp. 4559-4562, Nov. 2020. doi: 10.1109/TED.2020.3019019.

[16] G. Ghibaudo, "New method for the extraction of MOSFET parameters", Electronics Lett., vol. 24, no. 9, pp. 543-545, April 1988. doi: 10.1049/el:19880369.

[17] Y. Cheng, "Comparison of MOSFET threshold voltage extraction methods with temperature variation", Proc. Int. Conf. on Microelectronic Test Structures (ICMTS), pp. 126-131, Kitakytushu, Japan, 2019. doi: 10.1109/ICMTS.2019.8730978.

[18] N. Makris and M. Bucher, to be published.

[19] C. Enz and E. Vittoz, Charge-based MOS Transistor Modeling, The EKV model for low-power and RF IC deisgn, John Wiley and Sons, Chichester, 2006. doi: 10.1002/0470855460.

[20] J.-M. Sallese, M. Bucher, F. Krummenacher and P. Fazan, "Inversion charge linearization in MOSFET modeling and rigorous derivation of the EKV compact model", Solid-State Electron., vol. 47, no. 4, pp. 677–683, 2003. doi: 10.1016/S0038-1101(02)00336-2.

[21] A. S. Roy, S. P. Mudanai and M. Stettler, "Mechanism of long-channel drain-induced barrier lowering in halo MOSFETs", IEEE Trans. Electron Devices, vol. 58, no. 4, pp. 979-984, Apr. 2011. doi: 10.1109/TED.2011.2109387.

Digital twin of electrical motorcycle battery charger as AC Load in a Microgrid Based on Renewable Energy

Alcides Amado Herrera-Guerra
Dept. of Electrical Engineering
Universidad Tecnológica de Panamá
Panamá, Panamá
alcides.herrera@utp.ac.pa
ORCID: 0009-0000-2486-0388

Elkin Edilberto Henao-Bravo
Dept. of Mechatronics-Electromechanics
Instituto Tecnológico Metropolitano
Medellín, Colombia
elkinhenao@itm.edu.co
ORCID: 0000-0001-9663-1082

Juan Pablo Villegas-Ceballos
Dept. of Electronics-Telecommunications
Instituto Tecnológico Metropolitano
Medellín, Colombia
juanvillegas@itm.edu.co
ORCID: 0000-0002-2598-4486

Abstract—Advances in power electronics and improved battery autonomy have made electric vehicles (EVs) an increasingly feasible solution for mitigating greenhouse gases (GHG) levels. Due to increased energy demand, its massive arrival can impact traditional distribution and generation systems. Due to their characteristics, microgrids (MGs) are postulated as a critical element in solving this problem. To estimate the impact of electric vehicles connected to the microgrid of the "Electronics and Renewable Energy Laboratory" of the Metropolitan Technological Institute (ITM) and emulate different scenarios, this work presents the digital twin of an EV charger developed based on the modeling of the system and the control design. PSIM software allows the validation of the digital twin operation. Actual parameters of an EV battery charger allow for generating a database intended to propose different test scenarios that validate the operation of the MG.

Index Terms—electric vehicle, microgrid, battery charger, PFC, cascade control.

I. INTRODUCTION

Greenhouse gases (GHG) are chemical compounds in a gaseous state that is capable of absorbing infrared radiation from the sun, which increases the temperature of the atmosphere. The GHG are CO_2, CH_4, N_2O, H_2O, O and fluorinated gases. In the case of the Aburrá Valley, mobility vehicles based on internal combustion engines provide the 99.89% of methane, the 80.7% of carbon dioxide, and 65.5% of nitrous oxide, making them the main emitting source of GHG [1], [2].

Several researchers have reported that air pollution has become a problem linked to emissions from mobility sources. According to the Comptroller General of Medellin, in Aburrá Valley, there is an annual growth of 9.5% of vehicles based on internal combustion engines, especially motorcycles, which is 4% higher than the world average. This annual increase, together with poor controls and maintenance, causes an increase in the levels of CO_2 and other pollutants in the atmosphere of the region [3], [4].

Panama also presents serious problems related to air pollution in the country. According to the National Institute of Statistics and Census (INEC), the high vehicular flow in the country is one of the bigger sources of air pollution, which increases rapidly [5]. Official reports show that mobility sources represent the 95% of air pollutants emissions, with an annual increase of approximately 50.000 metric tons from 2013 to 2017. On the other hand, during this period, air pollution is composed of a 70% of carbon monoxide (CO), a 15.5% of oxides nitrous (NOx), a 9.3% of hydrocarbons (HC), a 2.8% of suspended particles (PST) and at 2.6% of sulfates (SOx), with CO being the country's largest atmospheric pollutant.

Electric vehicles (EVs) postulate as a mobility alternative to reduce GHG emissions, their massive incorporation requires an increase in the energy demand from the electrical distribution systems to recharge the battery of these vehicles.

To support the electrical grid due to the demand increase, some work such as [6]–[9] propose the electric microgrids (MGs) based on renewable energy sources (RES) to supply the energy demanded by the recharging stations for EVs. MGs provide efficient energy management and increase the participation of RES in electricity generation, providing a solution to the GHG problem [10].

Achieving optimal integration of EVs to MGs requires evaluating different management strategies at several power flow operating conditions; assessing the MG response when charging different electric vehicle models is also essential to achieve optimal integration of EVs in MGs.

This work presents a digital twin of an EV battery charger of an actual electric motorcycle to evaluate the incorporation of EV charging systems into a real microgrid available in the Electronics and Renewable Energies laboratory located in *Parque i* at the *Instituto Tecnológico Metropolitano (ITM)* from Medellin-Colombia. The digital twin allows for emulating any electric motorcycle charging system based on the charger model.

The organization of the document is as follows: Section

979-8-3503-1191-4/23 $31.00 © 2023 IEEE

2023 IEEE Latin American Electron Devices Conference (LAEDC)
Puebla, México, July 3-5, 2023

II presents the methodology used to develop the modeling, simulation, and charging data acquisition of the charger for EVs. Subsection II-A shows the results and discussions of the modeling of the charger for electric modeling EVs, presenting its respective averaged model of the converter and the state space model; later in Section III, the results of the simulation and the analysis of its behavior are shown. Completing section III are the results of the data acquisition obtained during this work and the results of the inclusion of EV chargers as a new charge to the MG. Section IV presents the conclusions of this work, and the Future work close the paper.

II. METHODS

The Electronics and Renewable Energies laboratory located in *Parque i* at the *Instituto Tecnológico Metropolitano (ITM)* from Medellin-Colombia, has a nonisolated MG of **5.2 kW**, with **3.6 kW** from photovoltaic panels; the MG has a monitoring system to record variables as power, voltage, among others, the storage data already available covers more than three years. Currently, the MG supplies the lighting demands of the laboratory. However, there is a need to generate new test environments that validate the MG's behavior in the face of new parameter variations, such as integrating other types of loads.

To improve the existing MG management system facing the addition of new loads is necessary to test changes in the MG management algorithms. Therefore, it is necessary to generate different scenarios to validate the correct functioning of the management algorithms. For testing proposes, the emulation based on digital twins allows modifying variables that affect the MG so that the management algorithm reacts and interacts with these changes; this procedure allows for the performance evaluation of the MG management algorithm.

This work focuses on implementing variations in the load of the MG by introducing a new load, which corresponds to a battery charger for electric motorcycles. Fig. 1 shows the schematic of the MG in the Electronics and Renewable Energies laboratory in conjunction with the digital twin (DT) that emulates the battery charger system.

The motorcycle STARKER is the best-selling electric motorcycle brand in Colombia. The Avanti 2.0 model has standard technical parameters for all models of this brand and even for other different brands [11], [12]. Therefore, the STARKER Avanti 2.0 electric motorcycle is the reference used to emulate its battery charger and connect to the MG. In Medellin-Colombia, the most used vehicle is the motorcycle, representing the **60%** vehicles in the city; therefore, an electric motorcycle is a suitable EV to reduce GHG emissions.

The manufacturer data highlights that the STARKER Avanti 2.0 model's charger must connect to an AC network of **110 V** at **60 Hz**. The output must supply DC capable of recharging a battery of **72 V**, and **32 Ah** [12]. In addition, the charger must fulfill the standards that regulate the power factor (PF) levels and the total harmonic distortion (THD) established in article 25 of CREG Resolution 108 of 1997 *"Power Factor*

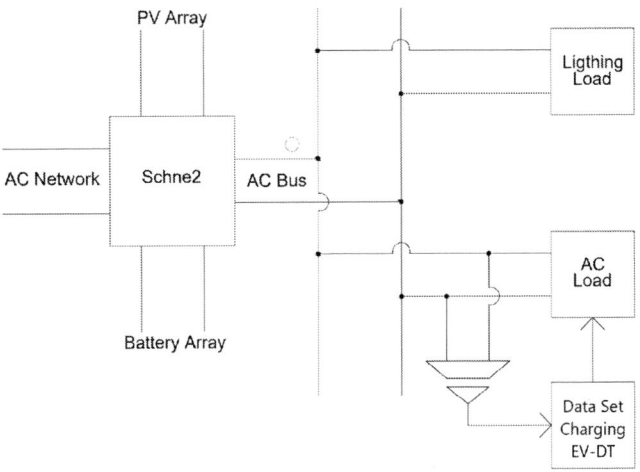

Figure 1. Microgrid including chargers for EVs.

Control in the Electric Power Service" and IEEE Standard 519-1992 respectively [13].

After choosing the EV and determining its nominal data, it is necessary to design the charger to include in the MG through a digital twin. This work selects the constant voltage model within the existing recharging methods recommended for slow EV recharging. This selection considers that EVs have different types of recharging; this depends on their power and charging time, classified as slow, semi-fast, and fast recharging [14]. According to the manufacturer's data, the motorcycle selected uses a slow type recharge; for this reason, the current supplied by the charger to the motorcycle battery (charging current) will be **10 %** of its nominal capacity; meanwhile, the voltage at the charger output terminals (charging voltage) will be a **10 %** greater than the nominal of the battery. [15].

Simulation of the EV battery's complete recharging process allows testing the charger's operation and storage power, voltage, current, and SOC data obtained during recharging. All the simulation results were stored to create a digital twin of the battery charger for further testing in the existing MG. The following subsections show the battery charger's modeling, control design, and validation. MATLAB software [16] helps in the modeling and control design process, while PSIM power electronics specialized software allows for validation in a simulation environment.

A. Modeling charger for electric motorcycles

Taking into account the input and output parameters of the charger, Fig. 2 shows the proposed topology. The circuit consists of a voltage reduction stage (transformer) that reduces the magnitude of the voltage in a ratio of $1 : \frac{1}{4}$. The next stage is a rectification (rectifier) responsible for converting the AC signal from the transformer to a DC. Finally, a stage sets the output voltage (boost converter) to provide a stable output voltage to the motorcycle battery. Finally, through the control stage, the converter maintains a constant output voltage and an input current in phase with the voltage to guarantee a unitary PF.

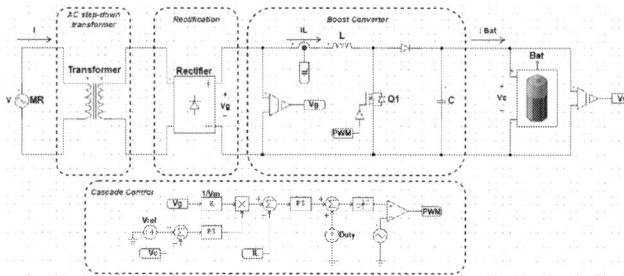

Figure 2. Proposed EV charger

Equations (1) and (2) show the averaged model that represents the behavior of the inductor current (L) and the capacitor voltage (C) of the boost converter, where: v_g is the average voltage at the input of the converter, v_c is the voltage across the capacitor, i_l is the current in the inductor, i_{Bat} is the battery charging current, d is the duty cycle and "D" is the duty cycle at the operating point.

$$L\frac{di}{dt} = v_g - v_c \cdot (1 - d) \qquad (1)$$

$$C\frac{dv}{dt} = i_l \cdot (1 - d) - i_{Bat} \qquad (2)$$

For control purposes, the load R replaces the battery; therefore, using Ohm's law [17], Equation (2) yields Equation (3).

$$C\frac{dv}{dt} = i_l \cdot (1 - d) - \frac{v_c}{R} \qquad (3)$$

Equations (1) and (3) allow to obtain the state-space model. The state space matrices are shown in Equations (4), (5). (6), (7).

$$Am \begin{bmatrix} \dot{i_l} \\ \dot{v_c} \end{bmatrix} = \begin{bmatrix} 0 & \frac{-(1-D)}{l} \\ \frac{1-D}{C} & \frac{-1}{R \cdot C} \end{bmatrix} \qquad (4)$$

$$Bm = \begin{bmatrix} \frac{V_C}{L} & \frac{1}{L} \\ \frac{-I_L}{C} & 0 \end{bmatrix} \begin{bmatrix} d \\ V_G \end{bmatrix} \qquad (5)$$

$$Cm = \begin{bmatrix} 1 & 0 \end{bmatrix} \qquad (6)$$

$$Dm = \begin{bmatrix} 0 & 0 \end{bmatrix} \qquad (7)$$

Two PI controllers in a cascade configuration regulate the output voltage and the input current for power factor correction. The inner loop controls i_l by defining d, the transfer function (TF) that relates i_l vs d has two poles and one zero on the left side of the complex plane; therefore the relationship i_L vs d is stable at all operating points. The outer loop keeps the voltage at the converter's output in the reference value by defining the i_l reference. In this way, the outer controller imposes a reference current on the inner controller allowing a dynamic system capable of maintaining

a constant output voltage and regulating the PF and THD at the circuit input. The outer controller was designed from the first-order TF that relates v vs i_l; this TF has a negative pole which guarantees a stable relationship between the mentioned variables. In this way, the complete stability of the controlled system is guaranteed.

B. Inclusion of digital twin of the EV battery charger to the MG

The intention of developing a digital twin of the battery charger for electric motorcycles is to include the charging profile within the data of the MG operating in the ITM's Electronics and Renewable Energies laboratory. Therefore, after validating the proper operation of the charger, two different charging scenarios will be simulated to include the data in the MG. The scenarios considered are charging one motorcycle starting from **0%** until reaching the **100%** starting at **9:00 am**. Later, at **7:00 pm**, charging is done on two motorcycles, with the difference that this time their EV batteries charge from **0%** at **50%**. The resulting data from the simulation scenarios can be included in the actual data of the MG through Matlab software to define the load profile, allowing future work to enhance the management algorithms in the MG.

III. RESULTS AND DISCUSSION

A. EV battery charger simulation and data acquisition

Before the modeling and control design, the system simulation in PSIM software shows in Fig. 3 that the EV charger maintains the average output voltage of **79.2V**, **10%** above the nominal battery voltage, as established in the design parameters. Regarding the input parameters to the charger, in Fig. 4, it can be seen that the the charger does not alter the waveform of the input current and voltage, and also maintains a PF of **0.99** and a THD of current at **60Hz** de **3.48681** $\times 10^{-2}$. Those simulation results confirm the proper operation of the EV battery charger.

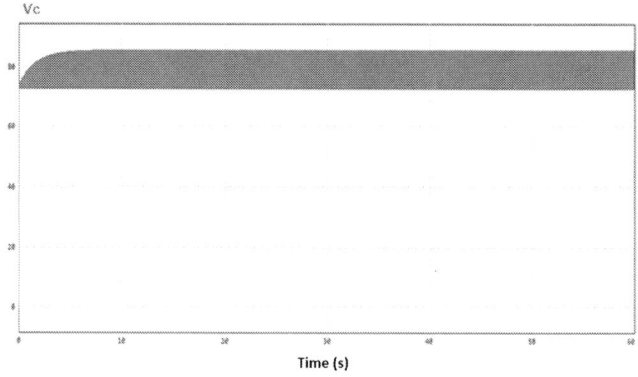

Figure 3. Charger's output voltage (V_c).

Once the correct operation of the charger was verified, the simulation of the complete battery recharge with the specifications of the Avanti 2.0 motorcycle was conducted. The simulation was carried out during **12 hours** and collected

2023 IEEE Latin American Electron Devices Conference (LAEDC)
Puebla, México, July 3-5, 2023

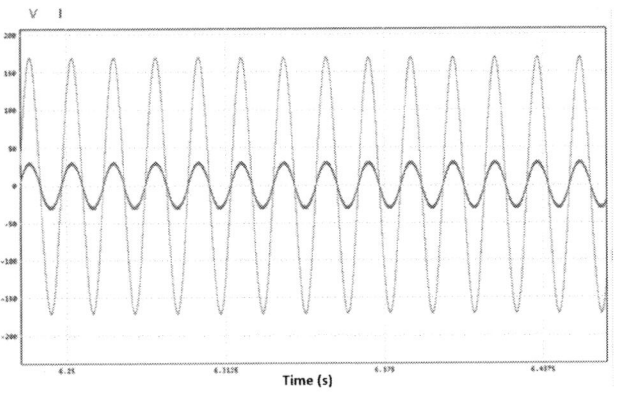

Figure 4. Waveform of charger's input voltage and current.

43200 data, i.e., one data per second. Recharging started with the battery charge at **10%** and ended by completing the **100%** of its capacity.

Fig. 5 shows the power curves demanded by the charger to the MG and the state of charge of the motorcycle battery (SOC). The average power demanded was **265W**, and the recharge takes **9** hours with **28** minutes; these results coincide with the power and charging time ranges established by the motorcycle manufacturer [12].

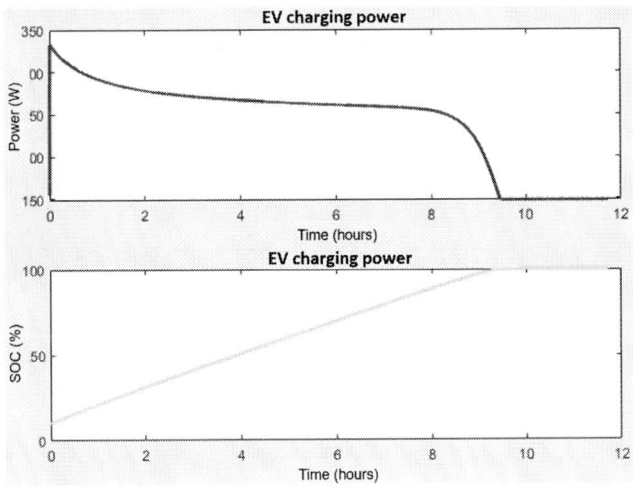

Figure 5. Recharge power and SOC obtained in the simulation.

B. Modified MG data because of the inclusion of the digital twin of the EV battery charger

As a final point of the results, in Fig. 6, the graphs of power demand to the MG are shown. The upper part of Fig. 6 is the power demand during a day without including EV chargers. Conversely, the lower graph shows the demand for the same day with chargers, according to Subsection II-B. Results show that the demand varies with the **9:00 am** and to the **7:00 pm** as expected; the graphs also reflect a correct inclusion of EV battery chargers in the demand curve.

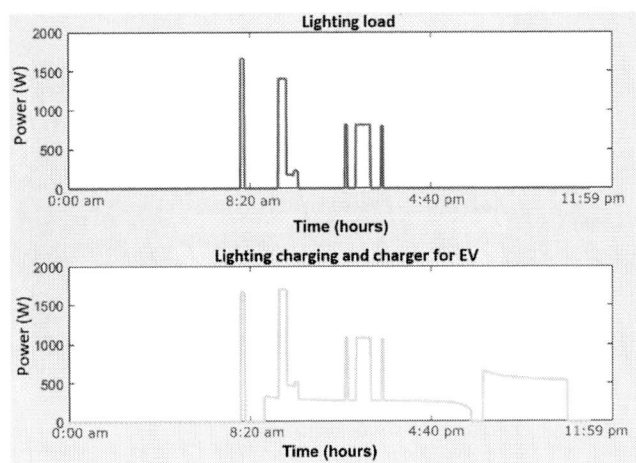

Figure 6. EV battery charger inclusion as load in the MG.

IV. CONCLUSIONS

In search of generating a database to emulate the MG of the ITM's Electronics and Renewable Energies laboratory, a digital twin of an EV battery charger for electric motorcycles was developed with the help of PSIM and MATLAB. The simulation of the complete recharge of the EV battery based on the proposed system shows that the recharge took **9** hours with **28** minutes at an average power of **265W**. The Waveform of the charger's input voltage and current shows a PF of **0.99** and a THD of $\mathbf{3.487224 \times 10^{-4}\%}$. This way, it is concluded that it meets the EV manufacturer's specifications and complies with international power quality standards. The recharging simulation allows the extraction of **43200** data on power, current, voltage, and state of charge, both at the input and output of the charger. Therefore, the acquired data was successfully included in a simulation environment as a new load to the MG. The future work is to obtain a digital twin of the MG to test management algorithms using DT, analyzing the current algorithm's performance in several scenarios and, if necessary, making modifications to the algorithm or proposing a new one in search of obtaining maximum efficiency in MG.

ACKNOWLEDGMENT

This work was supported by Minciencias, Universidad Nacional de Colombia, Universidad del Valle, and Instituto Tecnológico Metropolitano under the research project "Estrategias de dimensionamiento, planeación y gestión inteligente de energía a partir de la integración y la optimización de las fuentes no convencionales, los sistemas de almacenamiento y cargas eléctricas, que permitan la generación de soluciones energéticas confiables para los territorios urbanos y rurales de Colombia", (Minciencias code 71148), which belongs to the research program "Estrategias para el desarrollo de sistemas energéticos sostenibles, confiables, eficientes y accesibles para el futuro de Colombia", (Minciencias code 1150-852-70378, ITM code RC 80740-178-2021-1).

979-8-3503-1191-4/23 $31.00 © 2023 IEEE   43

REFERENCES

[1] C. G. de Medellín, "Estado anual de los recursos naturales y del ambiente del municipio de medellín," Aviable: https://www.cgm.gov.co/cgm/Paginaweb/IP/Informe%20Ambiental%202019/Informe%20Ambiental%20Vigencia%202019.pdf, 2019.

[2] U. P. Bolivariana, "Contrato de ciencia y tecnología n° 1179 de 2018," Aviable: https://www.metropol.gov.co/ambiental/calidad-del-aire/Documents/Inventario-de-emisiones/Inventario-de-Emisiones-2018.pdf, 2019.

[3] M. V. Toro, J. J. Ramírez, R. A. Quiceno, and C. A. Zuluaga, "Cálculo de la emisión vehicular de contaminantes atmosféricos en la ciudad de medellín mediante factores de emisión corinair," *Revista Acodal*, vol. 191, pp. 42–49, 2001.

[4] M. CÓMOVAMOS, "Informe de calidad de vida de medellín," Aviable: https://www.medellincomovamos.org/system/files/2020-04/docuprivados/Informe%20de%20indicadores%20objetivos%20sobre%20c%C3%B3mo%20vamos%20en%20movilidad%20y%20espacio%20p%C3%BAblico%2C%202018.pdf, 2018.

[5] I. N. de Estadística y Censo, "Estadísticas ambientales 2013-17," Aviable: https://www.inec.gob.pa/publicaciones/Default3.aspx?ID_PUBLICACION=938&ID_CATEGORIA=16&ID_SUBCATEGORIA=49, 2021.

[6] M. M. Llano, "La micro-red inteligente: una ciudad eficiente, en miniatura," *Universitas Científica*, vol. 18, no. 1, pp. 24–29, 2015.

[7] W. E. G. Muñoz, A. F. Rodríguez, L. M. Gómez, F. Santamaría, and C. Trujillo, "Desarrollo de un prototipo de micro-red residencial a baja escala," *TecnoLógicas*, vol. 21, no. 43, pp. 107–125, 2018.

[8] W. Cieslik, F. Szwajca, W. Golimowski, and A. Berger, "Experimental analysis of residential photovoltaic (pv) and electric vehicle (ev) systems in terms of annual energy utilization," *Energies*, vol. 14, no. 4, p. 1085, 2021.

[9] M. Z. Bernal Verdugo, "Dimensionamiento de una micro-red tipo estación de recarga para vehículos eléctricos conectados al sistema eléctrico." B.S. thesis, 2020.

[10] B. Hartono, Y. Budiyanto, and R. Setiabudy, "Review of microgrid technology," in *2013 international conference on QiR*. IEEE, 2013, pp. 127–132.

[11] AutoCRASH, "La actualidad de las motos eléctricas en colombia," Aviable: https://www.revistaautocrash.com/la-actualidad-de-las-motos-electricas-en-colombia/, 2021.

[12] A. Mobility, "Starker avanti 2.0," Aviable: https://www.autecomobility.com/moto-electrica-starker-avanti-2-0/p, 2020.

[13] J. W. Parra Betancur, H. R. Toro Bolívar *et al.*, "Estudio de distorsión armónica y factor de potencia en la red de alimentación eléctrica de una máquina papelera," 2009.

[14] J. E. Vallejo Ruiz, "Situación de la ciudad de medellín en cuanto a la capacidad que tiene en infraestructura de electrolineras para recargar de energía a los vehículos eléctricos," 2017.

[15] J. C. Ramírez García, C. C. Paipa Bocanegra *et al.*, "Análisis y evaluación del comportamiento del thdi a causa de la conexión masiva de vehículos eléctricos en una estación de carga."

[16] Z. Chen, W. Gao, J. Hu, and X. Ye, "Closed-loop analysis and cascade control of a nonminimum phase boost converter," *IEEE Transactions on power electronics*, vol. 26, no. 4, pp. 1237–1252, 2010.

[17] G. Mantilla Quijano and H. González, "Ley de ohm (aplicación)," 1977.

Non-iterative parameter extraction method based on the single diode model (SDM).

Manuel J Heredia-Rios, Luis Hernandez-Matinez, Monico Linares-Aranda, Mario Moreno-Moreno

Instituto Nacional de Astrofísica Optica y Electronica

dept. Electronics

Puebla, Mexico

manuel.heredia@susu.inaoep.mx, luish@inaoep.mx, mlinares@inaoep.mx, mmoreno@inaoep.mx

Abstract— **Currently, the analysis of photovoltaic devices that convert light energy into electrical energy is of great importance, as efforts are being made to reduce global pollution. In this regard, various tools such as software, models, and simulators have been developed that allow for the understanding and improvement of the efficiency of these devices. Such tools are a valuable resource for research and development in the field of photovoltaic solar energy. This paper presents a model that allows for the analysis of the I-V and P-V curves' behavior under different radiation and temperature conditions, as well as the influence of parameters that describe the single diode model. For this purpose, the commercial RTC France solar cell and the Kyocera 200GT solar panel, which are among the most reported in the literature, were used as study objects. The results obtained through the proposed model contribute to the understanding of the factors that affect the efficiency of photovoltaic devices and can be of great utility for the design and optimization of solar energy systems.**

I. INTRODUCTION

The model of a photovoltaic (PV) device is commonly represented by an equivalent circuit and a set of parameters that describe its electrical response and operation. The determination of these parameters is not a simple task since they are not available in the photovoltaic module information and their values vary according to the operating conditions. Research in recent decades has focused on the extraction of these parameters from the photovoltaic model, and there are several methods available that differ in their nature, reliability, complexity, and input data requirements. The extraction of parameters from the photovoltaic cell model is considered a research topic in itself and is known in the literature by similar names such as the "photovoltaic cell model parameter estimation problem" [1].

There are three main categories of methods used for parameter extraction in photovoltaic generator models: non-iterative, numerical (or iterative), and optimization. Non-iterative methods are characterized by symbolically or explicitly solving a set of equations, resulting in a simpler formulation and implementation [2-5]. In contrast, numerical (or iterative) methods use a system of equations that is numerically solved using trial and error or some other iterative algorithm. These equations are derived from the photovoltaic model equation applied to specific conditions such as short-circuit (SC), open-circuit (OC), or maximum power point (MPP). Although this category can achieve high accuracy, it often presents initialization and convergence difficulties, high computational

load, and non-optimal solutions [6-10].

Optimization methods, on the other hand, fit the model equation to a set of measurements using evolution, genetic, particle, swarm, or curve fitting algorithms, providing high accuracy and nearly global optimization but requiring computational complexity and difficulties in parameter tuning [11-17].

In [18], 52 research articles are reviewed, and it is determined that the most commonly used models are the Single Diode Model (SDM) and the Double Diode Model (DDM), with a 69.23% match. In contrast, one of the most used panels for model validation is the Kyocera KC200GT panel, present in 30% of the reported articles, followed by the RTC France cell with 24%.

In this paper, the methodology established to obtain the proposed model is based on the analysis of the I-V (current-voltage) curve image provided by the manufacturer. From this curve, key points are identified and lines are drawn to determine the slopes in the short-circuit current (Isc) and open-circuit voltage (Voc) branches. These slopes allow for a direct calculation of the Rs, Rp, and n. Using this approach a direct comparison is achieved between the I-V curve of the proposed model and the curve provided by the manufacturer. This provides real-time information about the behavior of the I-V and P-V (power-voltage) curves of the proposed model. Additionally, the influence of temperature and radiation on the efficiency of solar devices can be evaluated through the analysis of these parameters.

This proposed non-iterative model presents significant advantages in terms of computational efficiency and accuracy in determining the Rs, Rp, and n parameters. By avoiding the iterative process, the time and resources required to obtain accurate results are reduced. Furthermore, by directly calculating these parameters, potential errors introduced by approximate or iterative methods are minimized.

II. PROPOSED METHOD

The single diode model is shown in Figure 1, described by five parameters:

Ipv = photocurrent

IO1 = diode saturation current

n = diode ideality factor

Rs = series resistance

Rp/Rsh = parallel resistance

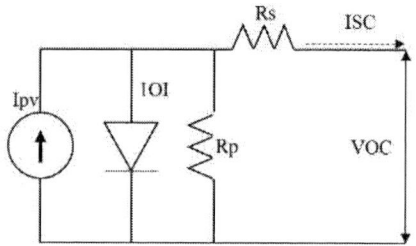

Fig. 1. Equivalent circuit for photovoltaic devices (single diode model).

The equation (1) presented in Figure 1 is known as the general equation for solar devices

$$I_{SC} = I_{PV} - I_{01}(e^{\frac{q(V + R_S I_{SC})}{nN_s kT}} - 1) - \frac{V + R_S I_{SC}}{R_P} \quad (1)$$

. In order to address other phenomena not considered in the single diode model, it is possible to incorporate additional diodes into the model, which results in the creation of the double diode model (DDM) and triple diode model (TDM) [19].

As previously mentioned, there are five parameters that describe the single diode model. However, calculating Rs, Rp, and n is still complex when attempting to find an optimal solution. Some authors [4,20,21] ignore one or even both resistances, while others propose an initial value of R_SO or R_PO [6] to subsequently calculate the other two.
This study adopts the method proposed by [2], where Rs and Rp are calculated directly using the principle of slope of lines passing through two points Ec. (2).

$$m = \frac{y_2 - y_1}{x_2 - x_1} \quad (2)$$

These slopes are calculated directly on the curve provided by the manufacturer, at STC (25°C and 1000 W/m^2), according to the initial points (Isc,0) and (0, Voc), where Isc and Voc are provided by the manufacturer. Each of the resistances, according to the initial point of their slope, are defined as follows:

$$Rp = \frac{1}{m_{sc}} \quad (3)$$

$$Rs = \frac{1}{m_{oc}} \quad (4)$$

To calculate Ipv, the equation proposed by [6 and 10] is used

$$Ipv = Ipv1 + (\mu*(T - Tn))G/Gn \quad (5)$$

Where

Tn = 298 K
Gn = 1000 W/m^2

To calculate Ipv1, the mathematical expression is given by the following equation

$$Ipv1 = I_{SC} \frac{R_P + R_S}{R_P} \quad (6)$$

In [6], the equation to calculate Rs is proposed. Since the proposed method calculates this parameter from the beginning, n (7) can be calculated from the coordinates of Pmax (Vmp, Imp) and using (5) and (6)

$$n = \frac{(Rs*Imp + Vmp - Voc)q}{kTN_s \ln\left(1 - \frac{Imp}{Ipv}\right)} \quad (7)$$

Up to this point, by applying equations (2)-(7), four of the five parameters that define the single diode model have been calculated. By applying the solution for Io in (1), it is possible to calculate this last parameter.

Having calculated the five parameters, applying a solution for Isc in (1), equation (8) is obtained, which plots the I-V curve of a solar device

$$I = -\left(\frac{V}{R_S R_P}\right) - \frac{\left(W\left(\frac{R_S R_P I_0 e^{\left(\frac{R_P(R_S I_{PV} + R_S I_0 + V)}{a(R_S + R_P)}\right)}}{a(R_S + R_P)}\right)a\right)}{R_S} + \frac{R_P(I_0 + I_{PV})}{R_S + R_P} \quad (8)$$

Where W refers to the Lambert W function.

III. RESUTLS

Equation (8) allows for a direct comparison with the manufacturer's curve or even an experimental curve if that were the case.

The commercial Kyocera KC200GT panel and the RTC France cell were analyzed, and a comparison was made with other models that have previously studied these devices. Table 1 shows the results obtained by the model named INAOE, which, compared in Io, is approximately the same as that calculated by [5,6,12,17, and 18]. For Ipv, a value of 8.21 A was obtained, which is within the trend marked by the other authors. For the parameters that present the conflict point in the calculations (Rs, Rp, and n), it was obtained that for n = 1.2935, which is close to [6], which indicates that for this parameter, values are assigned to the type of technology, so for a poly-Si panel, n = 1.3, which is very similar to what was obtained by [13 and 17].

The values of Rs are below 0.3 Ω, and for the proposed model, Rs = 0.1145 Ω, which is close to that obtained by [16], which applies optimization algorithms in its modeling. For Rp, Table 1 shows more variation in this parameter, ranging from values close to 100 Ω to the largest value obtained by the proposed model in this work, 801.5 Ω, since theory indicates that Rp → ∞ [26].

TABLE 1

PARAMETER EXTRACTION OF SOLAR PANEL KC200GT WITH SINGLE DIODE MODEL (SDM)

Algorithms	Authors/Parameters	Io (A)	Ipv (A)	n1	Rs (Ω)	Rp (Ω)
NR	Villalva (2009) [6]	9.8250E-08	8.2100	1.3000	0.22100	415.4050
SA	Saady (2013) [12]	9.8250E-08	8.2100	1.0000	0.22770	591.1123
GA	Echeverria (2014) [13]	8.5000E-09	8.2100	1.2931	0.20160	213.1306
AF	Cubas (2014) [5]	9.7600E-08	8.2130	1.3000	0.23000	597.3800
CSA	Jieming Ma (2016) [15]	5.1200E-10	8.1840	1.0170	0.25700	117.9220
WDO	Derick (2017) [16]	4.4230E-07	8.1812	1.4172	0.11320	747.4100
PS	Derick (2017) [16]	7.1836E-06	8.2014	1.7000	0.03390	624.3820
GA	Derick (2017) [16]	9.2200E-07	8.2112	1.4819	0.10670	728.5800
SA	Derick (2017) [16]	3.3484E-06	8.2109	1.6118	0.07690	713.1100
GA	Elezab (2018) [17]	8.7890E-08	8.2130	1.2900	0.23200	547.9100
WOA	Elezab (2018) [17]	8.5580E-08	8.2800	1.2900	0.28150	424.2200
NRMSE	Benahmida (2019) [20]	1.7500E-09	8.1900	1.0600	0.23000	113.9970
CBSA	Kadhim (2020) [18]	8.72100E-08	8.1810	1.2800	0.21500	332.6000
INAOE	INAOE (2023)	8.78900E-08	8.2120	1.2935	0.11470	801.5000

TABLE 2

PARAMETER EXTRACTION OF SOLAR PANEL RTC FRANCE WITH SINGLE DIODE MODEL (SDM)

Algorithms	Authors/Parameters	Io (A)	Ipv (A)	n1	Rs (Ω)	Rp (Ω)
PSO	Ye M (2009) [27]	3.2272E-07	0.7608	1.4838	0.03639	53.7965
GA	AlRashidi MR (2011) [28]	8.0870E-07	0.7619	1.5751	0.02990	42.3729
HS	Askarzadeh A (2012) [29]	3.0495E-07	0.7607	1.4754	0.03663	53.5946
BBO	Niu Q (2014) [30]	8.6100E-07	0.7610	1.5874	0.03214	78.8555
LI	lim LHI (2015) [31]	3.4566E-07	0.7609	1.4880	0.03614	49.4822
PPSO	Ma J (2016) [32]	3.2300E-07	0.7608	1.4812	0.03640	53.7185
GWO	Jordehi Ar (2016) [33]	2.4304E-07	0.7610	1.4512	0.03773	45.1163
TVACPSO	Jordehi Ar (2016) [33]	3.1068E-07	0.7608	1.4753	0.03655	52.8896
GOTLBO	Chen X (2016) [34]	3.3155E-07	0.7608	1.4838	0.03627	54.1154
ICA	Fathy A (2017) [35]	1.4650E-07	0.7603	1.4421	0.03890	41.1577
MSSO	Lin P (2017) [36]	3.2356E-07	0.7608	1.4812	0.03637	53.7425
ER-WCA	Kler D (2017) [37]	3.2270E-07	0.7608	1.4811	0.03638	53.6910
MPSO	Manel (2018) [38]	3.10683E-07	0.7608	1.4753	0.03655	52.8897
HISA	Dhruv (2019) [39]	3.10685E-07	0.7607	1.4773	0.03655	52.8898
NMA	Easwarakhanthan [40]	3.223E-7	0.7608	1.4837	0.0364	53.7634
ACPSBOP	Cubas (2017) [41]	2.92E-06	0.7606	1.7407	0.0459	246.78
INAOE	INAOE (2023)	1.03696E-08	0.7603	1.15456	0.075	333.33

Table 2 shows the results obtained from analyzing the RTC France cell (57 mm diameter), where it can be seen that the results obtained by
the proposed model differs significantly from those obtained by [27-39]. Io = 1.03696E-08 A is an order of magnitude lower than that obtained by other authors, while Ipv = 0.7603 A is the closest to the values obtained in previous works, and n = 1.15456 differs from the trend shown in Table 2, which is around n = 1.4. For Rs = 0.075 Ω, it can be said that it is twice the values presented in Table 2, as most values are around Rs = 0.036 Ω,

and finally, Rp = 333.33 Ω, which is a "high" value compared to those shown in this same table, but agrees with the theory [26]. In [40], the RTC France solar cell is studied for the first time, where it is indicated that the temperature at which the I-V curve was obtained is 33 °C. However, several models that have studied this device do not take into account temperature and radiation variations when formulating the equations. In the proposed model, the equations consider variations in G and T, hence the difference with previous works in calculating the reverse saturation current (Io).

The difference in Rs and Rp is due to the precision of the approximation for the Isc and Voc branches, where lines are drawn to calculate the slopes. The proposed model allows for adjusting the "step" of approximation.

Comparing with [41], a Rp approximately four times higher than the average shown in Table 2 is calculated, and for Rs, a value approximately 26.1% higher than the average obtained in other works is obtained in [41]. Thus, it is possible to obtain values outside the indicated trend.

Table 2 also presents the results obtained by [40], and as mentioned in previous paragraphs, the proposed models studying this device do not consider temperature and radiation variations, so they seek to approximate the results obtained by [40], suggesting that the values of T and G were held constant.

IV. CONCLUSIONS

The proposed model does not require prior measurements or initial values for the calculation of parameters that generate the most conflict when modeling (Rs, Rp, and n), as these are directly calculated from the manufacturer's curve and according to the results shown in Tables 1 and 2 for each of the case studies. It has good approximation compared to iterative and optimization models, and has the advantage over these two mentioned models of faster calculations, not falling into local minima (iterative), or having problems in setting up the search for the best solution (optimization).

The proposed model also allows for understanding the influence of each parameter of the single diode model (Io, Ipv, n, Rs, and Rp), as well as the influence of ambient conditions when varying radiant power (G) and temperature, all in a simple and real-time manner. It is also confirmed that the proposed model can perform calculations for both individual cells and solar arrays.

REFERENCES

[1] Efstratios Batzelis, Non-Iterative Methods for the Extraction of the Single-Diode Model Parameters of Photovoltaic Modules: A Review and Comparative Assessment, Energies 2019, 12, 358; doi:10.3390/en12030358.

[2] Phang, J.C.H.; Chan, D.S.H.; Phillips, J.R. Accurate analytical method for the extraction of solar cell model paramaters. Electron. Lett. 1984, 20, 406–408.

[3] Accarino, J.; Petrone, G.; Ramos-Paja, C.A.; Spagnuolo, G. Symbolic algebra for the calculation of the seriesand parallel resistances in PV module model. In Proceedings of the 2013 International Conference on Clean Electrical Power (ICCEP), Alghero, Italy, 11–13 June 2013; pp. 62–66.

[4] Sera, D.; Teodorescu, R.; Rodriguez, P. Photovoltaic module diagnostics by series resistance monitoring and temperature and rated power estimation. In Proceedings of the 2008 34th Annual Conference of IEEEIndustrial Electronics, Orlando, FL, USA, 10–13 November 2008; pp. 2195–2199.

[5] Cubas, J.; Pindado, S.; de Manuel, C. Explicit expressions for solar panel equivalent circuit parameters based on analytical formulation and the Lambert W-function. Energies 2014, 7, 4098–4115.

[6] Villalva, M.; Gazoli, J.; Filho, E. Comprehensive approach to modeling and simulation of photovoltaic arrays. IEEE Trans. Power Electron. 2009, 24, 1198–1208.

[7] Laudani, A.; Mancilla-David, F.; Riganti-Fulginei, F.; Salvini, A. Reduced-form of the photovoltaic five-parameter model for efficient computation of parameters. Sol. Energy 2013, 97, 122–127.

[8] ALQahtani, A.H. A simplified and accurate photovoltaic module parameters extraction approach using matlab. In Proceedings of the 2012 IEEE International Symposium on Industrial Electronics, Hangzhou, China, 28–31 May 2012; pp. 1748–1753.

[9] Chenni, R.; Makhlouf, M.; Kerbache, T.; Bouzid, A. A detailed modeling method for photovoltaic cells Energy 2007, 32, 1724–1730.

[10] Mohammed, S.S. Modeling and simulation of photovoltaic module using MATLAB/Simulink. Int. J. Chem. Environ. Eng. 2011, 2.

[11] Jordehi, A.R. Parameter estimation of solar photovoltaic (PV) cells: A review. Renew. Sustain. Energy Rev.2016, 61, 354–371.

[12] G.El-Saady, El-NobiA.Ibrahim, Mohamed EL-Hendawi, Simulated Annealing Modeling and Analog MPPT Simulation for Standalone Photovoltaic Arrays, January 2013.

[13] N. Echeverría, M.P. Cervellini, R. García Retegui, S.A. González, M. Funes y D. Carrica, Extracción de parámetros de un panel solar utilizando algoritmos genéticos, (2014).

[14] Jieming Maa, Ziqiang Bi, Tiew on Ting, Shiyuan Hao, Wanjun Hao, Comparative performance on photovoltaic model parameter identification via bio-inspired algorithms, Solar Energy 132 (2016) 606–616.

[15] Mathew, Derick & Rani, Christia & Farrag, Mohamed & Wang, Yue & Busawon, Krishna & Muthu, Rajesh. (2017). An improved optimization technique for estimation of solar photovoltaic parameters.

[16] Elazab, O.S.; Hasanien, H.M.; Elgendy, M.A.; Abdeen, A.M. Parameters estimation of single-and multiple-diode photovoltaicmodel using whale optimisation algorithm. IET Renew. Power Gener. 2018, 12, 1755–1761.

[17] Hassan, Kadhim & Rashid, Abdulmuttalib & Hani, Basil. (2021). Parameters estimation of solar photovoltaic module using camel behavior search algorithm. International Journal of Electrical and Computer Engineering. Vol.11, No.1. 788-793. 10.11591/ijece.v11i1.pp788-793.

[18] Radhakrishnan Venkateswari, Natarajan Rajasekar, Review on parameter estimation techniques of solar photovoltaic systems, Wiley, 2021. doi.org/10.1002/2050-7038.13113.

[19] Fahim, S.R.; Hasanien, H.M.; Turky, R.A.; Aleem, S.H.E.A.; Ćalasan, M. A Comprehensive Review of Photovoltaic Modules Models and Algorithms Used in Parameter Extraction. Energies 2022, 15, 8941. https://doi.org/10.3390/ en15238941

[20] Cannizzaro, S.; Di Piazza, M.C.; Luna, M.; Vitale, G. Generalized classification of PV modules by simplified single-diode models. In Proceedings of the 2014 IEEE 23rd International Symposium on Industrial Electronics (ISIE), Istanbul, Turkey, 1–4 June 2014; pp. 2266–2273.

[21] Saloux, E.; Teyssedou, A.; Sorin, M. Explicit model of photovoltaic panels to determine voltages and currents at the maximum power point. Sol. Energy 2011, 85, 713–722.

[22] De Blas, M.A.; Torres, J.L.; Prieto, E.; Garcia, A. Selecting a suitable model for characterizing photovoltaic devices. Renew. Energy 2002, 25, 371–380

[23] Cubas, Javier & Pindado, Santiago & Sorribes, Felix. (2017). Analytical Calculation of Photovoltaic Systems Maximum Power Point (MPP) Based on the Operation Point. Applied Sciences. 7. 870. 10.3390/app7090870.

[24] El Tayyan, Ahmed. (2013). A simple method to extract the parameters of the single-diode model of a PV system. Turkish Journal of Physics. 37. 121 – 131. 10.3906/fiz-1206-4.

[25] Arab, A.H.; Chenlo, F.; Benghanem, M. Loss-of-load probability of photovoltaic water pumping systems. Sol. Energy 2004, 76, 713–723.

[26] SM Sze and KK Ng. Physics of semiconductor devices John wiley & sons, 2006.

[27] Ye, M.; Wang, X.; Xu, Y. Parameter extraction of solar cells using particle swarm optimization. J. Appl. Phys. 2009, 105, 1–8.

[28] AlRashidi, M.R.; AlHajri, M.F.; El-Naggar, K.M.; Al-Othman, A.K. A new estimation approach for determining the I-V characteristics of solar cells. Sol. Energy 2011, 85, 1543–1550.

[29] Askarzadeh, A.; Rezazadeh, A. Parameter identification for solar cell models using harmony search-based algorithms. Sol. Energy 2012, 86, 3241–3249.

[30] Niu, Q.; Zhang, L.; Li, K. A biogeography-based optimization algorithm with mutation strategies for model parameter estimation of solar and fuel cells. Energy Convers. Manag. 2014, 86, 1173–1185.

[31] Lim, L.H.I.; Ye, Z.; Ye, J.; Yang, D.; Du, H. A linear identification of diode models from single I-V characteristics of PV panels. IEEE Trans. Ind. Electron. 2015, 62, 4181–4193.

[32] Ma, J.; Man, K.L.; Guan, S.U.; Ting, T.O.; Wong, P.W. Parameter estimation of photovoltaic model via parallel particle swarm optimization algorithm. Int. J. Energy Res. 2016, 40, 343–352.

[33] Jordehi, A.R. Time varying acceleration coefficients particle swarm optimisation (TVACPSO): A new optimisation algorithm for estimating parameters of PV cells and modules. Energy Convers. Manag. 2016, 129, 262–274.

[34] Chen, X.; Yu, K.; Du, W.; Zhao, W.; Liu, G. Parameters identification of solar cell models using generalized oppositional teaching learning-based optimization. Energy 2016, 99, 170–180.

979-8-3503-1191-4/23 $31.00 © 2023 IEEE

[35] Fathy, A.; Rezk, H. Parameter estimation of photovoltaic system using imperialist competitive algorithm. Renew. Energy 2017, 111, 307–320.

[36] Lin, P.; Cheng, S.; Yeh, W.; Chen, Y.; Wu, L. Parameters extraction of solar cell models using a modified simplified swarm optimization algorithm. Sol. Energy 2017, 144, 594–603.

[37] Kler, D.; Sharma, P.; Banerjee, A.; Rana, V.; Kumar, K.P.S. PV cell and module efficient parameters estimation using evaporation rate-based water cycle algorithm. Swarm Evolut. Comput. 2017, 35, 93–110.

[38] Manel, M.; Anis, S.; Faouzi, M.M. Particle swarm optimisation with adaptive mutation strategy for photovoltaic solar cell/module parameter extraction. Energy Convers. Manag. 2018, 175, 151–163.

[39] Dhruv, K.; Goswami, Y.; Kumar, R.V. A novel approach to parameter estimation of photovoltaic systems using hybridized optimizer. Energy Convers. Manag. 2019, 187, 486–511.

[40] Easwarakhanthan, T.; Bottin, J.; Bouhouch, I.; Boutrit, C. Nonlinear minimization algorithm for determiningthe solar cell parameters with microcomputers. Int. J. Sol. Energy 1986, 4, 1–12.

[41] Cubas, Javier & Pindado, Santiago & Sorribes, Felix. (2017). Analytical Calculation of Photovoltaic Systems Maximum Power Point (MPP) Based on the Operation Point. Applied Sciences. 7. 870. 10.3390/app7090870.

Development of Simplified Lumbar Spine Mechanism Implemented with Tendon-Driven Motion

1st Ruben Florez
LIECAR Laboratory
University of San Antonio Abad del Cusco
Cuzco, Peru
rubendfz2206@gmail.com

2nd Facundo Palomino-Quispe
LIECAR Laboratory
University of San Antonio Abad del Cusco
Cuzco, Peru
facundo.palomino@unsaac.edu.pe

3rd Thuanne Paixão
PAVIC Laboratory
University of Acre UFAC
Rio Branco, Brazil
thuannepaixao@gmail.com

4th Ana Beatriz Alvarez
PAVIC Laboratory
University of Acre UFAC
Rio Branco, Brazil
ana.alvarez@ufac.br

5th Lucas Angst
PAVIC Laboratory
University of Acre UFAC
Rio Branco, Brazil
llucasangst@gmail.com

6th Luis Maggi
PAVIC Laboratory
University of Acre UFAC
Rio Branco, Brazil
luis.maggi@gmail.com

Abstract—**The lumbar region of the spine has important characteristics for the investigation of pathological dysfunctions, where the major abnormalities present in the spine are caused. This paper presents the development and implementation of a controlled and simplified lumbar spine mechanism. The construction was done using 3D Slicer, SolidWorks and Ultimaker Cura software. The control action is based on a motion trajectory system realized with conventional, PID-type control. The physical implementation of the motion drive was performed and results show the correct execution of the planned movements, following the references specified for the movement of the structure.**

Index Terms—**Biomechanics, Motion control, Lumbar spine mechanism.**

I. INTRODUCTION

The construction of experimental models of the human spine, has as main motivation the design of computational and physical models, in order to provide the study, evaluation and demonstration of the specific structural behavior of these elements [1], [2]. Low back pain, can be characterized by several factors that can lead to injury, inflammation and degeneration of the lumbar vertebral structure [3], [4]. In the literature are found works and approaches that report the construction of models of the spine, in the work of Kakehashi et al. [5] a robotic mechanism capable of replicating the movement of the spine was built, applying a continuous robot technique with 9 motors and 24 wires. Similarly, in the work of Karadogan et al. [6] a dynamic robotic lumbar spine model with 15 degrees of freedom was developed, driven by 20 wires.

In this paper, the construction of an experimental model of the lumbar spine with movement analogous to human is presented, this model reproduces dynamic movements of the spine using only the bone geometry of the vertebral bodies

without considering the posterior structures of each vertebra. For the movement, a tendon drive will be considered whose control will be done using the PID (Proportional Integral Derivative) type, in order to realize the flexion and extension movement in the model. Considering that the mathematical model developed by the authors was for two degrees of freedom, as described in [7], the five vertebrae were grouped considering as criteria the movement of each vertebra and the initial position angle between each one of them. In Fig. 1, it is observed that, when joining the vertebrae L1, L2 and L3 by means of their centroids, they form a straight line. Thus, these vertebrae L1, L2 and L3 were grouped together to form the first group G1, leaving vertebrae L4 and L5 to form the second group G2. In the implementation and testing, movements of only two vertebrae, L1 and L4, which perform predefined movements, were considered.

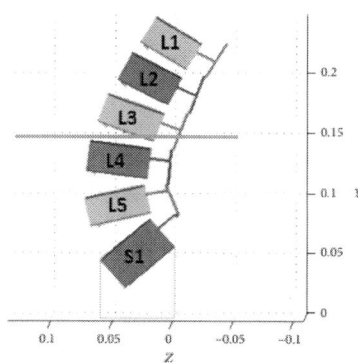

Fig. 1: Lumbar Spine Geometry - Side View. [6]

II. METHODOLOGY

A. Overview of lumbar spine design

The analysis of the development of the lumbar spine model was presented in [8]. This model is composed from anonymized CT scans made available by the CTSpine1K database. The construction of the vertebral bodies, occurred with the 3D Slicer program, using the tomographic image for the reconstruction of the physical object, resulting in an STL file. After editing the segmentation of the reconstruction, only the vertebral bodies belonging to the lumbar vertebrae were kept. To separate the meshes into 5 different files, the 3D Builder software was used. The files, also in STL format, represent the vertebral bodies of the lumbar vertebrae and can be used as a basis for simulating the flexion and extension movement of the lumbar vertebrae.

The construction of the parameterized models was carried out with the Solidworks software, the construction was carried out with the same print settings, which can be modified with the Ultimaker Cura software.

The lumbar spine design is 3D printed with premium acrylonitrile butadiene butadiene styrene (ABS) material printed on a GTMax A3 machine. The design consists of 5 lumbar vertebrae (L1 to L5) and the pelvis (S1) [8]. The spine is only limited to flexion (forward) and extension (backward) movements at maximum angles of 50° and 35° respectively [5].

Therefore, the Table I shows in summary the characteristics of the lumbar spine.

TABLE I: Characteristics of the lumbar spine model

6*General	Weight	50 [g]
	Height	12 [cm]
	Width	3 [cm]
	Depth	2.5 [cm]
	Material	ABS premium
	Degrees of freedom	5 DOF

B. Implementation

The complete design of the 3D parts of the lumbar spine can be seen in Fig. 2.

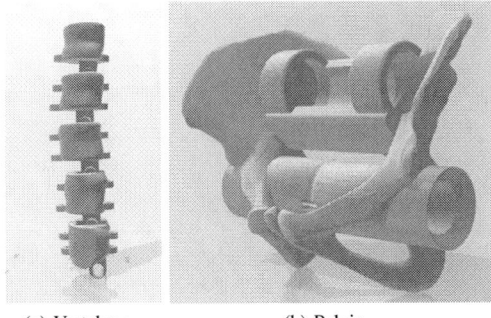

(a) Vertebrae.   (b) Pelvis.

Fig. 2: 3D design of the lumbar spine.

The vertebrae take the form of a semicircle which is a simplified representation of the sacrum and forms part of the base to join the pelvis, this arrangement of the vertebrae can be seen in the Fig. 3.

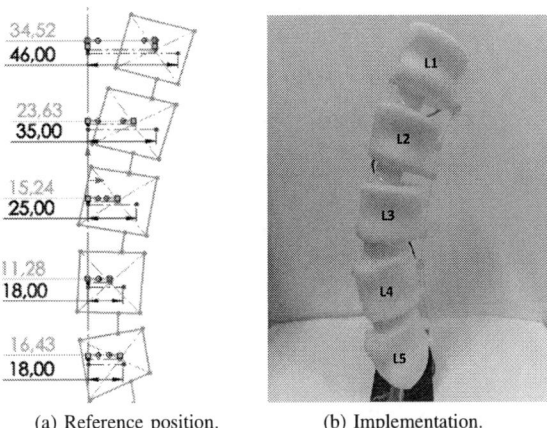

(a) Reference position.   (b) Implementation.

Fig. 3: Position of the lumbar vertebrae.

For the implementation of the tendon-actuated lumbar spine, the movement of two vertebrae were considered. Thus, for the actuation of the lumbar vertebrae, the L1-L2-L3 vertebrae were grouped forming G1 and the L4-L5 vertebrae forming the G2 group, to represent the flexion and extension movements. To simulate the tendons for G1 and G2 actuation, nylon threads were used, which are coupled to the servomotors. Two servomotors are used to reproduce the movements of G1 and two servomotors for G2, necessary to perform flexion and extension in each of the joints. The drive of the servomotors is adjusted by reading modules that measure the angular position at a certain degree of inclination. These modules are located in the L1 and L4 vertebrae, and the 4 servomotors are placed on a medium density fiberboard (MDF). This experiment does not consider the use of the 3D printed pelvis, which is being replaced by an acrylic shaft. Physically, the implementation consists of 4 MG995 servomotors, an Arduino Mega with ATmega2560 microcontroller, 2 MPU6050 modules (accelerometers), 1 12V power supply at 1.5A with 1 LM2596 step-down 3A DC-DC voltage converter, Table II.

TABLE II: Characteristics of the lumbar spine model

Actuator	Servomotor MG995	4 servomotores
4*Electronic	Arduino Mega 2560	1 unid
	Módulo MPU6050	2 unid
	DC-DC step-down 3A LM2596	1 unid
	Power supply	12V @ 1.5A

The complete implementation is shown in the Fig. 4.

For the realization of the movements, the plane of the MPU6050 sensors is considered to the X-axis of the sensor.

For the measurement of the orientation and inclination of G1 and G2, it was considered to find the angle of inclination using the accelerometer of the sensor, where gravity is taken into account as the only force acting on the sensor, assuming that the X-Z plane is in the inclined plane [9]. To find the tilt angle and orientation, equations (1) and (2) are used.

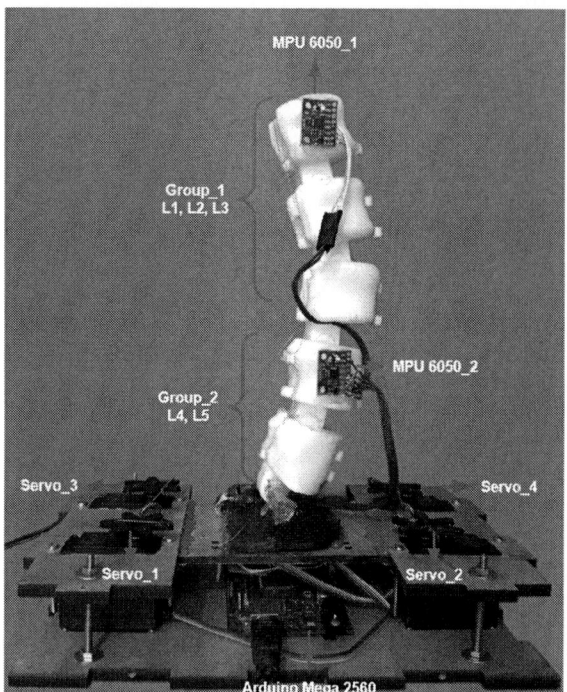

Fig. 4: Complete implementation of the lumbar spine.

$$\theta_x = \alpha \tan\left(\frac{a_x}{\sqrt{a_y^2 + a_z^2}}\right) \quad (1)$$

$$\theta_y = \alpha \tan\left(\frac{a_y}{\sqrt{a_x^2 + a_z^2}}\right) \quad (2)$$

Fig. 5 shows the reference planes of each sensor for the movements.

(a) Sensor axes.　　　　(b) Sensor drawings.

Fig. 5: Reference drawings of the sensors.

III. EXPERIMENTAL STUDY

For the experimental study, a simulation was performed in MATLAB/Simulink software, giving reference positions for L1 and L4 of the lumbar spine. The physical experiments consider the five vertebrae grouped together, L1-L2-L3 forming

group G1, and L4-L5 forming group G2. The experiments on the 3D structure are described in the following sections.

A. Simulation results

The mathematical model representing the simplified structure (two vertebrae) of lumbar spine uses the Lagrangian methodology, for 2DOF. For the simulation a linear behavior of the system is considered, and the PID controller used was tuned with the PID constants: $K_p = 10$, $K_i = 0.2$ and $K_d = 0.5$ for G1 and G2.

Two separate movements were simulated for G1 and G2. For G1 a reference of 40° representing flexion movement was given, returning then to its initial position of 0° and as a second movement a reference of -20°, representing extension, as shown in Fig. 6a and 6b, respectively. For G2 both movements were given at their 10° limits in flexion and extension, Fig. 7a and 7b, respectively.

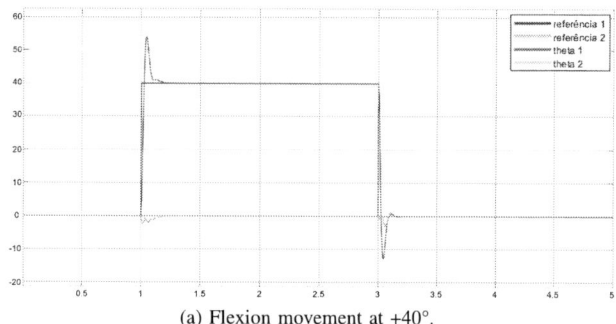

(a) Flexion movement at +40°.

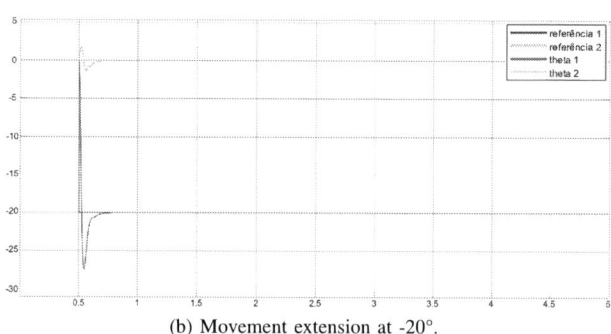

(b) Movement extension at -20°.

Fig. 6: Simulation of G1 movement.

B. Physical experiments

For the physical experiments of the lumbar spine structure, a filter for the sensor readings was implemented in the algorithm. Ten readings were taken and averaged to improve the sensitivity of the MPU6050 and reduce noise.

Fig. 8a and 8b show the movements of G1 with 40° of flexion and 20° of extension, respectively. Fig. 8c and 8d correspond to G2 movements with 10° of flexion and 10° of extension, respectively.

The movements performed generated the behavior shown in the Fig. 9, where the given references and the readings of the movement angles for G1 and G2 are observed.

979-8-3503-1191-4/23 $31.00 © 2023 IEEE

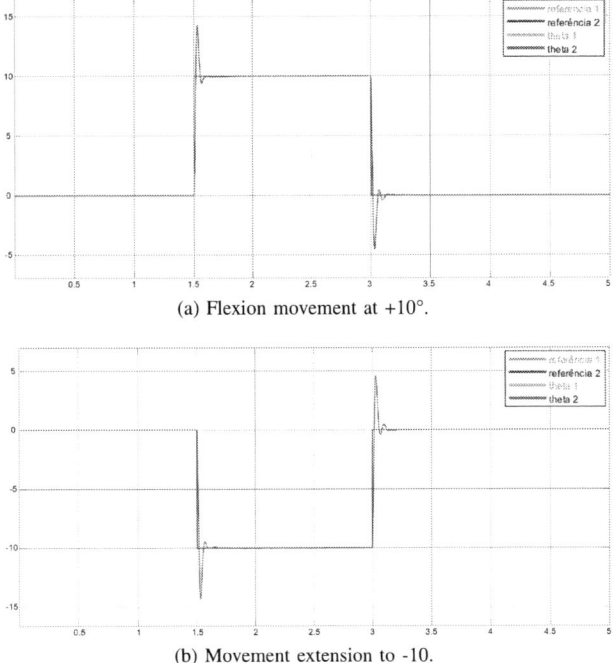

(a) Flexion movement at +10°.

(b) Movement extension to -10.

Fig. 7: G2 motion simulation.

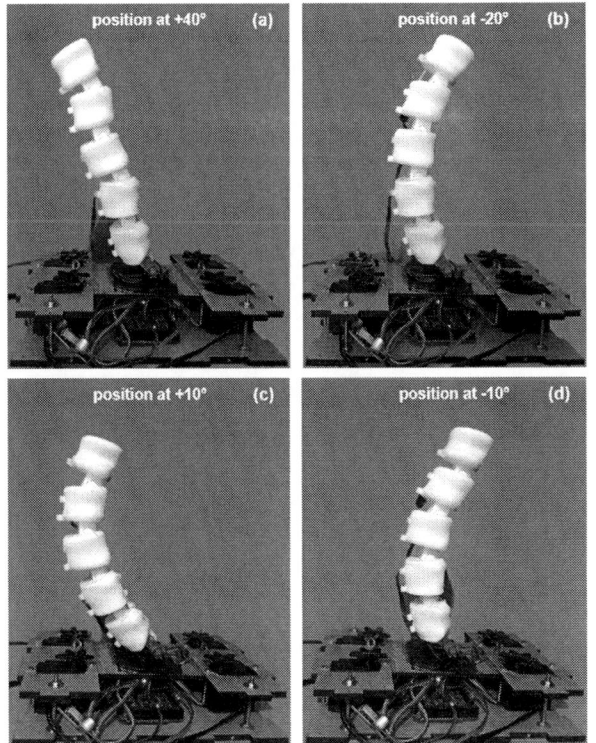

Fig. 8: Physical tests of the lumbar spine.

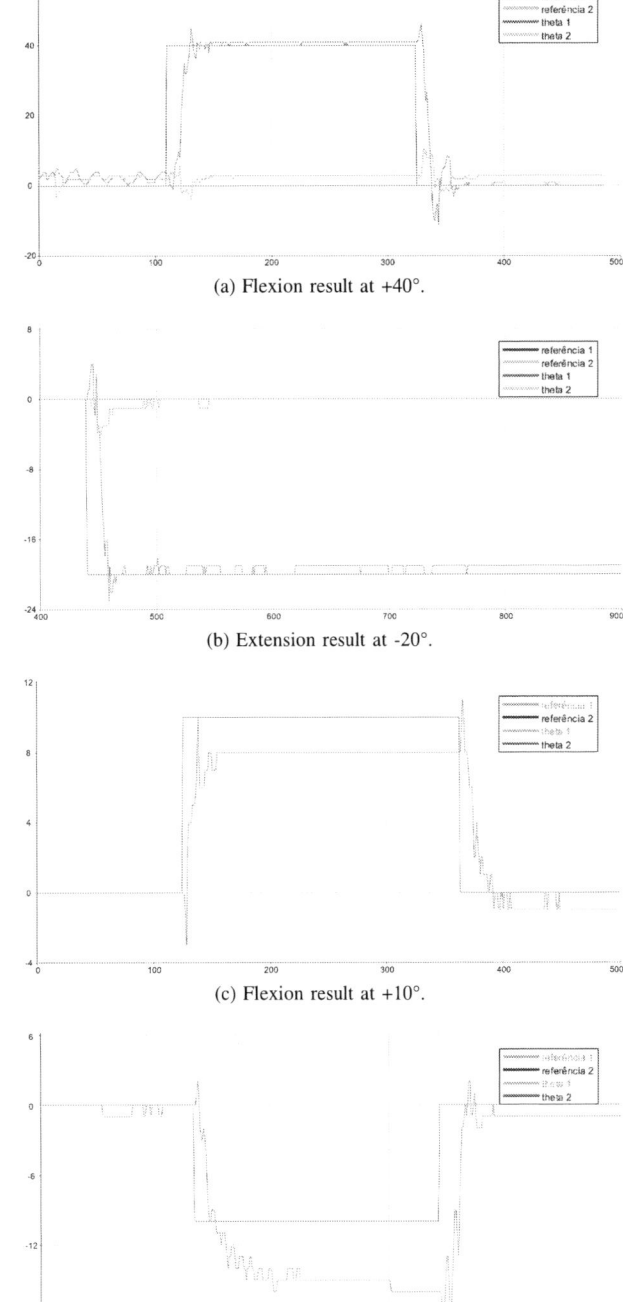

(a) Flexion result at +40°.

(b) Extension result at -20°.

(c) Flexion result at +10°.

(d) Extension result at -10°.

Fig. 9: Movement results: (a), (b) corresponding to G1, (c) and (d) G2.

When comparing the results of physical experiments with those of simulation, a similarity in the responses is observed. However, the physical responses of the G1 behavior present a slight delay, this may be due to the wires that drive the vertebrae group. In the physical responses of the G2 behavior, the presence of attenuation is observed, this may be due to the limitation of physical movement of the servomotors that only go from 0° to 180° because the G2 vertebrae are closer to the servomotor and therefore have smaller tendons.

IV. CONCLUSION

In the present paper the development of simplified lumbar spine mechanism implemented with tendon-driven motion was presented. The construction of the biomodel demonstrates the possibility of developing anatomical structures from computational resources and tools. The developed lumbar spine model represents the anatomical and kinesiological characteristics of the vertebrae, thus it is believed that it is possible to reproduce flexion and extension movements with high accuracy. In order to test the model, movements were planned to be performed by means of servomotor-driven tendons. The five vertebrae of the lumbar spine were organized in two groups, G1 and G2. The physical experiments considered the use of 4 servomotors to replicate flexion and extension movements in G1 and G2. In the physical experiments, flexion and extension movements analogous to those of human beings were reproduced, specifically for G1 a movement of $40°$ of flexion and $20°$ of extension was tested. While for G2, movements of $10°$ of flexion and $10°$ of extension were reproduced for evaluation. Considering that there are physical factors present such as friction, resistance and tendon tension, the angular position measurements of G1 and G2 obtained with the gyroscope are shown to be close to the simulation results, results that can be considered satisfactory. The results obtained and presented in this article are promising and make us think that with this lumbar spinal column model it is possible to replicate accurate natural movements of the human spine. In order to replicate the human movement of the five vertebrae of the lumbar spine, as future work we intend to use modern control techniques that can adapt with the perturbations present during the behavior. Also, for a more precise movement, continuous rotation servomotors ($360°$) can be used to ensure the best movement for each vertebra.

ACKNOWLEDGMENT

The work presented in this paper was supported by the *Pesquisa Aplicada em Visão e Inteligência Computacional* (PAVIC) project and the *Amazônia Legal* project at *Universidade Federal do Acre*, Brazil.

REFERENCES

[1] Z. Gao, I. Gibson, C. Ding, J. Wang, and J. Wang, "Virtual lumbar spine of multi-body model based on simbody," *Procedia Technology*, vol. 20, pp. 26–31, 2015.

[2] G. Eremina, A. Smolin, and I. Martyshina, "Convergence analysis and validation of a discrete element model of the human lumbar spine," *Reports in Mechanical Engineering*, vol. 3, no. 1, pp. 62–70, 2022.

[3] T. Mitchell, P. B. O'Sullivan, A. F. Burnett, L. Straker, and A. Smith, "Regional differences in lumbar spinal posture and the influence of low back pain," *BMC Musculoskeletal Disorders*, vol. 9, no. 1, pp. 1–11, 2008.

[4] C. T. Swain, F. Pan, P. J. Owen, H. Schmidt, and D. L. Belavy, "No consensus on causality of spine postures or physical exposure and low back pain: A systematic review of systematic reviews," *Journal of biomechanics*, vol. 102, p. 109312, 2020.

[5] Y. Kakehashi, K. Okada, and M. Inaba, "Development of continuum spine mechanism for humanoid robot: Biomimetic supple and curvilinear spine driven by tendon," in *2020 3rd IEEE International Conference on Soft Robotics (RoboSoft)*. IEEE, 2020, pp. 312–317.

[6] E. Karadogan and R. L. Williams, "The robotic lumbar spine: Dynamics and feedback linearization control," *Computational and mathematical methods in medicine*, vol. 2013, 2013.

[7] T. Paixão, A. B. Alvarez, R. Florez, and F. Palomino-Quispe, "Motion control of a robotic lumbar spine model," in *Bioinformatics and Biomedical Engineering, IWBBIO 2023, Gran Canaria, Spain, July 12–14, 2023, Proceedings, Part I*, vol. 13919. Springer, 2023, pp. 3–14.

[8] L. Angst, "Construction and validation of an experimental kinesiological model of the human lumbar spine," Master's thesis, Western Amazon Health Sciences Graduate Program. Federal University of Acre, 2022.

[9] J. Vazquez, C. Aavalos, J. Ontiveros, N. Galán, and G. Rubio, "Control pid y difuso de un seguidor solar fotovoltaico para obtención de potencia nominal con sensor mpu6050," *rotación*, vol. 1, p. 2.

2023 IEEE Latin American Electron Devices Conference (LAEDC)
Puebla, México, July 3-5, 2023

Precession motion of a CanSat Rover-Back prototype during its fall-to-earth descent

Luis Aiquipa Moreno
Facultad de Ciencias
Universidad Nacional de Ingeniería
Lima,Peru
luis.aiquipa.m@uni.pe

Roxana Pastrana Alta
Facultad de Ciencias
Universidad Nacional de Ingeniería
Lima,Peru
rpastranaa@uni.edu.pe

Walter Estrada Lopez
Facultad de Ciencias
Universidad Nacional de Ingeniería
Lima,Peru
westrada@uni.edu.pe

Abstract— This work presents the generation of the precession motion of a CanSat Rover-Back educational picosatellite falling to earth for future space exploration missions using the conventional design of reaction wheels, together with the basic concepts of electronics, mechanics, physical sciences and mathematics, being a technological innovation (STEM). For the modeling and operating parameters, are presented mainly the mass, the distance difference between the point where the structure hangs or pivots and the center of mass, the control of the angular velocity in the motors performing a change of direction of rotation. The results show the generation of a sinusoidal signal of the precession speed in revolutions per minute (RPM) of the prototype, demonstrating its validity, and the variation of its speeds in RPM of the two motors.

Keywords—Precession motion, CanSat Rover-Back, STEM

I. INTRODUCTION

Space exploration robotics that contribute to understanding the soils on extraterrestrial lands, such as the Moon and Mars, allow the creation of diverse designs that employ optimal methods to develop highly reliable, durable, and efficient systems. For this work, the parameters of environmental conditions such as gravity, pressure, temperature, and atmospheric composition are taken into account [1]. With each mission, new targets of research interest emerge. Nowadays, yaw rotation systems for cameras on a gimbal are used in flying robots such as drones for various applications, for example in the analysis of photovoltaic modules in a thermal image [2], however, solving problems of space within the design because the mission requires it and to economize for these prototypes are often complicated and more if we add the work of exploration on the ground, i.e the prototype must have wheels for mobilization or a mobility mechanism. The A-PUFFER robot developed at NASA for space exploration is able to move in small places using a flexible rigid body with the methodology of origami [3], in addition to the innovation systems can be allowed under the example given in this paper. In addition to the innovation systems, the use of the wheels themselves to generate the yaw movement during the descent can be allowed under the example set in drones and to avoid oversizing, from this can arise various applications such as for the detection of various objects or areas of interest, such as agriculture [4] and even

the search for minerals [5]. The CanSat picosatellite prototypes are a good basis for the use of various missions that may have a scientific purpose, where the sizing, mass and economic evaluation of the manufacture is at stake and must be able to successfully perform the mission [6].
This article shows the modeling of the system, the experimental bench under the concept of a CanSat Rover-Back picosatellite, to obtain the precession motion generated by the engines.

II. MECHANICAL DESIGN OF THE CANSAT PROTOTYPE

For the mechanical design, mainly four important points are taken into account: the mass, the moment of inertia in the x-axis, the pivot distance and the center mass. However, it is recommended that the rotation of the precession be oscillatory due to the generation of chord tension force, this would be considered an unwanted disturbance. In Fig. 1, the isometric view of the structure of a CanSat Rover-Back picosatellite is presented, in addition to the rotational chord, r is the vector from the center of mass (CM) to the pivot, and its x-axis component is r_x, with the reaction wheels A and wheel B.

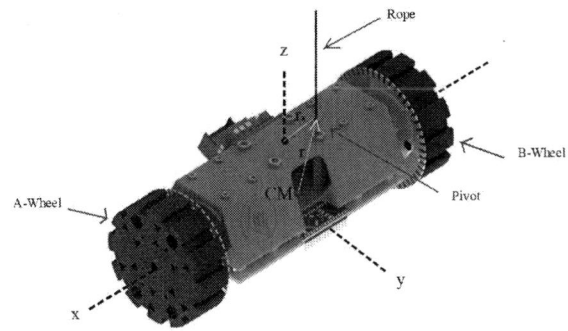

Fig. 1. Structure of the CanSat Rover-Back picosatellite for the generation of the precession motion.

III. ELECTRONIC DISPOSITION

The electronics for the generation of the precession movement for our prototype is mainly composed of a microcontroller, IMU, two DC motors, an H-bridge, a sensor capable of measuring the angular velocity of each wheel, an external power supply battery for the motors. The devices can vary and according to them it will be possible to obtain more or less precision in the results, in Fig.2 , the electronic

VRI-UNI FC-PF-10-2022

979-8-3503-1191-4/23 $31.00 © 2023 IEEE

disposition is presented by means of a block diagram. For the reading of our prototype, the ATMega328p microcontroller was used, the H-bridge used is the Lm298 module, the DC motors are composed of 37D 12V/1000 RPM, the reading of the angular velocity of the motors used the infrared sensor FC-03 and for the precession the use of the BNO-055.

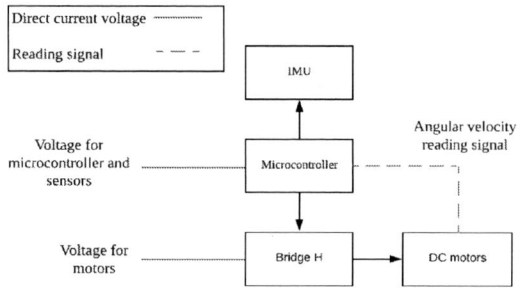

Fig. 2. Main electronic components of the CanSat Rover-Back picosatellite for precession motion.

IV. KINEMATICS

The system is taken into consideration the fulfillment of Newton's first law or in an approximate way. Thus the following equation for the precession velocity is fulfilled:

$$\frac{d(\Omega)}{dt} = \frac{Mgr_x}{I_x\omega} \qquad (1)$$

In order to show the results of the precession speed (Ω) it is necessary to have control of the angular velocity of the motors (ω) and to know the constant parameters defined by the design. For practical purposes it is possible to use a PID control system for DC motors [7].

V. EXPERIMENT

The experiment takes the specifications of Table I, this because it complies with the equations of precession motion (1). The experimental test bench used is composed of a universal support that allows holding a rope and attaching it to the pivot, a power supply for the system or battery, its position must be such that the cables do not generate any tension, a computer for data storage and the CanSat Rover-Back prototype.

TABLE I. LATEST PROTOTYPE SPECIFICATIONS

Specification	Value
Distance from pivot to CM (r)	30 mm
Moment of Inertia around x-axis	(4.5×10^5) g.mm^2
Mass	610.89 g
Dimensions	80 mm (Diameter) 200 mm (Long)

A. Generation of oscillatory motion

For the generation of the oscillatory motion, four rotations with change of angular velocity and direction of the wheels are taken. Table II indicates the series of stages to generate the change of direction of the precession motion, the change of stage is programmed to be executed in a time of 2 s.

TABLE II. STAGES OF THE DIRECTION OF ROTATION FOR THE GENERATION OF OSCILLATORY MOTION IN PRECESSION

Wheel	(1)	(2)	(3)	(4)
A-Wheel	Clockwise rotation	Counterclockwise rotation	No rotation	Clockwise rotation
B-Wheel	No rotation	Counterclockwise rotation	Counterclockwise rotation	Clockwise rotation

B. Monitoring and experimentation algorithm

The experiment shows the steps to follow to obtain the results, as shown in Fig. 3. First, it is necessary to carry out rotation tests beforehand to corroborate its operation, for this the electronic circuit must be prepared for the control of the angular velocity, a speed of 1300RPM is taken into consideration for each reaction wheel in average to a 12V/3A power supply, the disposition of the electronic circuit is placed so that it does not obstruct the movement of the motor and can correctly read the sensors, the change of the direction of rotation at these speeds produces a peak current of 1.1A. Second, corroborate the calculation and measurements of the specification tables. Third, perform the experimental setup. Fourth, compile the programming code inside the microcontroller. Fifth, corroborate the stability conditions and finally feed the system. In Fig. 4, visualize the rotation around the rope per frame, quick test of operation. If the desired results cannot be obtained, a calibration of the electronic components must be performed and the steps must be repeated.

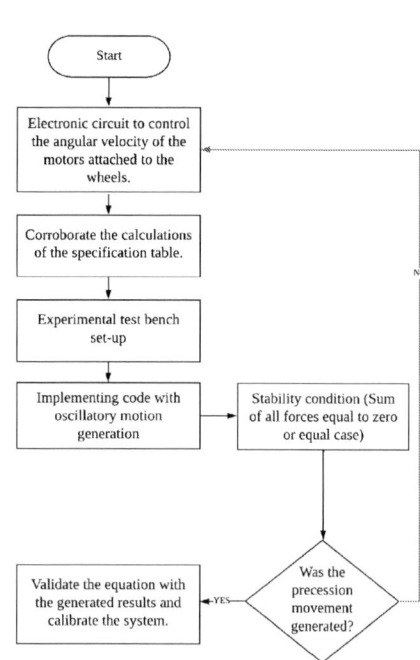

Fig. 3. Steps to follow to carry out the experiment.

979-8-3503-1191-4/23 $31.00 © 2023 IEEE

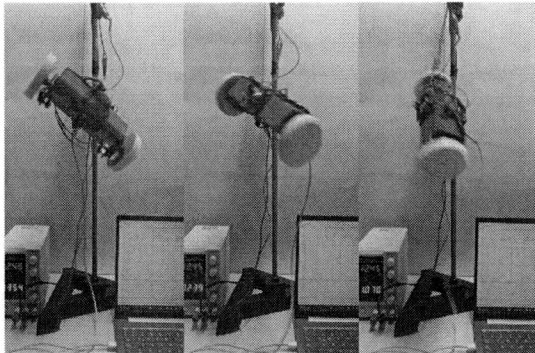

Fig. 4. Stills of the CanSat Rover-Back precession motion.

VI. RESULTS

The results of the experiment were carried out in an interval of 60 s, it was evidenced that the signals generated by the sensors are periodic in nature. A time interval of 16 s was taken for the data acquisition. Fig. 5A shows the oscillatory precession movement of the experiment with a maximum amplitude of 200 RPM, and an oscillation period of approximately 8 s per cycle, generated by the four stages of change in the rotation of the wheels. Fig. 5B represents the angular velocity changes in RPM for each wheel.

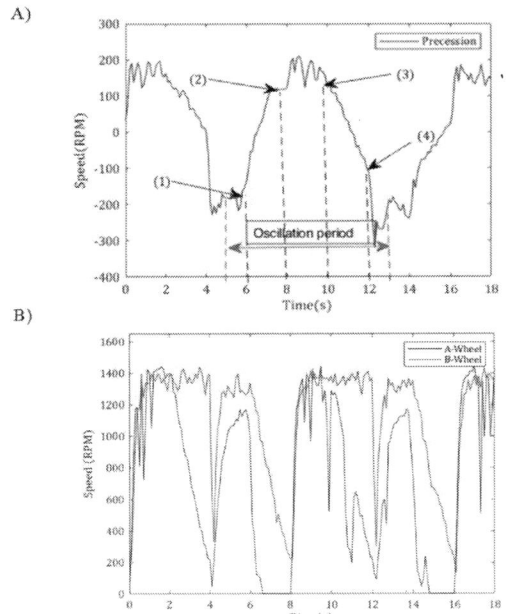

Fig. 5. Plots A) of the precession motion B) of the angular velocity of wheels A and B.

VII. CONCLUSIONS

The generation of the precession motion of a CanSat Rover-Back system presented in this paper provides a new approach for the implementation of new robots performing space or ground exploration that seeks to perform a tumbling rotation. The precession velocity presented in the experiment shows the generation of an oscillatory signal with a period of 8 s, due to the changes in the rotational stage of the A and B wheels driven by the electronic circuit.

VIII. FUTURE WORK

Future works for the implementation of these systems can be applied in conjunction with other works that may include a camera, such as the case of a robust control of a two degrees of freedom cardan [8], where the position control for the generation of precession must be evaluated, to perform various operations in the fields of engineering. The specifications made are of a basic nature applied to normal conditions, as under various temperature and pressure changes, the mechanical composites will need to be further evaluated. Finally, these applications will be presented in small size and low budget space exploration prototypes, which should perform this work within a region of the landing zone for search, detection or analysis of survey objects.

ACKNOWLEDGMENT

This work was carried out within the research laboratory of Bioinorganic Chemestry in Medicine, Environmental And Technology (BIOMET) of the Universidad Nacional de Ingeniería, under the economic support of the vice-rectorate of research whose project code is VRI-UNI FC-PF-10-2022. The authors would like to thank Victor Raul Quinde Savedra, Carlos Gonzales Lorenzo and Miguel Tarazona Tocto for their technical support for the elaboration of the experiment.

REFERENCES

[1] J. J. Zakrajse et al., "Exploration rover concepts and development challenges," A Collection of Technical Papers - 1st Space Exploration Conference: Continuing the Voyage of Discovery, vol. 1, pp. 258–280, 2005, doi: 10.2514/6.2005-2525.

[2] H.-C. Park, S.-W. Lee, and H. Jeong, "Image-Based Gimbal Control in a Drone for Centering Photovoltaic Modules in a Thermal Image," Applied Sciences, vol. 10, no. 13, p. 4646, Jul. 2020, doi: 10.3390/app10134646.

[3] J. T. Karras et al., "Pop-up mars rover with textile-enhanced rigid-flex PCB body," 2017 IEEE International Conference on Robotics and Automation (ICRA), Singapore, 2017, pp. 5459-5466, doi: 10.1109/ICRA.2017.7989642.

[4] J. J. Mejía González, S. A. Zapata Gil, S. León Serna, N. Buriticá Isaza, D. A. González Jaramillo, and J. M. Zamora Vélez, "Construcción de prototipo de CANSAT para toma de imágenes aéreas para detección de zonas de vegetación en agricultura de precisión," Ciencia y Poder Aéreo, vol. 16, no. 2, pp. 11–28, Nov. 2021, doi: 10.18667/CIENCIAYPODERAEREO.709.

[5] J. J. Nunez-Quispe, J. Lleren-Sernaque, and E. Lara-Chavez, "Mechanical Design of a ROVER prototype for Exploration tasks on Mars: Structural and Transient Dynamics simulation analysis," 2021 IEEE MIT Undergraduate Research Technology Conference, URTC 2021, 2021, doi: 10.1109/URTC54388.2021.9701623.

[6] A. Colin, "A pico-satellite assembled and tested during the 6th CanSat Leader Training Program," Journal of Applied Research and Technology. JART, vol. 15, no. 1, pp. 83–91, Feb. 2017, doi: 10.1016/J.JART.2016.10.003.

[7] S. K. Suman and V. K. Giri, "Speed control of DC motor using optimization techniques based PID Controller," 2016 IEEE International Conference on Engineering and Technology (ICETECH), Coimbatore, India, 2016, pp. 581-587, doi: 10.1109/ICETECH.2016.7569318.

[8] D. -H. Nguyen and V. -H. Nguyen, "Robust Control of Two-Axis Gimbal System," 2019 International Symposium on Electrical and Electronics Engineering (ISEE), Ho Chi Minh City, Vietnam, 2019, pp. 177-182, doi: 10.1109/ISEE2.2019.8921070.

4x1 Patch Antenna Array Using Taper-Based Coupling for 24 GHz Radar Applications

Arlen Gonzalez-Azuara
Department of electronics and telecommunications
CICESE
Ensenada, Mexico
arlen@cicese.edu.mx

Gabriela Méndez-Jerónimo
Department of electronics and telecommunicaions
CICESE
Ensenada, Mexico
ORCID: 0000-0001-8726-0814

Humberto Lobato-Morales
Department of electronics and telecommunicacions
CICESE
Ensenada, Mexico
hlobato@cicese.mx

Germán A. Álvarez-Botero
Institute of Smart Cities
Universidad Pública de Navarra
Pamplona, España
ORCID: 0000-0002-1143-6104

Abstract— Nowadays, microstrip patch antenna arrays are an important part of radar applications. In fact, due to their advantages such as small size, simple and economical fabrication, these structures are used as transmitters and receivers in radars operating at 24 GHz. At these frequencies, there are several considerations that designers must take into account, one of which is the appropriate antenna coupling feed. In this work, a 4x1 patch antenna array using conventional stepped impedance transitions is designed and compared to the proposed array using taper-based impedance transitions. Full-wave simulation data show how the proposal allows for a better coupling between the patch antennas and the feeding microstrip line. In fact, a reflection coefficient value of -43.60 dB was observed, compared to -18.55 dB and -22.23 dB for the single patch and stepped array designs, respectively. The above points highlight the simplicity and convenience of taper-based transitions at microwave operating frequencies.

Keywords—patch antenna, coupling feed, impedance transformer, taper transition, S-parameters.

I. INTRODUCTION

Microstrip patch antennas are devices widely used in several applications, due to the design and economic advantages offered by the materials and fabrication processes [1]. Currently, radar systems are one of the areas where microstrip patch antennas are receiving more attention [2]. This is because they allow the design of small devices with a relatively simple and economical design and fabrication process. In the field of radars, different operating frequencies of interest exist according to their application, and due to their small wavelength (λ), radars operating at 24 GHz are used as short-range radar, useful for detecting small objects. Since antennas are considered the most important element of wireless communication systems, the design of new architectures, the improvement of fabrication techniques, and the reduction of their losses are relevant contributions to the state of the art. In works such as the one reported in [2], a novel 1x16 series-fed microstrip patch array at 24 GHz is designed for automotive radar systems to improve the gain and directivity. Similarly, in [3] the design of the 24 GHz vehicle radar antenna method of a low sidelobe antenna array is proposed.

On the other hand, at lower frequencies, it is possible to find very interesting designs such as those presented in [4] and [5], where a 4x1 and 4x2 array show an increase in the gain and directivity. However, these types of designs consider the use of quarter wave transformers, which introduce abrupt impedance transitions and for applications at 24 GHz, they introduce multiple reflections due to the abrupt impedance transitions. This effect has been studied for years in high-frequency devices such as the one presented in [6], where it was proposed to use taper-based transitions instead of stepped impedance changes for coupling a microstrip to a substrate-integrated waveguide.

In this regard, the present work proposes a 4x1 microstrip patch antenna array operating at 24 GHz with taper-based impedance couplers, which allows to reduce the reflection losses of the antenna. Section II presents the design of the single patch operating at 24 GHz, the 4x1 design with abrupt impedance changes and the 4x1 array with the new taper-based transitions. Section III, explains the full-wave simulations carried out to validate the proposals and the results and discussion is presented in Section IV. Finally, Section V corresponds to the conclusions of this work.

II. MICROSTRIP PATCH ANTENNAS DESIGN

One of the most important parameters in the design of radio frequency and millimeter wave devices is the selection of appropriate materials. In this case, all the designs were carried out considering Rogers RO4350B as substrate, it has a low loss tangent ($\tan\delta = 0.0037$) and its relative permittivity ($\varepsilon_r = 3.66$) allows for a small size of the antenna. Besides this, substrate height (h) of 0.254 mm and copper thickness (t) of 35 μm were considered.

A. Single Microstrip Patch Antenna

To design the 4x1 microstrip patch antenna array, it is necessary to start with the design of a single patch. Fig. 1 shows the geometry of this single patch and it indicates the main dimensions. The width of the patch (W) can be calculated using [7]:

$$W = \frac{c}{2f_r}\sqrt{\frac{2}{\varepsilon_r + 1}} \tag{1}$$

This work was partially supported by CONACyT-Mexico through scholarship 1136109

979-8-3503-1191-4/23 $31.00 © 2023 IEEE

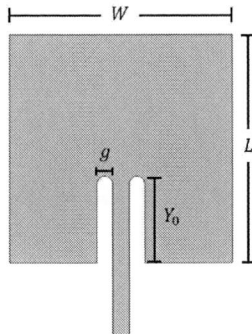

Fig. 1. Sketch of a single microstrip patch antenna with microstrip line feed. The main dimensions are indicated.

where f_r is the center frequency of the antenna. On the other hand, the patch length (L) is determined by means of [7]:

$$L = L_{eff} - 2\Delta L \qquad (2)$$

$$L_{eff} = \frac{c}{2f_r\sqrt{\varepsilon_{eff}}} \qquad (3)$$

$$\frac{\Delta L}{h} = 0.412 \frac{(\varepsilon_{eff} + 0.3)(W/h + 0.264)}{(\varepsilon_{eff} + 0.258)(W/h + 0.8)} \qquad (4)$$

$$\varepsilon_{eff} = 0.5(\varepsilon_r + 1) + 0.5(\varepsilon_r - 1)\left(1 + \frac{12h}{W}\right)^{-1/2} \qquad (5)$$

In (2), L_{eff} is the effective length of the patch and ΔL is a term known as the extension of the length, and they are given by (3) and (4) where ε_{eff} is the effective permittivity of the medium. For microstrips lines, it can be calculated from (5).

From Fig. 1, it is possible to see that Y_0 and g are dimensions related to the feeding of the antenna. They are defined by the following expressions:

$$Y_0 = \frac{L}{\pi}\cos^{-1}\left(\frac{Z_{in}}{R_{in}}\right) \qquad (6)$$

$$g = \frac{c}{f}\left(\frac{4.65 \times 10^{-12}}{\sqrt{2\,\varepsilon_{eff}}}\right) \qquad (7)$$

In (6), Y_0 is the inset feed depth, L is given by (2), Z_{in} and R_{in} are the resonant input impedance (taken as 50 Ω) and resonant input resistance respectively. Finally, in (7) g is the inset feed gap [8-9].

All the resulting calculated dimensions for $f_r = 24$ GHz considering the previously described substrate, are summarized in Table I. It should be noted that the inset feed gap has been modified in order to provide a better response.

TABLE I.   GEOMETRIC DIMENSIONS CALCULATED FOR A SINGLE PATCH

Parameter	Dimension (mm)
W	4.20
L	3.18
Y_0	1.16
g	0.23

B. Array Design

Using the single patch antenna above designed as the radiating element, the 4x1 patch antenna array shown in Fig. 2 is designed. As it is shown in Fig. 2, the array consists of a bifurcating network where the distance between the patches is $\lambda/2$. To achieve a uniform feed, a division of two lines with a characteristic impedance of 100 are used to divide the power and phase (see Fig. 2). Since a microstrip feed of 50 is used, a quarter wave transformer of $\lambda/4$ length is used to match the 50 and 100 lines. The calculated impedance of these transformers is 70.71. The corresponding widths and lengths of the microstrip lines to obtain the desired characteristic impedance and electrical length of 90° were calculated for the frequency and material considered [10], and the results are indicated in Table II.

TABLE II.   DIMENSIONS OF THE MICROSTRIP FEED LINE

Impedance (Z)	Dimensions	
	Width (mm)	Length (mm)
50 Ω	0.53	1.86
70.71 Ω	0.27	1.92
100 Ω	0.11	1.99

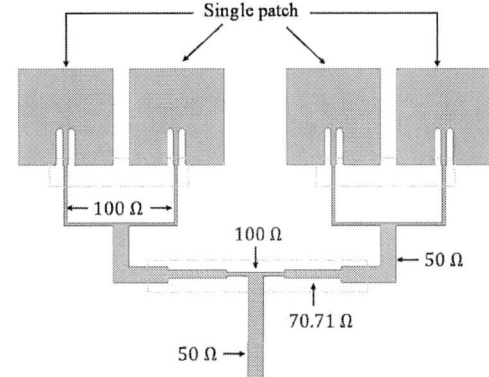

Fig. 2. Illustration of the 4x1 patch antenna array designed using quarter lambda transformers.

C. Taper-Based Transitions Design

The array shown in Fig. 2, is most used for applications at frequencies below 5 GHz. However, as it is known that abrupt

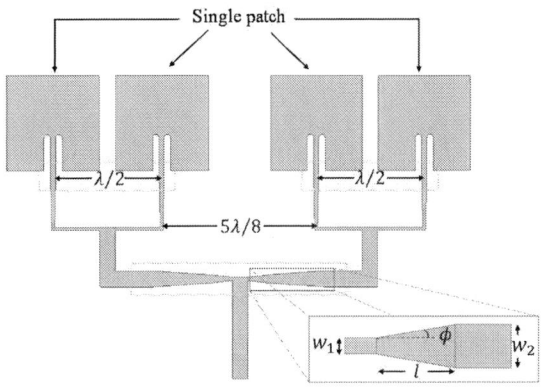

Fig. 3. Proposed 4x1 patch antenna array using taper-based transitions to reduce the reflections due to impedance changes.

impedance transitions can introduce multiple reflections as the operating frequency increases. In fact, at 24 GHz the effect of stepped transitions is expected to be more accentuated. For this reason, the configuration for the 4x1 array shown in Fig. 3 is proposed.

This proposal includes taper transitions in the feeding network. A taper transition is a microstrip line with variable width, beginning with a width w_1 and finishing with a width w_2 used to match the corresponding impedances. In this case, taper transformers realize the impedance matching between the 100 Ω impedance resulting from the split of the main feeding line, with the 50 Ω impedance of each line feeding an antenna pair. As shown in Fig. 3, the physical dimensions calculated for w_1 and w_2 are 0.11 mm and 0.53 mm, respectively. The length of the taper (l) is selected close to $\lambda/4$, because at the millimeter frequency range this is enough to minimize reflection loss, and it allows to maintain the distance between the patches [6]. Then, considering the wave propagation velocity in the substrate, the value obtained for the taper length is $l = 2.65$ mm. Finally, the taper angle obtained for calculated dimensions is $\phi = 4.53$ degrees.

III. FULL-WAVE SIMULATIONS

In order to validate the proposed designs and to systematically identify the effect of the main elements of the array, full-wave simulations were carried out using the Ansys HFSS software. The three designs described in Section II were implemented in HFSS considering Rogers RO4350B as the substrate, with the characteristics mentioned in the previous section. In addition, adaptive meshing and copper with a root mean square roughness (Rq) of 0.4 μm were considered for simulations up to 30 GHz.

It is important to note that all simulations were performed including the connector's effect for a more realistic result. For this purpose, the Southwest 1092-04A-5 End Launch Connector was selected; materials and dimensions were obtained from its data sheet [11]. Fig. 4 shows the 3D model implemented in Ansys HFSS for the single patch antenna, including the connector. The results of these simulations are presented in the following section.

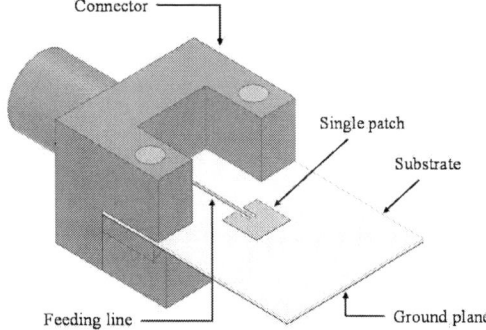

Fig. 4. 3D model implemented in Ansys HFSS for the full-wave simulation of the single patch antenna including the effect of the connector. The main elements of the model are indicated.

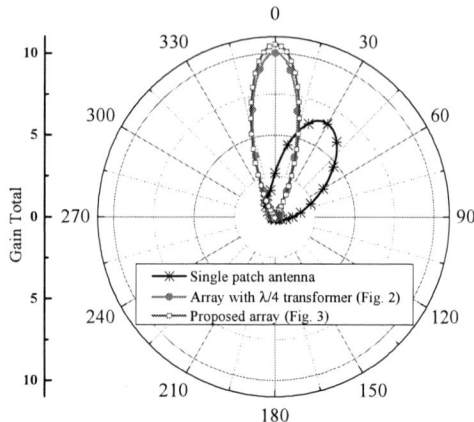

Fig. 5. Total gain radiation pattern obtained for the single patch antenna and the 4x1 arrays with stepped (Fig. 2) and taper-based (Fig. 3) impedance transformers.

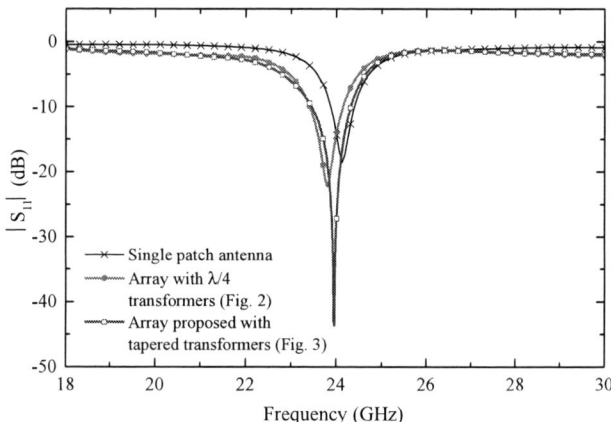

Fig. 6. Comparison between the S_{11} parameter of the single patch antenna and the 4x1 arrays with stepped and taper-based impedance transformers.

IV. RESULTS

Full-wave simulations were implemented for all the designs proposed in this work. Fig. 5 shows the comparison between the total gain radiation pattern obtained for the single patch antenna, and the 4x1 arrays with stepped and taper-based impedance transformers (see Fig. 2 and 3) at 24 GHz, where magnitudes of 6.56 to 30 degrees, 10.03 and 10.54 to 0 degrees, respectively, were obtained. As we can see, when we use an array of antennas like the proposed here, it is possible to obtain more gain as well as better directivity. From Fig. 5 it can be seen that the proposed array (Fig. 3) has a higher gain compared to the array corresponding to Fig. 2. This improvement in the performance of the antenna array can also be observed in the reflection parameters of Fig. 6. In this figure, it is clear how the addition of the tapers allows a better coupling between the microstrip lines and the patch antennas. Also, it is interesting to observe that when considering the 4x1 array of Fig. 2, the center frequency of the response is shifted to a lower frequency, due to the presence of power dividers and the high reflections in the regions of impedance changes. Comparing the three designs, we can observe that the arrays present high bandwidths with respect to the single patch. In fact, bandwidths of 3.07% and 3.59% were determined for the designs of Fig. 2 and Fig. 3, respectively.

979-8-3503-1191-4/23 $31.00 © 2023 IEEE

On the other hand, observing the reflection level of the designs of Fig. 2 and the proposal (Fig. 3), it is clear that the proposal allows for a better coupling between the microstrip lines used as feed and the radiating element. It is worth mentioning that a better coupling allows a higher power at the input of the antenna, and then, it contributes to reduce the power losses of the whole system.

V. CONCLUSIONS

This paper shows that it is possible to reduce the reflection losses of array antennas at 24 GHz when taper-based couplings are used instead of stepped transformer impedances. For this purpose, a 4x1 patch antenna array was proposed, using tapered impedance transformers to couple the 50 and 100 ohms feeding microstrip lines. The proposal was compared with a similar geometry but without taper-based couplings. A significant reduction in mismatch losses was observed. In this way, the proposal presents the advantage of being a simple way to contribute to the reduction of the noise/signal ratio of antenna systems at 24 GHz used in radar applications. Finally, this work allows to point out the simplicity and convenience of using the taper-based transitions at microwave operating frequencies, so that, it opens the possibility of extending the benefits of using these transitions in other applications.

REFERENCES

[1] I. Singh, and V. S. Tripathi. "Microstrip patch antenna and its applications: a survey," *Int. J. Comp. Tech*, vol. 2, no. 5, 2011, pp. 1595–1599.

[2] F. Suliman and A. Yazgan, "24 GHz Patch Antenna Array Design with Reduced Side Lobe Level for Automotive Radar System," in *Proc. 28th Signal Processing and Communications Applications Conference (SIU)*, Gaziantep, Turkey, Oct. 2020, pp. 1–4.

[3] Z. Xiang, B. Wang and X. Tian, "Design of Low Sidelobe Antenna Array for 24GHz Vehicular Radar," in *Electromagnetics*, vol. 41, no. 7, Dec. 2021, pp. 533–543

[4] N. Ab Wahab, Z. B. Maslan, W. N W. Muhamad and N. Hamzah, "Microstrip rectangular 4x1 patch array antenna at 2.5 GHz for WiMax application," in *Proc. 2nd International Conference on Computational Intelligence, Communication Systems and Networks*, Liverpool, United Kingdom, Nov. 2010, pp. 164–168.

[5] A. R. Nair, A. S. Bharati, and S. S. Thakur, "Design of Rectangular Microstrip 4x2 Patch Array Antenna at 2.4 GHz for WLAN Application," in *Proc. Second International Conference on Advances in Computing and Communication Engineering (ICACCE)*, Dehradun, India, May. 2015, pp. 53–56.

[6] D. Deslandes, "Design equations for tapered microstrip-to-substrate integrated waveguide transitions," in *Proc. IEEE MTT-S International Microwave Symposium*, Anaheim, CA, USA, May. 2010, pp. 704–707.

[7] C. A. Balanis, "Microstrip and Mobile Communications Antennas," in Antenna theory: analysis and design, 4th ed. Hoboken, USA: Wiley, 2016, ch. 14, sec. 2, pp. 788–798.

[8] A. M. Abdulhussein, A. H. Khidhir, and A. A. Naser. "Design and Implementation of Microstrip Patch Antenna Using Inset Feed Technique for 2.4 GHz Applications," in International Journal of Microwave and Optical Technology, vol. 16, no 4, May. 2021, pp. 355–361.

[9] M. A. Matin and A. I. Sayeed, "A design rule for inset-fedrectangular microstrip patch antenna", in *WSEAS Transactions on Communications*, vol. 9, no. 1, Jan. 2010, pp. 63–72.

[10] D. M. Pozar, "Transmission line theory," in Microwave engineering, 4th ed. USA: Wiley, 2011, ch. 2, sec. 5, pp. 72–75.

[11] Centric RF, "1092-04A-5 2.92mm End Launch Conn 40ghz .007" Pin .039" Dielectric Dia PCB to 0.1" Block 0.5" Wide" https://www.centricrf.com/connectors/end-launch-sma-2-92-2-4-1-85-1-0-connectors/1092-04a-5-2-92mm-end-launch-connector/ (accessed Feb 13, 2023).

979-8-3503-1191-4/23 $31.00 © 2023 IEEE

IoT-Based Neonatal Incubator Monitoring System

Alisson Waleska Martínez
Engineering Faculty
Universidad Tecnológica
Centroamericana (UNITEC)
Tegucigalpa, Honduras
aliwalem@unitec.edu

Fernanda de Lourdes Cáceres
Engineering Faculty
Universidad Tecnológica
Centroamericana (UNITEC)
Tegucigalpa, Honduras
fernanda_lcl@unitec.edu

Kevin Fabricio Martínez
Engineering Faculty
Universidad Tecnológica
Centroamericana (UNITEC)
Tegucigalpa, Honduras
kevin.cruz@unitec.edu

Abstract—Neonatal incubators are essential in neonatology rooms and, as they are life support equipment, they require a maintenance frequency of at least every six months to guarantee their performance and safety. To analyze them, several devices exist designed to measure parameters such as temperature, humidity, and noise level inside the baby´s chamber. A complete analysis consists of 9 tests with a duration of 5 hours and 56 minutes overall. During this time, the workflow of the incubator is interrupted. The objective of this research is to develop an IoT-based prototype capable of monitoring the parameters inside the chamber; without stopping the operation of the equipment and that provides the security of not affecting the health of the premature baby. The IoT architecture was designed and the safety criteria for each component used to build the prototype were determined. Experimentation was used to perform tests in simulated environments and in a real environment. The results measured by the prototype were compared with those obtained by the certified measurement instrument Fluke INCU II. The values were compared using non-parametric statistic analysis to determine points for improvement.

Index Terms—biomedical engineering, IoT, monitoring system, neonatal incubators, prototype

I. INTRODUCTION

Neonatal incubators are the medical devices that create an ideal environment for premature babies where temperature, humidity, noise, and oxygen flow are regulated [1]. This devices provide a protected environment for low birth weight infants [2]. These are used to increase the chances of survival of premature babies in the extrauterine environment [3]. It allows the baby's body reduce the heat loss [4].

A neonatal incubator has a chamber to place the baby covered by a transparent canopy. Several access ports allow the medical team access to the baby for feeding, examination and treatment. The baby is laid on a mattress inside located on top of a fan and heater. These are responsible for providing a forced circulation of hot air within the compartment [5].

The malfunction of neonatal incubators can cause serious injuries and even death of premature babies. The most frequent factors that cause these events are: the lack of inspection, the lack of an analysis protocol, errors in the production of said equipment and environmental noise [4] [6].

An incubator analyzer measures parameters such as temperature, humidity, sound, and air circulation inside the incubator. It´s main objective is to verify that it is operating within the

ranges defined by both the manufacturer and the institution that owns the equipment [7].

There are neonatal incubator analyzers from different manufacturers and different models. However, one of the most widely used analyzers worldwide is the Fluke INCU II. It is easy to use and gives reliable results ensuring the safety of babies [7].

The UNE-EN 60601-2-19 standard describes the essential safety requirements for the proper operation of neonatal incubators, ensuring the health of the newborn and the operator. This standard contains the mechanical and electrical safety guidelines, as well as regulations related to protection against excessive temperatures and alarm systems [8].

The Fluke analyzer allows you to verify all international standards: IEC 60601-2-19, IEC 60601-2-20 and IEC 60601-2-21. This device allows to check incubators and radiant heaters. It performs 9 different tests [7].

It is important to mention that each test has a different duration time. The analyzer needs 5 hours and 56 minutes to run all tests. It is not possible to analyze an incubator without stopping it's use and altering the workflow within a hospital.

This research project proposes the development of an initial prototype capable of performing certain tests of a neonatal incubator analyzer that does not interrupt workflow and is safe for the baby. This device allows real-time monitoring of certain environmental parameters created by the incubator.

This paper is divided in chapters. Chapter II shows the state of the art. Chapter III the methodology used for the development of the prototype, IV shows the results and finally chapter V the conclusions.

II. STATE OF THE ART

Multiple medical devices, sensors, diagnostic and imaging devices can be seen as smart objects. IoT-based healthcare services are expected to reduce costs, increase quality of life and enrich the user experience [9]. The IoT healthcare network facilitates the transmission and reception of medical data, and allowing the use of communications [15]. IoT in healthcare has multiple applications like remote monitoring; this includes patients and clinical settings [11] [16].

979-8-3503-1191-4/23 $31.00 © 2023 IEEE

A. Temperature And Relative Humidity Verification Device Inside Infant Incubator Through IoT in Thailand

The authors of reference [6] developed a device that measures the temperature and humidity inside the chamber of the neonatal incubator remotely through IoT. 5 sensors are positioned on the baby's mattress that transmit the data to Thing Speak platform via Wi-Fi.

B. Autonomous Incubation System for the Monitoring of Premature Infants in Romania

The authors in [17] designed an incubation system in which temperature and humidity are controlled autonomously. Different software and platforms were used, which are: IBM Cloud, the IBM Watson IoT platform and MIT App Inventor. The system collects information about the infant, sets the optimal conditions based on that, and displays the LCD screen.

C. Emergency Alert System for Neonatal Unit using IoT

In reference [18] the author created a system that can monitor several incubators to solve the problem of not receiving early alerts to treat premature babies. If the system detects an alteration, an alert will be sent to the medical professionals through the Blynk IoT software application.

III. METHODOLOGY

The research was developed with a mixed approach. Information from standards was used to select the safest components for the prototype. Operational tests were done both in a controlled environment and in a real one, the parameters measured by the prototype and an incubator analyzer were recorded. They were then compared to determine the effectiveness of the prototype.

According to [12] IoT architecture for medical applications follow the three-tier framework that includes application, network and perception layers. Fig. 1 shows the architecture proposed as a baseline for design. The device's sensors connect to a microcontroller with a Wi-Fi connection.

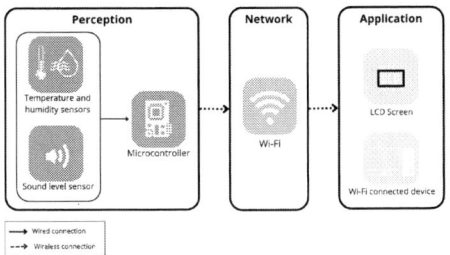

Fig. 1. Proposed IoT architecture

IV. RESULTS AND DISCUSSION

A. A. Criteria for the Prototype's Design

From information obtained in bibliographic databases, the importance of monitoring temperature, relative humidity and noise was determined. The UNE-EN 60601-2-19 /A1:2016 standard was used to determine the safety requirements inside neonatal incubators. The criteria taken into consideration for the development of the prototype are:

- The device casing must be able to withstand temperatures of at least 38°C.
- The temperature sensor must have an accuracy of ± 0.8 °C.
- The humidity sensor must have an accuracy of ± 10
- There must be a microphone amplifier module to measure the noise inside the incubator room.
- The device must be located so that it is difficult for the baby to touch some of its surfaces.

A 38-pin ESP32 was used as a microcontroller, due to the ease with which it can transfer data via Wi-Fi. It can operate normally in a temperature range of -40°C to 85°C, ideal for use inside the neonatal incubator compartment. It is also characterized by it´s ability to operate continuously for long periods of time, their small size and the possibility of connecting different sensors and actuators that work together.

An LCD screen shows the registered values, making it easier for the user to read them. A 20x4 LCD screen was used with the purpose of displaying the data in an orderly manner. The I2C module was used to connect with the ESP32 with only 4 pins. The screen can operate normally in temperatures from 0 to 50°C.

A DHT22 sensor was used to measure the temperature and relative humidity (RH) inside the chamber. It's accuracy of ±0.8°C and <10% RH matches the standard's requirements.

For the noise leve detection the MAX4466 microphone module with amplifier was selected, which is capable of detecting sounds and can work with the ESP32 microcontroller.

A copper plated Bakelite PCB on one side was used to build the electronic circuit. A Risk Analysis Matrix was done for the material selection of the device's casing. PMMA acrylic was used for the prototype casing because it supports temperatures of up to 75°C and has good stability to ultraviolet rays, ideal for situations in which the baby is receiving photo-therapy. It was decided not to use any type of acrylic glue because when exposed to high temperatures it could emit toxic gases, so screws were used to join the casing.

The interior of the device was covered with Expanded Polystyrene (EPS) due to its ability to insulate heat and it supports temperatures of up to 85°C withput emitting toxic gases [13]. This material does not release harmful gases when exposed to high temperatures and is one of the most resistant to the adverse effects of moisture [14].

A 10000mAh power bank was used as an external power supply for the microcontroller. It was decided to use a long micro-USB to USB cable to locate the power bank outside the incubator and avoid exposing it to high temperatures. The battery allows continuous operation for more than 10 hours.

Tests were carried out on the components operating individually with the microcontroller and then tests on all the components working together. Fritzing software was used to perform the schematic design and the PCB diagram as seen in Fig. 2.

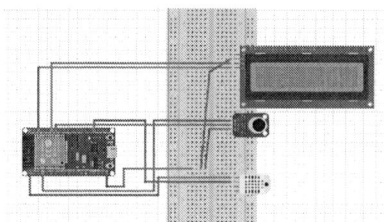

Fig. 2. Circuit schematics on Fritzing

B. Prototype´s Case

The size of the device were designed so that does not interfere with the access to the chamber. The casing has four suction cups on its front face to firmly hold the device in place and prevent it from falling.

Fig. 3 shows an isometric view of the prototype casing in SolidWorks with it´s size in cm. A laser cutter was used to make the acrylic cuts precisely. The size of the casing is enough to keep all components safely inside and small to be placed inside the incubator, as seen in Fig. 4.

The prototype design provides the facility to open its casing for maintenance in case any component fails. It also provides the opportunity of cleaning the device to prevent cross-contamination.

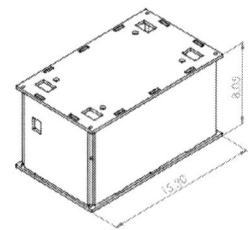

Fig. 3. Isometric view of casing

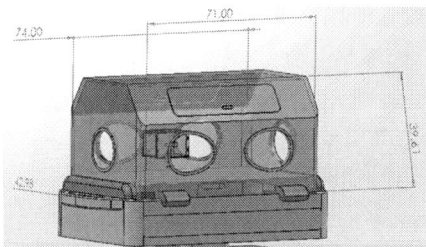

Fig. 4. Simulation inside incubator

C. Final IoT architecture

Fig. 5 shows the final architecture, which includes every component included in the prototype after testing. The DHT22 temperature and humidity sensor are connected via GPIO25 to the ESP32, and the MAX4466 is connected via GPIO39.

The next block is made up of the LCD screen that displays the data when connected to the ESP32. Simultaneously, the ESP32 connects to a Wi-Fi signal and transfers the measured data to the UBIDOTS IoT platform linked with a dashboard token number. This platform receives the information from the sensors and displays them on a dashboard. The last block consists of the display, any electronic device with internet access and that has the display link can monitor data transferred to UBIDOTS.

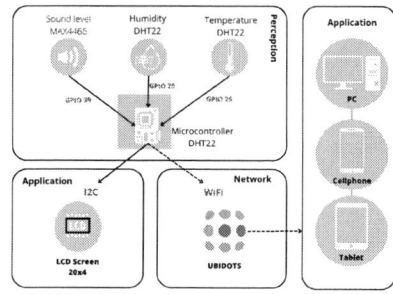

Fig. 5. Prototype´s IoT architecture

D. Prototype testing in controlled environment

Pilot tests of the prototype were carried out in a controlled environment, for which the control booths of the Industrial Engineering laboratory of UNITEC, Tegucigalpa were used. These cabins have sound, high temperature, low temperature, and humidity simulators.

The prototype's data was compared to a hygrometer and decibel meter. The prototype was fixed to the wall of the controlled cabin to see if the suction cups can keep it in place. It was observed that data was transferred simultaneously to UBIDOTS.

E. Prototype testing in real environment

Functional tests of the prototype were carried out in a neonatal incubator at Hospital María in Tegucigalpa. A Fluke INCU II was used simultaneously to compare the data. Fig. 6 shows both devices measuring the 3 parameters.

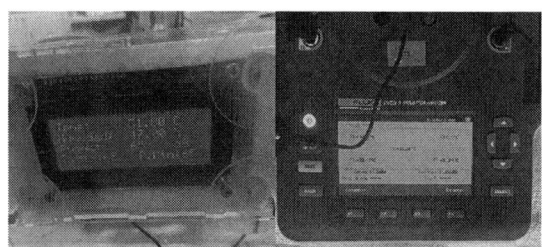

Fig. 6. Performing test with Fluke INCU II

The tests lasted for 6 hours, the incubator was turned on at a control temperature of 36°C and humidity of 60% RH. The Fluke analyzer was placed inside the incubator enclosure, general test was selected as the test type.

The values were manually recorded every five minutes for the first hour, then every 10 minutes until the end of the second hour and every hour until completing 6 hours. The results for the temperature sensor are shown in Table I with the difference between both devices. The same was done for the humidity and noise levels.

TABLE I
TEMPERATURE READINGS IN 6 HOURS

Time(min)	Prototype(°C)	Fluke(°C)	Difference(°C)
0	29.7	32.1	-2.4
5	30	32	-2
10	31.2	32.1	-0.9
15	31.4	32.2	-0.8
20	31.6	32.2	-0.6
25	31.8	32.28	-0.48
30	31.9	32.3	-0.4
35	31.9	32.2	-0.3
40	32	32.3	-0.3
45	32.2	32.37	-0.17
50	32.2	32.35	-0.15
60	33.6	33	0.6
70	33.6	33	0.6
80	33.4	32.71	0.69
90	33.3	32.57	0.73
100	33.1	32.7	0.4
110	32.7	32.7	0
120	32.7	32.8	-0.1
180	33.3	32.81	0.49
240	33	32.75	0.25
300	33	32.86	0.14
360	33	32.82	0.18

To compare the data from both devices, SPSS was used to determine if there is a significant difference between the 3 parameters. First, a normality test using the Shapiro-Wilk method was carried out due to the sample size being less than 50. The results show that non of the three parameters have a normal behavior and non-parametric tests were needed.

A non-parametric sign test was carried out to determine if the measurements of the prototype were significantly different from the Fluke. According to the results from SPSS the significance between the temperature readings is 0.664, which is greater than 0.05, which means the data is significantly similar. For humidity and noise this was not the case, being 0 the significance value; which means those two parameters are significantly different. This is verified showing that humidity values vary by approximately 5 to 6% and the noise values by 5 to 6 dB.

V. CONCLUSIONS

A prototype was developed capable of emulating the general test of temperature sensing, accuracy of relative humidity and the sound level inside the baby´s chamber of an incubator. The prototype was designed considering certain safety criteria of the UNE-EN 60601-2-19 /A1:2016 standard and is capable of monitoring parameters inside it´s compartment without altering the workflow. The measured data is displayed on an LCD screen and on the UBIDOTS digital platform, which allows monitoring the parameters in real time remotely.

The values measured by the prototype were compared with those of the Fluke INCU II analyzer inside the chamber.

Temperature measurements are significantly similar between the prototype and the Fluke analyzer, which indicates this parameter shows the real temperature inside the incubator and can be used as an analyzer.

For the humidity and noise level was not the case; measurements are significantly different, which indicates those parameters are not correctly displayed by the prototype. This error could be solved using a correction factor on the code o selecting more accurate sensors.

The next step for this research is improve the code to be able to emulate more tests performed by a certified analyzer.This prototype is the first step to develop an incubator analyzer that can perform each test indicated by the standards.

ACKNOWLEDGMENT

Thanks to Hospital María de Especialidades Pediatricas, for its openness and willingness to perform the last tests for the prototype. Specially, to Engineer Gracia Girón, for her time and willingness to collaborate.

This work was supported by the Research Fund of Universidad Tecnológica Centroamericana (UNITEC), Honduras.

REFERENCES

[1] P. Tiam Kapen, M. Youssoufa, M. Foutse, J. Dongmeza Koudjou, y F. de P. Mkankam Kamga, "A multi-function neonatal incubator for low-income countries: Implementation and ab initio social impact", Med. Eng. Phys., vol. 77, pp. 114–117, mar. 2020, doi: 10.1016/j.medengphy.2019.10.021.

[2] D. B. Zimmer, A. A. P. Inks, N. Clark, y C. Sendi, "Design, Control, and Simulation of a Neonatal Incubator", en 2020 42nd Annual International Conference of the IEEE Engineering in Medicine and Biology Society (EMBC), 2020, pp. 6018–6023. doi: 10.1109/embc44109.2020.9175407.

[3] M. Koli, P. Ladge, B. Prasad, R. Boria, y N. J. Balur, "Intelligent Baby Incubator", en 2018 Second International Conference on Electronics, Communication and Aerospace Technology (ICECA), mar. 2018, pp. 1036–1042. doi: 10.1109/ICECA.2018.8474763.

[4] A. Ili Flores, H. J. Konno, A. M. Massafra, y L. Schiaffino, "Simultaneous Humidity and Temperature Fuzzy Logic Control in Neonatal Incubators", en 2018 Argentine Conference on Automatic Control (AADECA), nov. 2018, pp. 1–6. doi: 10.23919/AADECA.2018.8577290.

[5] Draeger, "Neonatal Incubators–NICU Incubators for Premature Babies Draeger", 2021. https://www.draeger.com/en-us_ us/Hospital/Neonatal-Care/Incubators-NICU.

[6] J. Prinyakupt y K. Roongprasert, "Verification Device for Temperature and Relative Humidity Inside the Infant Incubator via IoT", en 2019 12th Biomedical Engineering International Conference (BMEiCON), nov. 2019, pp. 1–6. doi: 10.1109/BMEiCON47515.2019.8990351

[7] A. Y. K. Chan, Biomedical Device Technology: Principles and Design. Springfield, Illinois, 2008.

[8] Fluke, "INCU II Infant Incubator Tester Fluke Biomedical". 2016. [Online]. Available: https://www.flukebiomedical.com/products/biomedical-test-equipment/incubator-radiant-warmer-analyzers/incu-ii-incubator-radiant-warmer-analyzer.

[9] G. E. Valdés-de la Torre, M. Martina Luna, A. Braverman Bronstein, J. Iglesias Leboreiro, y I. Bernárdez Zapata, "Medición comparativa de la intensidad de ruido dentro y fuera de incubadoras cerradas", Perinatol. Reprod. Humana, vol. 32, núm. 2, pp. 65–69, jun. 2018, doi: 10.1016/j.rprh.2018.06.002.

[10] NE, "UNE-EN 60601-2-19:2009/A1:2016 (Ratificada) Equipos electroméd...", 2016. Available: https://www.une.org/encuentra-tu-norma/busca-tu-norma/norma?c=N0057743.

[11] P. P. Ray, D. Dash, y N. Kumar, "Sensors for internet of medical things: State-of-the-art, security and privacy issues, challenges and future directions", Comput. Commun., vol. 160, pp. 111–131, jul. 2020, doi: 10.1016/j.comcom.2020.05.029.

[12] M. Aledhari, R. Razzak, B. Qolomany, A. Al-Fuqaha and F. Saeed, "Biomedical IoT: Enabling Technologies, Architectural Elements, Challenges, and Future Directions," in IEEE Access, vol. 10, pp. 31306-31339, 2022, doi: 10.1109/ACCESS.2022.3159235.

[13] S. M. R. Islam, D. Kwak, MD. H. Kabir, M. Hossain, y K.-S. Kwak, "The Internet of Things for Health Care: A Comprehensive Survey", IEEE Access, vol. 3, pp. 678–708, 2015, doi: 10.1109/ACCESS.2015.2437951.

[14] IDAE, "Guía Técnica para la Rehabilitación de la Envolvente Térmica de los Edificios". 2007. [Online]. Available: https://www.idae.es/uploads/documentos/documentos_GUIA\ _TECNICA_EPS_Poliestireno_Expandido_v06_972d8feb.pdf

[15] S. Bhattacharya y M. Pandey, "Significance of IoT in India's E-Medical Framework: A study", en 2020 First International Conference on Power, Control and Computing Technologies (ICPC2T), ene. 2020, pp. 321–324. doi: 10.1109/ICPC2T48082.2020.9071513.

[16] M. M. Dhanvijay y S. C. Patil, "Internet of Things: A survey of enabling technologies in healthcare and its applications", Comput. Netw., vol. 153, pp. 113–131, abr. 2019, doi: 10.1016/j.comnet.2019.03.006.

[17] H. Alqaheri, S. Radha, J. Chatterjee, S. SHOORIYA, S. J., y N. SATISH, "Toward an Autonomous Incubation System for Monitoring Premature Infants", Stud. Inform. Control, vol. 30, pp. 121–131, dic. 2021, doi: 10.24846/v30i4y202111.

[18] V. Govindaraj, A. Thiagarajan, S. Uthirapathi, A. S. Murugiah and D. D. Jeevagan, "Emergency Alert System for Neonatal Unit using IoT," 2021 3rd International Conference on Advances in Computing, Communication Control and Networking (ICAC3N), Greater Noida, India, 2021, pp. 700-703, doi: 10.1109/ICAC3N53548.2021.9725481.

A Fast Transient Full Class AB LDO Regulator

A. Serrano-Reyes
INAOE
Puebla, Mexico
anferse@inaoep.mx

M.T. Sanz-Pascual
INAOE
Puebla, Mexico
materesa@inaoep.mx

B. Calvo-López
Universidad de Zaragoza
Zaragoza, Spain
becalvo@unizar.es

Abstract—**A novel fully integrated Low Dropout Regulator (LDO) using a class AB error amplifier, class AB indirect compensation and a QFG-based dynamic transient enhancement circuit is presented in this paper. The LDO was designed in a 180nm CMOS process with a 2.1-3V input supply and 1.8V regulated output voltage. Post-layout simulations show a line regulation of 3mV/V and a load regulation of 77μV/mA. For large load transients from 0 to 50mA, the overshoot and undershoot values are less than 330mV and 211mV, with a recovery time of 2μs while the quiescent current is below 14μA, overall achieving competitive state-of-art performance.**

Index Terms—**Voltage regulator, low-dropout regulator (LDO), CMOS design, indirect compensation, class AB amplifiers, fast-transient.**

I. INTRODUCTION

THE growing demand for Systems on a Chip (SoCs) for application in standalone devices has led current research to focus on fully integrated, power-size, efficient systems. These SoCs require precise and regulated supply voltages, therefore causing low-dropout (LDO) voltage regulators to be a central axis in their design. An LDO voltage regulator converts an input supply voltage, such as that from batteries which has variations over time, into an accurate and stable regulated voltage suitable for the standalone SoC under large variations of load and input voltage.

Conventional LDOs require a relatively large output capacitor, in the microfarad range, for stability. However, this does not make them suitable for SoC systems. Removing the large off-chip output capacitor reduces area and the overall cost of the design, but makes compensation more complex. Besides, operation with low quiescent currents to optimize power efficiency critically slows down the LDO transient performance, mainly determined by its frequency response and the slew-rate at the critical system nodes. Therefore, the design of on-chip capacitor-less LDO regulators requires alternative internal compensation schemes [1] [2] and transient enhancement techniques [3] [4] that do not degrade the overall performance in terms of regulation, size and power efficiency, a real challenge since these parameters are interrelated.

This paper is mainly focused on the improvement of the transient response of low power, fully integrated LDOs. The

simplest high-gain one stage error amplifier, a telescopic architecture, is adopted to achieve good regulating performance while easing the LDO compensation, reduced to a two-pole topology. To enhance transient performance, a class AB telescopic amplifier is designed, while a hybrid cascode-indirect Miller compensation with class AB current injection is introduced to improve both stability and dynamic response. A Time Response Enhancement Circuit (TREC) which requires minimal additional hardware and becomes active only during the transient stages but not in steady state further improves dynamic performance without increasing power consumption, thus achieving very competitive state-of-the-art performance with a simple and compact approach.

The paper is organized as follows. Section II presents the proposed LDO. Section III discusses the post-layout simulation results for static and dynamic response. Finally, the conclusions are drawn in Section IV.

II. PROPOSED LDO

Fig. 1a shows the topology of a classic CMOS LDO regulator. It consists of an error amplifier (EA), a PMOS pass transistor connected in series between the unregulated V_{IN} and the regulated output V_{OUT}, and a negative feedback resistive sampling circuit R_{F1}-R_{F2}. The EA senses the voltage difference between a reference voltage V_{REF} and the scaled output voltage $V_{FB}=[R_{F2}/(R_{F1}+R_{F2})]V_{OUT}$ and adjusts the gate of the PMOS pass transistor to deliver the demanded current to the load (modelled by C_L) over the operating V_{IN} range, keeping V_{OUT} stable according to:

$$V_{OUT} = \left(1 + \frac{R_{F1}}{R_{F2}}\right) V_{REF} \qquad (1)$$

Fig. 1b shows the scheme of the proposed LDO regulator with hybrid compensation and Fig. 2 shows its schematic. The LDO core, the compensation scheme and the TREC are next described.

A. LDO core

The M_P pass transistor is designed to handle load currents up to 50mA preserving a drop-out voltage –defined as the minimum voltage across the pass device to maintain regulation– of

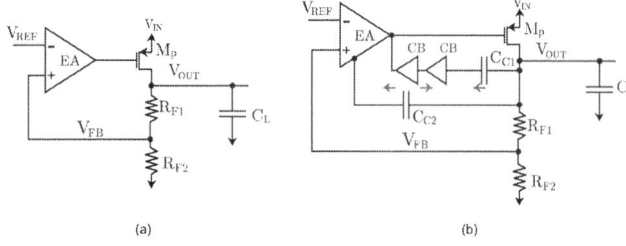

Fig. 1: LDO Regulator (a) classic compensation (b) proposed compensation

300mV, for which its length and width were set to L=0.34μm and W=4.5mm. Minimum transistor length (for 3.3V MOS transistors) is used to minimize the parasitic capacitance at the gate of the pass transistor, which is C=12pF, 20pF (for I_{LOAD}=0, 50mA).

Taking into account that V_{REF}=1.2V is an external bandgap reference, according to (1), to obtain V_{OUT}=1.8V, R_{F2}=2R_{F1} is required. These feedback resistors are implemented with three identical PMOS diode transistors (W_{Rf}=1.6μm, L=500nm) to minimize area. Values are set to R_{F1}=200kΩ and R_{F2}=400kΩ, so the static consumption in that branch is I_{QFB}=3μA (for I_{LOAD}=0), as a trade-off between power consumption and feedback resistance values.

The EA is a single-stage class AB Telescopic OTA that provides high gain -and therefore good regulating performance- with low current consumption thanks to the use of adaptive bias to obtain class AB operation [5] and quasi-floating gate (QFG) cascode transistors for enhanced performance. Briefly, transistors M_9-M_{12} and the bias current sources I_B implement a maximum circuit that sets the voltage at the common source node V_2 to V_2=max(V_{FB},V_{REF})+V_B, where V_B is the DC gate to source voltage V_{GS} of M_{11}, M_{12}, so either M_1 or M_2 experiences the full swing of the input signal and a large output dynamic current not limited by I_B is generated. The EA transistor sizes are given in Table I. It features a 61dB gain over the range (2.1V - 3V), with gain-bandwidth product >1.02MHz and a phase margin PM=89° for a load capacitor equal to C_P (12pF – 20pF), 9.8V/μs slew-rate, and only 3.8μA current consumption under static conditions.

TABLE I: Transistor sizes in the EA

	W(μm)/L(μm)
M_1, M_2	2.5/0.4
M_3, M_4	2.5/0.5
M_5, M_6, M_7, M_8	4/0.5
M_9, M_{10}, M_{11}, M_{12}	1.3/0.5

B. Compensation

For the resulting LDO to be stable, the dominant pole of the system must be located at the gate of the PMOS pass transistor over all the V_{IN}, I_{LOAD} operating conditions, being the major problem that for low load currents the non dominant pole at the output gets closer to the dominant pole, reducing the phase margin for the uncompensated LDO down to 11°, below the 60° margin that guarantees stability.

The simplest compensation strategy that does not require additional quiescent current is the cascode one, implemented through compensation capacitor C_{C2} connected from the output to the EA low-impedance node at the cascode transistor M_6. This choice, however, does not enhance the dynamic performance. Therefore, a hybrid conventional cascode-class AB indirect Miller approach has been adopted as compensation and dynamic performance enhancement strategy. Capacitor C_{C1} senses the changes in V_{OUT} in the form of a current. This current is injected into the gate of M_P through the network formed by two cascaded class AB current buffers. In this way, besides contributing to compensation, this path further enhances the driving capability at the gate of the pass transistor [1]. Note that as the current buffers invert the sense of the current, two of them are required for proper operation.

Fig. 2 also shows the class AB current buffer CB, based on the QFG technique [6]. Under dynamic operation, the change in voltage at the gates of M_1-M_2 is transferred through the three floating capacitances to the gates of M_7-M_8, providing class AB operation, and to the gates of M_3-M_4, M_5-M_6 to preserve the voltage drops V_{DS} constant, achieving an increase in linearity and dynamic range. As for the EA, reverse PMOS diodes and 0.8pF capacitances were used for the QFG implementation. The transistor dimensions are given in Table II. The CB total consumption is I_B=1.8μA, and it can handle current amplitudes up to 1mA.

TABLE II: Transistor sizes in the CB

	W(μm)/L(μm)
M_1	10/0.5
M_2	20/0.5
M_3, M_4, M_8,M_9	50/0.5
M_5, M_6, M_7, M_{10}	100/0.5

This combined compensation solution, with C_{C1}=4pF and C_{C2}=2pF, stabilizes the LDO keeping the phase margin >60° (Fig. 3) with better figures than single cascode compensation or single class AB indirect Miller compensation, at the same total compensation capacitance.

C. Time Response Enhancement Circuit (TREC)

The TREC is shown in Fig. 2 for overshoot and undershoot enhancement. The undershoot TREC operates as follows: transistor QFG M_{U3} (60μm/0.34μm) is biased in steady state with $V_{SG,U3} = V_{DD} - V_P = 0.5 < |V_{THP}|$ in such a way that it remains off; when the load current suddenly increases, capacitor C_{D2}=0.8pF senses the voltage drop in V_{OUT} making $V_{SG,U3} > |V_{THP}|$. The current generated by M_{U3} is copied by the current mirror M_{U1}-M_{U2} (6μm/0.34μm, 3μm/0.34μm) and injected into the EA V_2 node to decrease the undershoot. When V_{OUT} is regulated back to its nominal value, M_{U3} returns to the off region.

979-8-3503-1191-4/23 $31.00 © 2023 IEEE

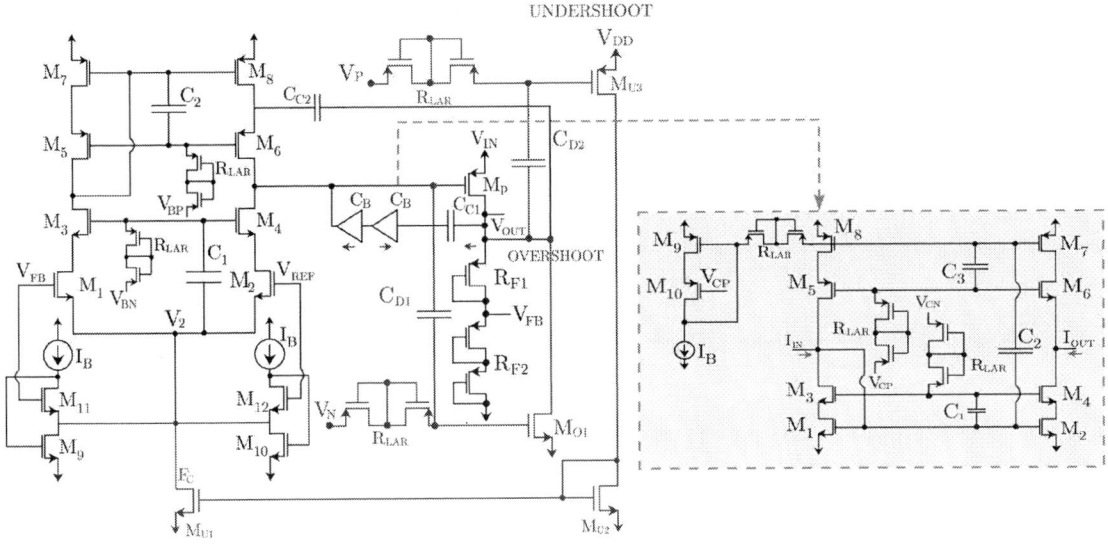

Fig. 2: Schematic of the proposed LDO Regulator

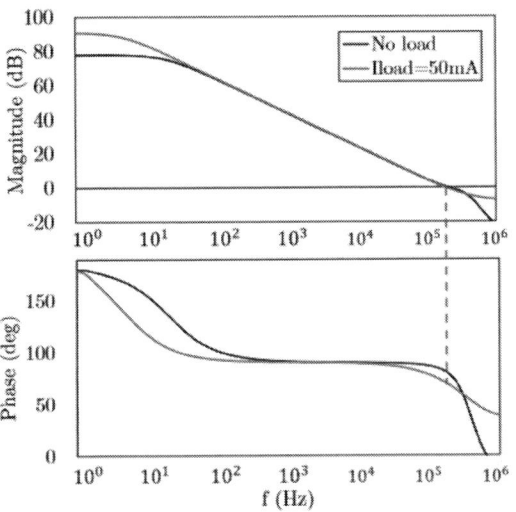

Fig. 3: Open-loop frequency response for the proposed LDO with different I_{LOAD}, V_{IN}=3V

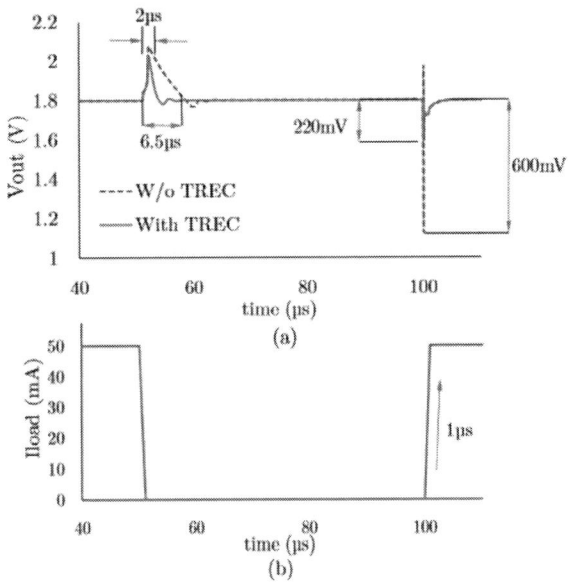

Fig. 4: Time response in (a) output voltage V_{OUT} due to (b) full load variations

Similarly, the overshoot TREC consists of QFG transistor M_{O1}, biased in such a way that it is off in steady state. When an overshoot occurs, the voltage at the EA output node increases, this change is transmitted through C_{D1}, turning on the transistor M_{O1} and generating a discharge current path at the output node. When V_{OUT} returns to its nominal value, M_{O1} returns to the off region.

Fig. 4 shows the LDO regulator response time with and without the TREC block for a full change in load current 0-50mA. Undershoots are reduced from 600mV to 220mV whereas overshoot settling times are reduced by up to 70%. Therefore, the dynamic TREC effectively improves the transient response at minimal hardware and consumption penalty.

III. POST-LAYOUT SIMULATION RESULTS

The proposed LDO regulator was designed in a 0.18μm CMOS process, using MOS transistors with 3.3V nominal voltage supplies (V_{THP}=-0.72V, V_{THN}=0.59V). It provides a regulated output voltage V_{OUT}=1.8V from an unregulated input voltage V_{IN} varying from 3 to 2.1V, for a load I_{LOAD}=0-50mA and C_L=0-100pF, with a total quiescent current consumption of only 14μA. The layout of the proposed regulator is shown in Fig. 5. The total area of the chip is equal to 0.12mm^2.

Fig. 6 shows the DC $V_{in}(V_{dd})$-V_{out} characteristic of the

979-8-3503-1191-4/23 $31.00 © 2023 IEEE   69

TABLE III: PERFORMANCE SUMMARY AND COMPARISON

	Maity (exp) [2]	Vahideh (post) [3]	Khan (sim) [7]	This work (post)
Technology	180nm	180nm	180nm	180nm
V_{IN}	1.4V	1.2V	3.3V	2.1V-3V
V_{OUT}	1.2V	1V	2.8V	1.8V
V_{DO}	200mV	200mV	83mV	300mV
I_Q	0.61-141μA	3.1μA	37.7μA	14μA
I_{LOAD}	0.01-100mA	100mA	0-50mA	0-50mA
C_{COMP}	9pF	12.5pF	25pF	6pF
C_{LOAD}	100pF	10pF	10pF	100pF
LNR=$\Delta V_{OUT}/\Delta V_{IN}$	0.6mV/V	2.23mV/V	23.4mV/V	3mV/V (worst case)
LDR=$\Delta V_{OUT}/\Delta I_{LOAD}$	270μV/mA	23μV/mA	0.31μV/mA	77μV/mA (worst case)
$\Delta V_{OUT}@T_{EDGE}$	110mV@1μs	220mV@1μs	800mV@1μs	330mV@1μs
Tsettle ts	5μs	3.6μs	2μs	2.3μs
FOM(ps)[a]	634	141	3700	38

[a]$FOM=(t_s I_Q C_C)/(I_{LOAD} C_L)$

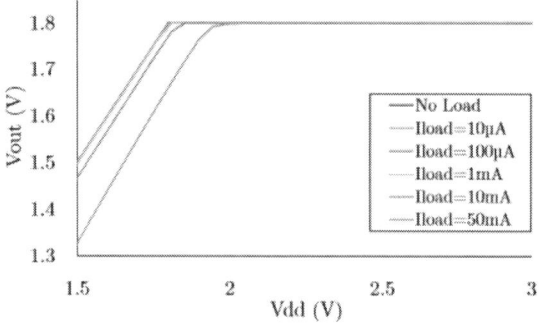

Fig. 5: Layout of the proposed LDO regulator

proposed LDO regulator. The LDO output voltage is set to 1.8V in a regulation region ranging from 3 to 2.1V, so the dropout is 300mV (Fig. 6) at the maximum output current 50mA. The load regulation for V_{DD}=3V (worst case) is 0.077mV/mA and the line regulation for I_{LOAD}=0mA (worst case) is 3mV/V. In the transient response of the proposed LDO at V_{IN}=2.1V and CL=100pF, the measured overshoot and undershoot are less than 330mV and 211mV under maximum load current changes (with an edge time of 1μs), and the recovery time is 2.3μs.

Fig. 6: DC characteristic of the proposed LDO regulator for different I_{LOAD}

The performance summary and comparison with other recent fully integrated designs is shown in Table III. The proposed design actually presents the lowest FOM, i.e., the best trade-off between power consumption, dynamic performance, and capacitive compensation/load ratio load. Furthermore, it can work under no-load conditions, as it is properly compensated in all cases.

IV. CONCLUSION

A compact low quiescent current 1.8V CMOS LDO voltage regulator has been proposed, which relies on a high slew-rate error amplifier together with a combined stability/speed circuit strategy and a dynamic TREC to achieve a stable LDO over full load conditions with fast transient performance. Based on the achieved figures, the proposed LDO regulator is suitable to be integrated into SoC power management modules.

ACKNOWLEDGMENT

This work was supported by CONACYT through the Doctoral Grant 787982 and the Ministerio de Ciencia e Innovación (Spain) through the PID2019-106570 RB-I00 Research Project.

REFERENCES

[1] R. J. Milliken, J. Silva-Martinez and E. Sanchez-Sinencio, "Full On-Chip CMOS Low-Dropout Voltage Regulator," in IEEE Transactions on Circuits and Systems I: Regular Papers, vol. 54, no. 9, pp. 1879-1890, Sept. 2007, doi: 10.1109/TCSI.2007.902615.

[2] A. Maity and A. Patra, "Tradeoffs Aware Design Procedure for an Adaptively Biased Capacitorless Low Dropout Regulator Using Nested Miller Compensation," in IEEE Transactions on Power Electronics, vol. 31, no. 1, pp. 369-380, Jan. 2016, doi: 10.1109/TPEL.2015.2398868.

[3] V. Shirmohammadli, A. Saberkari, H. Martinez-Garcia and E. Alarcón-Cot, "Low power output-capacitorless class-AB CMOS LDO regulator," 2017 IEEE International Symposium on Circuits and Systems (ISCAS), Baltimore, MD, USA, 2017, pp. 1-4, doi: 10.1109/IS-CAS.2017.8050958.

[4] Y. Kim and S. Lee, "A Capacitorless LDO Regulator With Fast Feedback Technique and Low-Quiescent Current Error Amplifier," in IEEE Transactions on Circuits and Systems II: Express Briefs, vol. 60, no. 6, pp. 326-330, June 2013, doi: 10.1109/TCSII.2013.2258250.

[5] M. P. Garde, A. Lopez-Martin and J. Ramirez-Angulo, "Power-efficient class AB telescopic cascode opamp", Electronics Letters, vol. 51, no. 10, pp- 620-622, May 2018.

[6] F. Esparza-Alfaro, A. J. Lopez-Martin, J. Ramirez-Angulo and R. G. Carvajal, "Low-voltage highly-linear class AB current mirror with dynamic cascode biasing," Electronics Letters, vol. 48, no. 31, pp. 1336-1338, 2012.

[7] M. Khan and M. H. Chowdhury, "Capacitor-less Low-Dropout Regulator (LDO) with Improved PSRR and Enhanced Slew-Rate," 2018 IEEE International Symposium on Circuits and Systems (ISCAS), Florence, Italy, 2018, pp. 1-5, doi: 10.1109/ISCAS.2018.8351039.

2023 IEEE Latin American Electron Devices Conference (LAEDC)
Puebla, México, July 3-5, 2023

A Multi-Stage CTLE Design and Optimization for PCI Express Gen6.0 Link Equalization

Karla G. López-Araiza[#*1], Francisco E. Rangel-Patiño[#*2], Jorge E. Ascencio-Blancarte[#*3],
Edgar A. Vega-Ochoa[#4], José E. Rayas-Sánchez[*5], and Omar Longoria-Gandara[*6]

[#] Intel Corp. Zapopan, Jalisco, 45019 Mexico

[*] Department of Electronics, Systems, and Informatics, ITESO – The Jesuit University of Guadalajara,
Tlaquepaque, Jalisco, 45604 Mexico

[1]karla.lopez.araiza, [2]francisco.rangel, [3]jorge.e.ascencio.blancarte, [4]edgar.vega.ochoa, {@intel.com},
[5]erayas, [6]olongoria {@iteso.mx}

Abstract—The continuously increasing bandwidth demand from new applications has led to the development of the new peripheral component interconnect express (PCIe) Gen6, reaching data rates of 64 giga-transfers per second (GT/s) and adopting the pulse amplitude modulation 4-level (PAM4) signaling scheme. While PAM4 solves the bandwidth requirements, it brings new challenges for the physical channel design. PAM4 is more susceptible to errors due to various noise sources caused by reduced voltage (and timing) ranges, yielding a higher bit error rate (BER). It also introduces new challenges in slicers, transition jitter, and equalizers, making of equalization (EQ) a critical process for PAM4 signaling. In this paper, we propose a multi-stage continuous-time linear equalizer (CTLE) with high-band, mid-band, and low-band frequency boost stages to deal with highly lossy channels. Given the complexity of EQ of multi-level signals, optimization techniques are used, including an efficient optimization of the transmitter finite impulse response (FIR) filter and the receiver CTLE tuning.

Keywords—channel, CTLE, equalization, eye-diagram, FIR, ISI, jitter, optimization, PAM4, PCIe, receiver, transmitter.

I. INTRODUCTION

The ever-increasing bandwidth required by new applications has deployed the peripheral component interconnect express (PCIe) Gen6, reaching data rates of 64 giga-transfers per second (GT/s) and adopting the pulse amplitude modulation 4-level (PAM4) signaling scheme. By contrast to the conventional non-return-zero (NRZ) signaling, the design of PAM-4 transceivers brings many new challenges for the physical channel analysis and design. The intrinsic 1/3 eye amplitudes of PAM-4 lead to a signal-to-noise ratio (SNR) penalty, and the transitions between non-adjacent levels with finite rise and fall times reduce the horizontal eye openings. Additionally, many undesired channel effects worsen with higher data rates (*e.g.*, noise and attenuation in the received signal).

An intense industry effort is presently ongoing regarding the development of PAM-4 receiver (Rx) architectures featuring high bandwidths, high gain, low noise, and high linearity [1]. In addition, equalizers are used to cancel many undesired physical channel effects, including inter-symbol interference (ISI), making PAM-4 equalization (EQ) more demanding [2]. A combination of continuous-time linear equalizer (CTLE) and decision feedback equalization (DFE) is widely used to eliminate ISI. However, due to the higher transmission rates, the conventional CTLE is no longer able to meet the

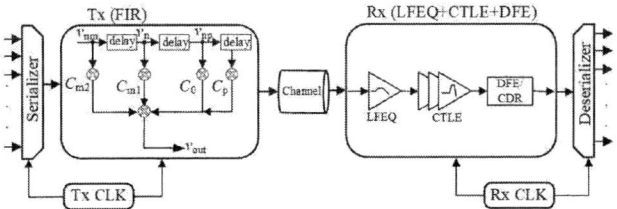

Fig. 1. Block diagram of the PCIe Gen6 serial link transceiver.

requirements in a wide range of channel losses [3],[4].

In this paper, we propose a 3-stage CTLE and a low-frequency equalizer (LFEQ) designed to compensate for highly lossy channels with high, mid and low frequency bands boosting stages. We also propose an efficient optimization methodology to determine the optimal coefficients for the transmitter (Tx) feed-forward equalizer (FFE) and the Rx CTLE. The procedure implies defining a new objective function as a figure of merit (FOM) suitable for PAM4, and then applying a numerical optimization method using a combination of pattern search [5] and Nelder-Mead [6] methods. We validate our proposed methodology by using MATLAB SerDes Toolbox with realistic parameters.

II. PCI EXPRESS EQUALIZATION

PCIe Gen6 specification defines the requirements to perform on-chip EQ at the Tx and at the Rx to mitigate undesired channel effects and minimize the bit error rate (BER). The Tx EQ coefficients for 64 GT/s are based on a FFE 4-tap finite impulse response (FIR) filter (C_{m2}, C_{m1}, C_0, and C_p) as illustrated in Fig. 1. The serial data output is obtained by the superposition of four consecutive received pulses (v_{nm2}, v_{nm1}, v_n, v_{np}) that are weighted with the four different filter coefficients [7]. The FIR filter output, v_{out}, can be adjusted by varying the coefficient values, since

$$v_{out} = v_{nm2}C_{m2} + v_{nm1}C_{m1} + v_nC_0 + v_{np}C_p \quad (1)$$

PCIe specification defines some predefined set of values for the Tx coefficients, referred to as presets, which are adaptively changed during the on-chip EQ. The Tx EQ coefficients are computed at the upstream port by the coefficient adaptation algorithm using the received signal. These coefficients are communicated to the downstream port by using the PCIe protocol. The Tx at the downstream port then applies the

This work was supported in part by ITESO (*The Jesuit University of Guadalajara, Mexico)* and Intel Corp. through a scholarship granted to K. G. Lopez-Araiza.

979-8-3503-1191-4/23 $31.00 © 2023 IEEE

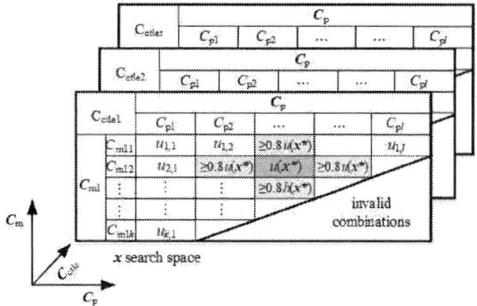

Fig. 2. EQ map coefficients search space for optimization. From [8].

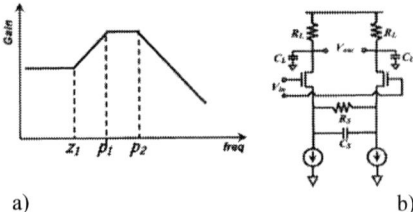

Fig. 3. a) CTLE bode plot, b) CTLE circuit using capacitive degeneration. From [9].

received coefficients setting to its Tx EQ circuitry. This process of computing the coefficients, communicating them to the Tx, and checking the signal quality can be repeated multiple times until the required BER is achieved [7],[8].

To have a unity gain for the Tx equalizer, the Tx coefficients are subject to the following protocol constraints (as per the PCIe specification [6]):

$$|C_{m2}| + |C_{m1}| + |C_0| + |C_p| = 1$$
$$\text{subject to } C_{m2} \geq 0, C_{m1} \leq 0, C_p \leq 0 \tag{2}$$

These constraints are implemented by determining only C_{m1} and C_p to fully define v_{out} from (1), being $C_{m2} = 1/24$ (per specification) and C_0 implied by (2). The coefficients must support all eleven values for the presets, and their respective tolerances, as defined by the PCIe specification [7].

When all the PCIe specification constraints are applied, the resulting coefficients space may be mapped onto a triangular matrix, as shown in Fig. 2. C_{m1} and C_p coefficients are mapped onto the y-axis and x-axis, respectively. Each matrix cell corresponds to a valid combination of C_{m1} and C_p coefficients, and $u(x^*)$ correspond to a combination of C_{m1}, C_p that results in an eye diagram qualified as optimum (see Section IV).

III. CTLE DESIGN

At higher data rates, several EQ techniques can be used to compensate ISI impairments, and then maximizing the eye diagram before the Rx sampling process fails to satisfy the required BER. Tx pre-emphasis suffers from peak power constraints, while Rx equalizer performance is limited by the amplifier bandwidth, therefore, design trade-offs are required between Tx and Rx implementations or a combination of both. However, many times the perfect channel state information is unknown, and they can change due to PCB manufacturing process, voltage, and temperature conditions. Continuous-time adaptive equalizers can be used to overcome these challenges.

A. Continuous Time Linear Equalization

A CTLE is a continuous-time circuit with high-frequency gain boosting, whose transfer function can flatten the channel frequency response. One of the most common types of CTLE is a source-coupled differential-pair circuit with source degeneration, whose basic topology is shown in Fig. 3 [9]. The differential-pair source resistor attenuates the low-frequency signals while the source capacitor allows the high-frequency signal content, resulting in high frequency gain boosting [10].

The transfer function of this circuit can be represented by one zero and two poles, where the zero provides +20dB/decade slope and a pole gives −20dB/decade giving a total of −40dB/decade. This topology can be modeled by

$$H(s) = w_{p2} \frac{s + w_{z1}}{(s + w_{p1})(s + w_{p2})} \tag{3}$$

where $w_{z1} = w_{p1}A_{DC}$, $w_{p1} = 2\pi f_{p1}$, and $w_{p2} = 2\pi f_{p2}$, with $w_{z\#}$ representing a zero location, $w_{p\#}$ a pole location, $f_{p\#}$ a pole frequency, and A_{DC} the DC gain. By placing $w_{p2} > w_{p1} > w_{z1}$, the CTLE provides high-frequency gain boosting [3].

The PCIe Gen6 specification [7] defines the requirements to support 64 GT/s, defining a CTLE with six poles and three zeros, and an adjustable DC gain, so the system transfer function can be modeled as

$$H(s) = \sigma \frac{(s + w_{z1})(s + w_{p2}ADC)}{(s + w_{p1})(s + w_{p2})(s + w_{p3})} \cdot$$
$$\cdot \frac{(s + w_{z3})}{(s + w_{p4})(s + w_{p5})(s + w_{p6})} \tag{4}$$

where σ is defined by,

$$\sigma = \frac{w_{p1}w_{p3}w_{p4}w_{p5}w_{p6}}{w_{z1}w_{z3}} \tag{5}$$

Considering that the CTLE must support a wide frequency range of channel loss, the proposed CTLE consists of three stages to cover the overall transfer function at low-, mid-, and high-frequency ranges, respectively. Henceforth (4) can be described as

$$H(s) = \sigma G_1(s) G_2(s) G_3(s) \tag{6}$$

where,

$$G_1(s) = (s + w_{z1}) / [(s + w_{p1})(s + w_{p6})] \tag{7}$$

$$G_2(s) = (s + w_{p2}A_{DC}) / [(s + w_{p2})(s + w_{p4})] \tag{8}$$

$$G_3(s) = (s + w_{z3}) / [(s + w_{p3})(s + w_{p5})] \tag{9}$$

Consequently, the EQ topology at the Rx is a combination of a 3-stage CTLE and a DFE, as shown in Fig. 1.

B. Low Frequency Equalizer

A conventional CTLE cannot compensate for the small amount of low-frequency channel loss since its primary objective is to compensate for high-frequency channel losses [11]. Since the slope of the low-frequency loss is quite flat (<3dB/dec), an extra circuit is required.

The uncompensated low-frequency loss causes nonnegligible

long-term residual ISI that results in data dependent jitter (DDJ) that is difficult to reduce further by enhancing a CTLE, unless a LFEQ is added [3]. The LFEQ is based on a negative feedback topology. The objective is to minimize the small slope of low-frequency loss by placing together w_{z1} and w_{p1} pairs to achieve a small amount of low frequency gain (0 to 4dB) [10]. The transfer function of the LFEQ can be defined by (7), where w_{p1} is tuned to provide the expected DC gain in the low frequency range.

IV. Optimization of the PCIe Gen6.0 Link Equalization

We aim at finding the optimal set of Tx and Rx EQ settings to maximize the eye diagram margins. Let $R_m \in \Re^2$ denote the electrical system margins response,

$$R_m = R_m(x) = \begin{bmatrix} e_w(x) & e_h(x) \end{bmatrix}^T \quad (10)$$

where $e_h \in \Re^1$ is the smallest of the three PAM4 eye height measurements and $e_w \in \Re^1$ is the smallest of the three PAM4 eye width measurements, which are functions of the Tx FFE and Rx CTLE EQ settings contained in vector x.

We need to ensure the optimal system margin response also meets an eye linearity, $e_{linearity}$, larger than 0.85, and a vertical eye closure (VEC) below 6 dB. An initial optimization problem can be defined through a constrained formulation,

$$x^* = \arg \min_x u(x) \quad (11)$$

subject to $e_{linearity}(x) > 0.85$ and $VEC(x) < 6\text{dB}$, where $u(x)$ is the total area of the PAM4 eye diagram,

$$u(x) = -e_w(x)e_h(x) \quad (12)$$

$e_{linearity}$ is the measure of the vertical linearity defined by the variance of amplitude separation among the different PAM4 levels, and VEC is the smallest of the ratios of voltage swing to eye height.

A more convenient unconstrained objective function is

$$u'(x) = -w_1 u(x)\rho(x) + w_2 \| \lambda(x) \|_2^2 \quad (13)$$

where $\rho(x)$ is a vertical eye closure penalty function defined as

$$\rho(x) = 10^{\frac{VEC(x)}{6}} \quad (14)$$

and $\lambda(x)$ is eye linearity penalty function defined as

$$\lambda = \max\left\{0, 0.85 - e_{linearity}(x)\right\} \quad (15)$$

Both terms in (13) are scaled by weighting factors w_1, $w_2 \in \Re^1$ such that they become comparable. The initial unconstrained formulation can then be defined as

$$x^* = \arg \min_x u'(x) \quad (16)$$

Additionally, we need to ensure the optimal system response is within a suitable area in the coefficients search space of the EQ map. Here we follow our work in [8] and [11] to redefine the corresponding objective function. The four responses around $u'(x^*)$ must be at least 80% of the value of $u'(x^*)$, as shown in Fig. 2, where $u'_{i,j}$ are the objective function values per (13) for the i-th C_{m1} and j-th C_p values.

The new optimization problem can be defined through a constrained formulation, such that the optimal set of

$e_{h\_up} = 63\text{mV}$
$e_{w\_up} = 8.6\text{ps}$
$e_{h\_mid} = 62\text{mV}$
$e_{w\_mid} = 9.0\text{ps}$
$e_{h\_low} = 63\text{mV}$
$e_{w\_low} = 8.6\text{ps}$
a)
$e_{h\_up} = 44\text{mV}$
$e_{w\_up} = 9.1\text{ps}$
$e_{h\_mid} = 44\text{mV}$
$e_{w\_mid} = 9.8\text{ps}$
$e_{h\_low} = 44\text{mV}$
$e_{w\_low} = 9.1\text{ps}$
b)

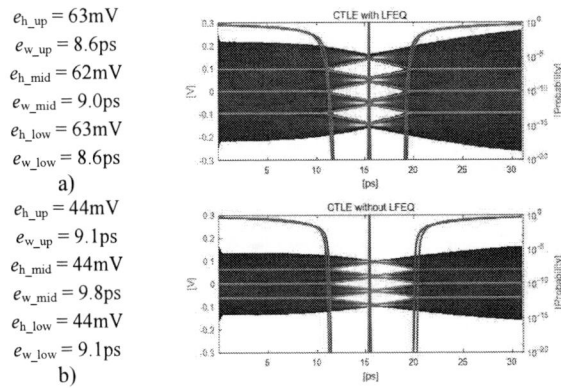

Fig. 4. CTLE performance a) with LFEQ and b) without LFEQ.

coefficients maximizes the system response without violating the lower bound of $0.8u'(x^*)$ in the vicinity,

$$x^* = \arg \min_x u'(x)$$

subject to $l_{11}(x) \leq 0,\ l_{12}(x) \leq 0,\ l_{21}(x) \leq 0,\ l_{22}(x) \leq 0$ (17)

with

$$l(x) = \begin{bmatrix} u(C_{m1i^*+1}, C_{ctle}, C_{pj^*}) & u(C_{m1i^*-1}, C_{ctle}, C_{pj^*}) \\ u(C_{m1i^*}, C_{ctle}, C_{pj^*+1}) & u(C_{m1i^*}, C_{ctle}, C_{pj^*-1}) \end{bmatrix} - $$
$$0.8u(C_{m1i^*}, C_{ctle}, C_{pj^*}) \begin{bmatrix} 1 & 1 \\ 1 & 1 \end{bmatrix} \quad (18)$$

where C_{m1i^*} and C_{pj^*} are the Tx set of coefficients that minimize (13) for each of the Rx CTLE setting values (C_{ctle}).

Similarly, a more convenient unconstrained objective function can be defined by adding a penalty term,

$$U(x) = u'(x) + w_3 \| L(x) \|_F \quad (19)$$

where $\| L(x) \|_F$ is the Frobenius norm of matrix $L(x)$ defined as

$$L(x) = \max\{0, l(x)\} \quad (20)$$

and w_3 is a weighting factor.

Our final unconstrained formulation is

$$x^* = \arg \min_x U(x) \quad (21)$$

with $U(x)$ defined by (13), (19) and (20).

We find the optimal set of coefficients x^* by solving (21). To avoid estimating gradients and considering that the objective function has many local minima, we use a combination of pattern search [5] and Nelder-Mead [6] methods. We start the optimization with pattern search, to explore the design space until finding a potential region where the global minimum is located. Then, the solution found by pattern search is used as seed for the Nelder-Mead method, which further minimizes the objective function for a more precise solution.

V. Simulation Results

To validate our methodology, we use MATLAB SerDes Toolbox considering a short, medium, and long-reach channels (CEI-56G serial links) of 10dB, 20dB and 27dB losses, respectively, in a 64 GT/s PCIe Gen6 link, where the pass/fail criteria is defined in terms of a time domain eye diagram at

979-8-3503-1191-4/23 $31.00 © 2023 IEEE

$e_{h\_up} = 68mV$
$e_{w\_up} = 6.1ps$
$e_{h\_mid} = 68mV$
$e_{w\_mid} = 6.4ps$
$e_{h\_low} = 68mV$
$e_{w\_low} = 6.1ps$
a)

$e_{h\_up} = 39mV$
$e_{w\_up} = 5.4ps$
$e_{h\_mid} = 39mV$
$e_{w\_mid} = 5.8ps$
$e_{h\_low} = 39mV$
$e_{w\_low} = 5.4ps$
b)

$e_{h\_up} = 49mV$
$e_{w\_up} = 5.9ps$
$e_{h\_mid} = 49mV$
$e_{w\_mid} = 6.3ps$
$e_{h\_low} = 49mV$
$e_{w\_low} = 5.9ps$
c)

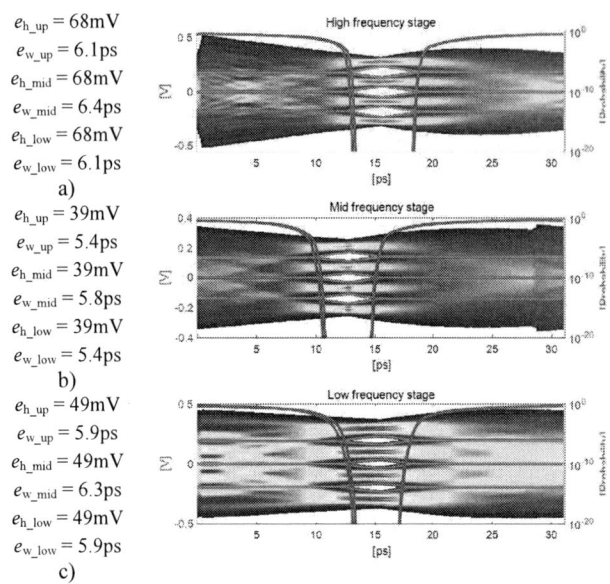

Fig. 5. Eye diagrams at different stages of the CTLE: a) high-frequency stage, b) mid-frequency stage, and c) low-frequency stage.

BER=10^{-6}. The link is simulated with the corresponding Tx jitter parameters (deterministic and sinusoidal) based on [7], and Rx jitter parameters from a common reference clock Rx architecture. The simulator generates an output containing the three statistical eye heights and widths.

The simulation results within a medium-reach channel in Fig. 4 demonstrate how the LFEQ-CTLE combination enhances overall performance in the Rx equalization scheme, yielding an eye area improvement of 35.3%.

The 3-stage CTLE equalization effects within a short-reach channel as reference are shown in Fig. 5. It is seen how each CTLE stage target a range of frequencies boosting the DC Gain. The high-frequency stage results in an improved eye opening.

To validate the proposed design within worst-case conditions, we added Tx and Rx deterministic and sinusoidal jitter parameters to the system and proceed to a link equalization optimization in a long-reach channel as reference. The eye diagrams at the receiver, before and after applying the optimization process in Section IV, are shown in Fig. 6. Additionally, Table I confirms that the resultant top eye width and height amply satisfy the channel tolerancing eye mask defined in the PCIe Gen6 Spec [7]. The optimized eye-diagram under worst-case channel conditions confirm the effectiveness of the proposed optimization approach.

VI. CONCLUSION

We proposed in this paper a 3-stage CTLE and a LFEQ designed to compensate for PAM-4 PCIe Gen6 highly lossy channels considering high, mid and low frequency bands boosting stages. We also proposed an efficient optimization

Table I. 64 GT/s Eye margins. Specification versus simulation.

Eye diagram parameter	PCIe Gen6 spec (min)	27dB channel simulation - worst-case Tx/Rx jitter parameters
top eye height	6.0 mV	20.0 mV
top eye width	0.1 UI	0.26 UI

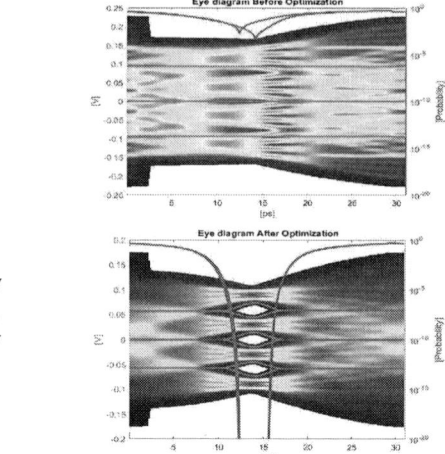

$e_{h\_up} = 20mV$
$e_{w\_up} = 8.1ps$
$e_{h\_mid} = 20mV$
$e_{w\_mid} = 8.3ps$
$e_{h\_low} = 20mV$
$e_{w\_low} = 8.1ps$
$VEC = 5.84$
$e_{linearity} = 0.99$

Fig. 6. Eye diagram before and after the optimization process.

approach to find the optimal coefficients for the Tx FFE and Rx CTLE. We validated the proposed method by using MATLAB SerDes Toolbox. The optimized EQ coefficients were tested by measuring the eye diagrams at the receiver, confirming a significant improvement on eye height, eye width, eye linearity, and vertical eye closure.

REFERENCES

[1] H. Wang, Y. Chen, Y. Gao, N. Li, Z. Zhang, C. Guo and J. Li, "A quad linear 56Gbaud PAM4 transimpedance amplifier in 0.18 μm SiGe BiCMOS technology," in *IEEE Int. System-on-Chip Conf. (SOCC)*, Singapore, Sep. 2019, pp. 165-170.

[2] J. L. Zerbe, C. W. Werner, V. Stojanovic, F. Chen, J. Wei, G. Tsang, D. Kim, W. F. Stonecypher, A. Ho, T. P. Thrush, R. T. Kollipara, M. A. Horowitz and K. S. Donnelly, "Equalization and clock recovery for a 2.5-10-Gb/s 2-PAM/4-PAM backplane transceiver cell," *IEEE J. Solid-State Circuits*, vol. 38, no. 12, pp. 2121–2130, Dec. 2003.

[3] J. He, N. Qi, N. Yu, L. Wu, P. Yin Chiang, X. Xiao, and N. Wu., "A 2nd-order CTLE in 130nm SiGe BiCMOS for a 50GBaud PAM4 optical driver," in *IEEE Int. Conf. Integr. Circ. Tech. Applic.*, Beijing, China, Nov. 2018, p. 151.

[4] S. Parikh, T. Kao, Y. Hidaka, J. Jiang, A. Toda, S. Mcleod, W. Walker, Y. Koyanagi, T. Shibuya and J. Yamada, "A 32Gb/s wireline receiver with a low-frequency equalizer, CTLE and 2-tap DFE in 28nm CMOS," in *IEEE Int. Solid-State Circ. Conf.*, CA, USA, Feb. 2013, pp. 28-29.

[5] R. Hooke and T. A. Jeeves, "Direct search solution of numerical and statistical problems," *J. of the ACM*, vol. 8, no. 2, pp. 212-229, Apr. 1961.

[6] J. C. Lagarias, J. A. Reeds, M. H. Wright, and P. E. Wright, "Convergence properties of the Nelder-Mead simplex method in low dimensions," *SIAM J. Opt.*, vol. 9, no. 1, Jan 1998, pp. 112–147.

[7] PCI SIG Org. (2022), *PCI Express® Base Specification Revision 6.0.1* [Online]. Available: https://pcisig.com/specifications.

[8] F. E. Rangel-Patiño, J. E. Rayas-Sánchez, E. A. Vega-Ochoa, and N. Hakim, "Direct optimization of a PCI Express link equalization in industrial post-silicon validation," in *IEEE Latin American Test Symp. (LATS 2018)*, Sao Paulo, Brazil, Mar. 2018, pp. 1-6.

[9] W. T. Beyene, "The design of continuous-time linear equalizers using model order reduction techniques," in *IEEE EPEP Elec. Perform. Electron. Packag.*, San Jose, CA, USA, Oct. 2008, pp. 187-190.

[10] R. Farjad-Rad, H.-T. Ng, M.-J. E. Lee, R. Senthinathan, W. J. Dally, A. Nguyen, R. Rathi, J. Poulton, J. Edmondson, J. Tran and H. Yazdanmehr, "0.622-8.0 Gbps 150 mW serial 10 macrocell with fully flexible preemphasis and equalization," in *Symp. VLSI Circuits Dig. Tech. Papers*, Kyoto, Japan, Jun. 2003, pp. 63-66.

[11] R. J. Ruiz-Urbina, F. E. Rangel-Patiño, J. E. Rayas-Sánchez, E. A. Vega-Ochoa, and O. Longoria-Gándara, "Transmitter and receiver equalizers optimization for PCI Express Gen6.0 based on PAM4," in *IEEE MTT-S Latin America Microw. Conf. (LAMC)*, Cali, Colombia, May 2021, pp.1-4.

2023 IEEE Latin American Electron Devices Conference (LAEDC)
Puebla, Mexico, July 3-5, 2023

A New Microstrip Directional Filter Configuration Composed by Hybrid-Mode Resonators

Humberto Lobato-Morales
Electronics & Telecommunications
CICESE
Ensenada, Mexico
humbertolm@ieee.org

Gabriela Méndez-Jerónimo
Electronics & Telecommunications
CICESE
Ensenada, Mexico
gmendez@cicese.mx

Germán Álvarez-Botero
Electrical & Electronic Engineering
Public University of Navarra
Navarra, Spain
germanandres.alvarez@unavarra.es

Abstract—The design, fabrication and experimental test of a microstrip two-pole directional filter based on a new approach consisting of hybrid-mode resonators is presented in this paper. The use of hybrid *even*-mode and *odd*-mode resonators allows the structure to be composed by two stages having bandstop filter behavior, and following a symmetrical configuration, which in turn makes it easier to design and adjust while keeping low dimensions. The proposed filter is designed to operate over the uplink channel of the 1.8 GHz band used for mobile communications with a 4% bandwidth.

Keywords—directional filter, microstrip, odd-mode, even-mode, resonators.

I. INTRODUCTION

A directional filter DF is a four-port circuit having the capability to produce a bandpass response at one port and the complementary bandstop at other output; one of the ports remains isolated [1], [2]. A DF seen as a single block is shown in Fig. 1(a). The relevance of this type of filters relies on the fact that they can be used to design multiplexing networks by cascading several DFs operating at different bands, following a modular concept [3]; thus, they are highly useful in communications systems. Moreover, due to their directional capabilities, DFs have been proposed for new applications such as the coupling of harvesting circuits composed by rectifying stages inside a communication system [4].

Two main approaches have been proposed for the design of DFs: 1) based on travelling-wave resonators; and 2) based on standing-wave-resonators. The first approach, shown in

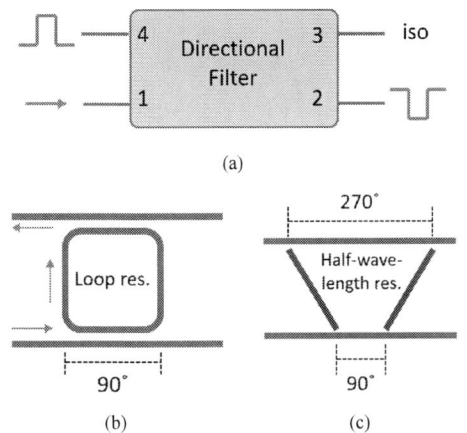

Fig. 1. Directional filter DF (a) as a basic block; (b) formed by traveling-wave resonators; and (c) formed by standing-wave resonators.

Fig. 1(b), makes use of lines coupled to loop resonators allowing the wave to travel in a defined direction in order to provide signal directivity [5]-[7]; multi-layer [5], [6] and substrate-integrated-waveguide SIW [7] technologies are required for a proper coupling of the loops, which in turn adds complexity in the design and fabrication, and bulky structures are obtained. On the other hand, the second approach for DFs consists of stages with coupled standing-wave-resonators and connected by lines of specific electrical lengths, as shown in Fig. 1(c), in order to produce the directivity of the signal; single-layer microstrip DFs can be designed with this configuration [8]-[10] allowing easier designs and fabrication processes. Independently, by following either the first or second approach, they have been evolved from single-pole responses [1], [2] to the implementation of multi-pole DFs by properly coupling more resonators, increasing selectivity of the bands [7]-[10].

Particularly, in order to design an *N*-pole DF following the second afore-mentioned approach, the structure requires the coupling of 2*N* resonators [8]-[10]; therefore, compactness of these filters is somewhat compromised. Some important efforts in DF miniaturization include the use of composite-right/left-hand CRLH lines, obtaining shorter distances between stages [10], [11]. However, complexity in the design, tuning, and fabrication of the DF tends to increase, as more elements (capacitors and inductors) are involved.

In this paper, a new and simple approach in the design of DFs based on the use of hybrid-mode resonators is presented; the use of *even*- and *odd*-mode resonators introduces a longitudinal symmetry in the DF, which in turn makes it simpler to design and adds a level of compactness. Hybrid mode-resonators have been introduced in the design of conventional bandpass and bandstop filter topologies [12], [13]; *even*- and *odd*-mode resonators can be placed adjacently with no cross couplings. Half-wavelength $\lambda_g/2$ resonators are used here for the design of a two-pole DF following the proposed approach, which is based on two bandstop stages operating under *even*- and *odd*-modes, and properly coupled for the directivity capability. The paper is organized as follow: Section II presents the proposed design approach; Section III covers results and discussion; and conclusions are addressed in Section IV.

II. MICROSTRIP DF WITH HYBRID RESONATORS

A. Proposed design approach

By analyzing the conventional configuration of a DF composed by standing-wave resonators [Fig. 1(c)], it can be observed that the first resonator section acts as a bandstop filter operating under the *odd*-mode [8], [10], while the second

979-8-3503-1191-4/23 $31.00 © 2023 IEEE

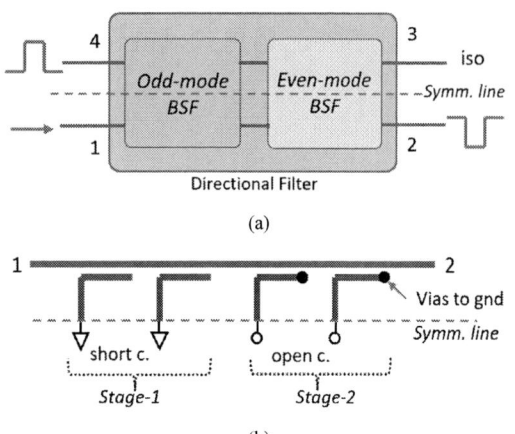

(a)

(b)

Fig. 2. (a) Proposed block diagram for the design of a multi-pole DF; and (b) equivalent single-ended circuit.

resonator section, including the additional 180° length of the upper line, transforms the *odd*-mode structure into an *even*-mode bandstop filter. Based on this statement, the proposed block diagram of a DF is a simplified scheme composed by a first stage corresponding to an *odd*-mode bandstop filter and a cascaded second stage formed by an *even*-mode bandstop filter, as shown in Fig. 2(a). Both stages, although they operate under different modes, must present the same operating frequency, number of poles, and bandwidth. A longitudinal symmetry is then obtained, which is an important factor for a simpler design of each stage and final adjustment of the filter, as will be commented below.

Particularly, the first stage can be composed by open-ended $\lambda_g/2$ resonators which naturally operate under the *odd*-mode. Thanks to the symmetry of the stage, the bandstop filter design can be simplified using the single-ended equivalent-circuit and applying conventional bandstop filter theory [14]; the resonators become then of $\lambda_g/4$-length and short-circuited at the end point (symmetrical plane). On the other hand, the second stage can be composed by short-ended $\lambda_g/2$ resonators for an *even*-mode operation; the same way as for the first

stage, the bandstop filter can be designed from the simplified equivalent circuit of the *even*-mode structure, where each complete resonator become an open-terminated $\lambda_g/4$ resonator at the symmetrical plane, as observed in Fig. 2(b).

Finally, after properly tuning both stages independently, they are cascaded in order to form the complete DF; the electrical length of the lines connecting both stages plays the main role in generating the directivity of the structure, and its adjustment can be simply performed, as will be shown.

B. Design of the microstrip DF

In order to demonstrate the concept, a microstrip two-pole DF is designed following the presented approach. The selected band for operation of the prototype is the *FDD*-uplink channel of the 1.8-GHz mobile communications band, having a 4.04% fractional bandwidth *FBW* centered at 1.71 GHz [14]. A *Rogerscorp* RO4003C® substrate is used for the structure design, having 1-oz. copper conductive layers, $\varepsilon_r = 3.55$, $\tan\delta = 0.0021$, and height $h = 0.813$ mm. The stages are designed using the conventional bandstop filter theory having a *Butterworth* response [14].

The first stage is composed by two resonators coupled to upper and lower 50-Ω transmission lines by means of interdigital fingers in order to obtain the required electrical *E*-coupling; the two identical resonators are configured in a "*C*" shape and aligned towards the same direction, in order to form the two-pole filter response. For the second stage, the resonators are configured short-ended at their extremes and following a hairpin shape in order to enable a parallel-section magnetic *H*-coupling to the lines. Fig. 3 shows the geometry of the stages and their simulated responses.

Finally, both stages are connected in cascade by 50-Ω transmission lines, as observed in Fig. 4(a), to form the complete structure. An additional and final process is done by varying the length of such lines in order to maximize the directivity level $D = |S_{41}| - |S_{31}|$. Conventionally, it has been stated that lines connecting the stages must present specific lengths of 90° and 270° between resonator coupling points [Fig. 1(c)] [2], [8]; however, their exact positions are no longer identified because couplings of the resonators are distributed

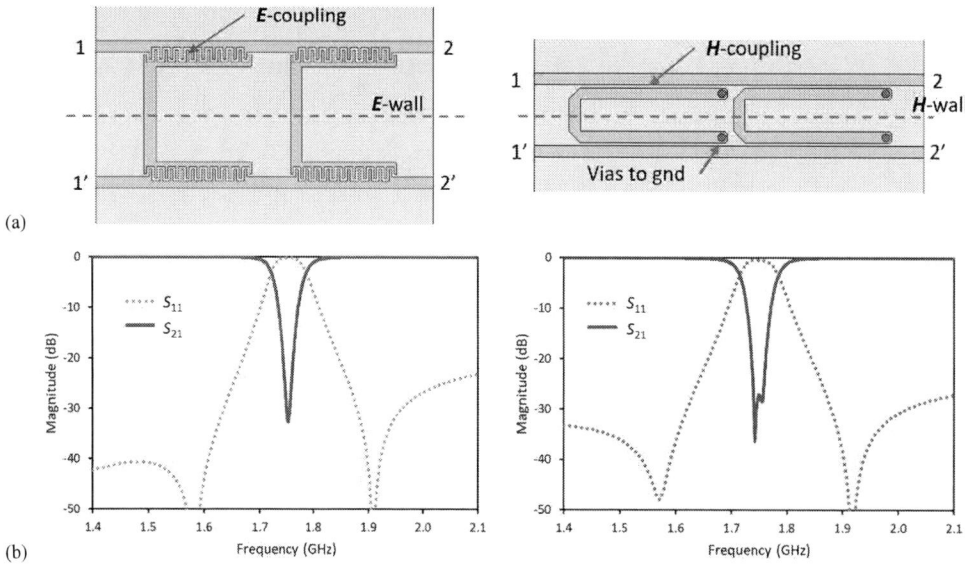

Fig. 3. Stages for the DF implementation, (a) *odd*-mode and *even*-mode bandstop filters; (b) correspondent frequency responses.

979-8-3503-1191-4/23 $31.00 © 2023 IEEE

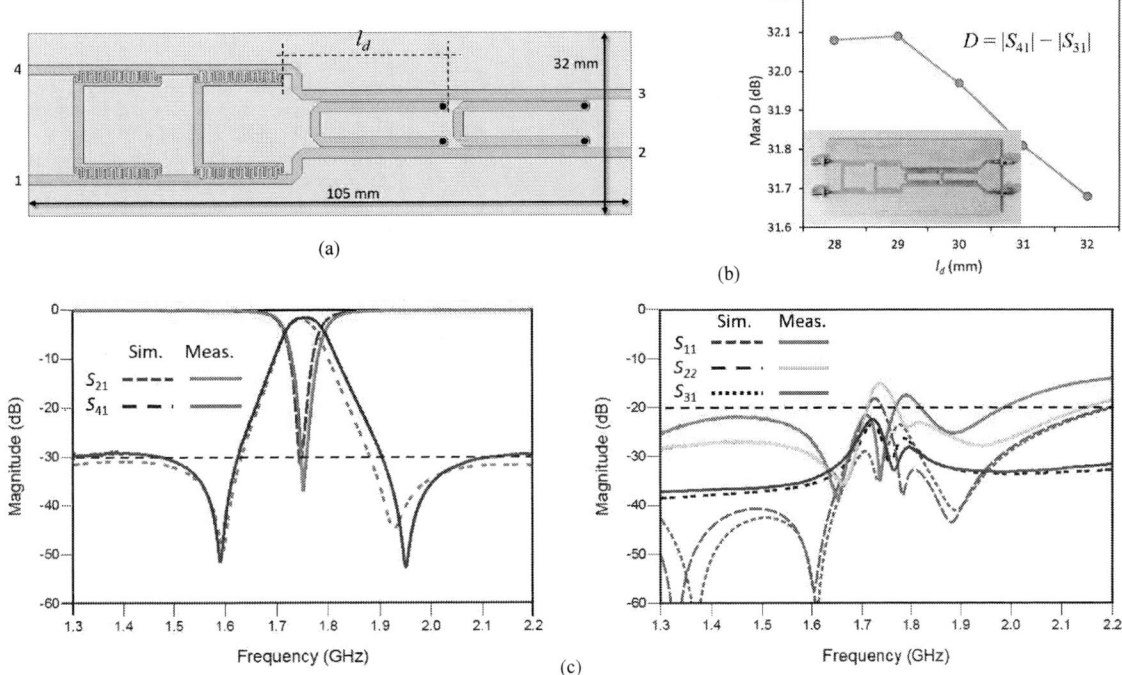

Fig. 4. Complete DF: (a) geometry; (b) max. D vs. l_d; (c) $|S_{21}|$ and $|S_{41}|$ responses; (d) $|S_{11}|$, $|S_{22}|$, and $|S_{31}|$ responses.

over a certain line length instead of being at a single point. In this design, the stages are configured having the resonators aligned towards the same direction (right side) and connected by transmission lines having an electrical length close to 90° from the termination of the last *odd*-mode resonator to the termination of the first *even*-mode resonator. Small variations in length of these lines are configured in simulations under a simple parametric analysis. Fig. 4(b) shows the result of varying the distance l_d from 28 mm to 32 mm in steps of 1 mm, looking for the maximum D level inside the passband, which is obtained for $l_d = 29$ mm ($D = 32.1$ dB).

III. RESULTS AND DISCUSSION

The designed DF is fabricated using an *LPKF* milling machine, and SMA-type connectors are soldered at the correspondent ports; a photograph of the structure is shown in the inset of Fig. 4(b). Responses of the filter are shown in Fig. 4(c). The simulated behavior (including material losses) is obtained utilizing *Sonnet-Software®*, and the measured data is obtained using an *Anritsu MS2038C*-series vector network analyzer VNA; good agreement is appreciated between both simulations and experiments. The simulated and measured *FBW* values are of 3.83% and 4.33%, respectively; a small variation of 0.5% is obtained; insertion losses of the passband S_{41} are 2.1 dB and 1.7 dB, simulated and measured

respectively. Transmission zeros in S_{41} appear close to the passband at 1.6 GHz and 1.95 GHz along with out-of-band lobes around -30 dB. In both, simulations and measurements, reflection S_{11}, S_{22}, and isolation levels S_{31} are generally found to be below -20 dB within a frequency range going from 1.3 GHz up to 2.2 GHz.

The overall dimensions of the structure are 105 mm x 32mm. Table I tabulates relevant characteristics of other microstrip multi-pole filters for comparison purposes. As seen, only ref. [9] presents a smaller DF structure due to the open-loop configuration of the used resonators.

It is worth to mention that, as the presented approach introduces symmetry along the structure, a faster design of the stages can be carried out by configuring them as simplified equivalent single-ended circuits (also reducing memory space and simulating time by configuring half of the structure). The final design step of properly cascading both stages is carried out in a single adjusting process maximizing D, in contrast to previous methods where upper and lower lines have different lengths, having to adjust them in two iterative processes.

IV. CONCLUSION

A new configuration of microstrip DF composed by cascaded stages consisting of *odd*-mode and *even*-mode bandstop filters has been presented. Each stage has been designed utilizing conventional filter theory. The final adjustment has been done with a single and simple parametric analysis by varying the distance between both stages and looking for the maximum directivity level. Good agreement was obtained between simulations and measurements. The proposed DF configuration does not introduce additional dimensions, keeping the design under a good level of compactness. The characteristics of the obtained bandstop, bandpass, isolation, and out-of-band responses evidence that this design can be used to form multiplexing systems by connecting other DFs.

TABLE I. COMPARISON WITH OTHER MICROSTRIP DFs.

Ref.	Dimensions*	Tot. area*	Poles	FBW @ f_0	Bands
[4]	0.95 x 0.49	0.47	2	7.3% @ 1.575	1
[8]	1.00 x 0.5	0.50	2	2.5% @ 2.5	1
[9]	0.78 x 0.28	0.22	2	8.6% @ 0.85	1
[10]	1.12 x 0.37	0.41	3	5% @ 1.5	1
[11]	0.69 x 0.44	0.30	2	8% @ 0.8	2
This work	1.00 x 0.30	**0.30**	2	4.3% @ 1.8	1

*Note: dimensions and area values are in terms of the guided wavelength λ_g at the operating frequency for a fair comparison (lowest band in case of multi-band operation); f_0 units are GHz.

ACKNOWLEDGMENT

The authors express their gratitude to technician Diana Soto-Castañeda for her contribution in the fabrication of the filter, and *Conacyt* project 269927.

REFERENCES

[1] S. B. Cohn, F. S. Coale, "Directional channel-separation filters," *Proc. IRE*, Aug. 1956, pp. 1018-1024.

[2] G. Matthaei, L. Young, and E.M.T. Jones, *Microwave Filters, Impedance Matching Networks and Coupling Structures*, Norwell, MA: Artech House, 1985.

[3] R. J. Cameron and M. Yu, "Design of manifold coupled multiplexers," *IEEE Microw. Mag.*, pp. 46-59, Oct. 2007.

[4] L. Mizquez-Corona, H. Lobato-Morales, R. Chávez-Pérez, J. Medina-Monroy, "RF/microwave energy harvesting system in a front-end using a directional filter," in *Proc. Int. Conf. Electr. Commun. Comput.*, Puebla, Mexico, Feb. 2019, pp. 27-30.

[5] S. Uysal, "Microstrip loop directional filter," *Electron. Lett.*, vol. 33, no. 6, pp. 475-476, 1997.

[6] A. Petrovich-Gorbachev, "The reentrant wide-band directional filter," *IEEE Trans. Microw. Theory Techn.*, vol. 50, no. 8, Aug. 2002.

[7] Y. Cheng, W. Hong, and K. Wu, "Half mode substrate integrated waveguide (HMSIW) directional filter," *IEEE Microw. Wirel. Compon. Lett.*, vol. 17, no. 7, Jul. 2007.

[8] J. P. Kim, "Improved design of single-section and cascaded planar directional filters," *IEEE Trans. Microw. Theory Techn.*, vol. 59, no. 9, Sep. 2011.

[9] H. Lobato-Morales, A. Corona-Chávez, J. Olvera-Cervantes, D.V.B. Murthy, "Multi-pole microstrip directional filters for multiplexing applications," in *Proc. Int. Conf. Electr. Commun. Comput.*, Puebla, Mexico, Feb. 2011, pp. 344-349.

[10] J. S. Sun, H. Lobato-Morales, J. H. Choi, A. Corona-Chávez, and T. Itoh, "Multistage directional filter based on band-reject filter with isolation improvement using composite right/left-handed transmission lines," *IEEE Trans. Microw. Theory Techn.*, vol. 60, no. 12, Dec. 2012.

[11] H. Lobato-Morales, A. Corona-Chávez, T. Itoh, and J. Olvera-Cervantes, "Dual-band multi-pole directional filter for microwave multiplexing applications," *IEEE Microw. Wirel. Compon. Lett.*, vol. 21, no. 12, Dec. 2011.

[12] M. Ohira, T. Kato, and Z. Ma, "Novel microstrip realization and straightforward design of fully canonical Cul-de-Sac coupling bandpass filters," in *IEEE Intl. Microw. Symp.*, San Francisco, CA, USA, May 2016.

[13] M. Ohira, M. Kanomata, Z. Ma, and X. Wang, "A new microstrip bandstop filter for fully canonical Cul-de-Sac coupling configuration," in *IEEE Intl. Microw. Symp.*, Philadelphia, PA, USA, Jun. 2018, pp. 708-711.

[14] J. S. Hong, and M. J. Lancaster, *Microstrip Filters for RF/Microwave Applications*, New York: Wiley, 2001.

2023 IEEE Latin American Electron Devices Conference (LAEDC)
Puebla, México, July 3-5, 2023

Fabrication and characterization of Al-based integrated MIS capacitors

Daniel Rocha-Aguilera
Electronics Department, Microelectronics Group
National Institute of Astrophysics, Optics and Electronics
Tonantzintla, Mexico
daniel.rocha@inaoep.mx

Daniel Ferrusca-Rodríguez
Astrophysics Department, Milimeter Wavelength Astronomical Instrumentation Laboratory
National Institute of Astrophysics, Optics and Electronics
Tonantzintla, Mexico
dferrus@inaoep.mx

Joel Molina-Reyes
Electronics Department, Microelectronics Group
National Institute of Astrophysics, Optics and Electronics
Tonantzintla, Mexico
jmolina@inaoep.mx

Abstract—In this work, the fabrication and characterization of Al/Al_2O_3/Si capacitors which uses atomic layer deposition to deposit ultra-thin Al_2O_3 films is presented. Results of materials' characterization of the capacitors are included, as well as electrical characterization at room and cryogenic temperatures, a brief analysis of conduction mechanisms through the dielectric and the obtention of electrical properties of the materials.

Keywords—metal-insulator-semiconductor capacitor, ultra-thin metal oxide, atomic layer deposition, conduction mechanisms in dielectrics, cryoelectronics.

I. INTRODUCTION

In recent years, quantum computing has become a promising alternative to classical or conventional computation to solve certain types of problems which result too complicated for conventional computers. Some of the most relevant capabilities of quantum computing, such as performing parallel operations in superposition states or entangling quantum states that are physically apart from one another, give place to applications in computational chemistry, cryptography, search algorithms and machine learning. [1-3]

Quantum bits or *qubits* are the basic units of quantum information. They are quantum systems with two accessible and well-defined states. Their physical implementation is one of the main challenges in the creation of functional quantum computers. There are multiple types of qubits, and one that has been exploited with great success because of its great scalability is superconducting qubits, which are superconducting integrated circuits.

The central element of superconducting qubits is a metal-insulator-metal (MIM) structure called Josephson junction (JJ), comprised of two superconducting electrodes separated by an insulator with a thickness of 1 or 2 nm. When in a superconducting state, a quantum phenomenon known as the Josephson effect takes place, in which a tunneling current appears through the insulator even in absence of applied voltage because of the quantum tunnelling of Cooper pairs. [4]

Typical fabrication of JJs involves the deposition of the metallic electrodes through e-beam evaporation in two deposition steps with a thermal oxidation in between to grow the insulator from the bottom electrode. A resist bilayer mask with and undercut and e-beam lithography are used to define the

junction area and electrode geometry. The most used materials are Al as superconductor and Al_2O_3 as insulator. [5]

In this work, a fabrication process for Al-based MIS capacitors using atomic layer deposition (ALD) to deposit the Al_2O_3 instead of thermal oxidation is proposed. The ALD technique allows a very precise control of the insulator film thickness and control of the deposition parameters can be used to optimize the process, minimizing defects in the interfaces between metals and insulator as well as defects in the bulk of the insulator. Process optimization to minimize defects is very important, as one of the main problems for an adequate performance of superconducting qubits is precisely defects in amorphous dielectrics that manifest as two-level quantum systems that cause decrease the lifetime of quantum information through energetic relaxation and dephasing. [6]

To develop this process, Al-based integrated metal-insulator-semiconductor (MIS) capacitors were fabricated on Si substrates with insulators of various thicknesses. The MIS capacitor is a useful reference structure to characterize the deposited thin films, since the Al_2O_3 is deposited directly on the high-quality surface of the silicon wafer, which has a very low surface roughness and defect density. This work presents the proposed fabrication process as well as materials and electrical characterization of the resulting devices.

II. EXPERIMENTAL

Samples with two different MIS capacitor structures were processed, as shown in Fig. 1, where a) is a simple MIS capacitor and b) is a MIS capacitor which includes a thick oxide layer which was used as a protection layer for the thin oxide to use a wire bonding technique on the capacitors.

For all the fabricated samples, prime grade n-type silicon wafers with 2-5 $\Omega\cdot$cm resistivity and a (100) plane surface were used as substrates. Initially, the silicon wafers were cleaned (RCA clean and HF-last treatment) to remove contaminants and obtain an H-terminated surface.

Only for the capacitors with thick oxide, thermal oxidation of the silicon wafers was used to grow a 60 nm oxide using dry oxygen + TCE at 1000 °C followed by a 30 min inert anneal in N2 atmosphere to decrease the fixed charge density near the SiO2-Si interface. Photolithography and wet etching with a buffer-HF solution were used to open 70x70 um windows in the oxide which define the active area of the capacitors.

979-8-3503-1191-4/23 $31.00 © 2023 IEEE

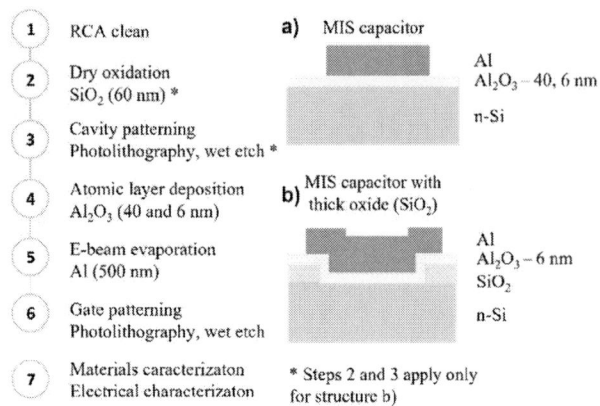

Fig. 1. Fabrication process sequence for integrated MIS capacitors (left). Schematics for a) normal MIS capacitor, b) MIS capacitor with thick oxide.

The Al_2O_3 films were deposited by thermal ALD (Savannah S100 from Cambridge NanoTech) at 200 °C and 200 mTorr using H_2O and TMA (trimethylaluminum) as precursors. After the ALD deposit, the samples were immediately introduced to the evaporation chamber to minimize the exposure of the Al_2O_3 to ambient contamination. All the aluminum layers were deposited by e-beam evaporation (Temescal BJD-1800 from Edwards) in ultrahigh vacuum (1E-7 Torr) with a deposition rate of 10 Å/s to obtain a 500 nm thickness. Photolithography and wet etching steps were used to pattern the top aluminum layer in the form of gate electrodes with areas of 40x40 and 80x80 um².

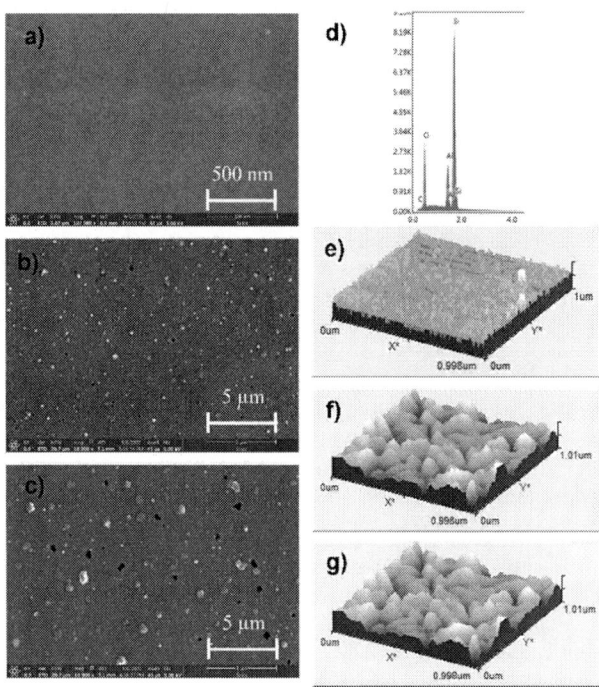

Fig. 2. SEM micrographies for a) Al_2O_3/Si, b) Al/Si, c) Al_2O_3/Al/Si samples. d) EDS spectrum for Al_2O_3/Si sample. AFM measurements of e) Al_2O_3/Si, f) Al/Si, g) Al_2O_3/Al/Si samples.

After wafer-level characterization, dies from the MIS capacitors with thick oxide samples were cut and packaged in a ceramic dual in-line package with 24 pins (CDIP-24) to introduce the samples to a cryostat and perform electrical characterization at cryogenic temperatures. Wire bonding with gold wire was used to connect the devices to the CDIP.

MIS capacitors with various Al_2O_3 thicknesses were fabricated. Fig. 1 summarizes the fabrication process.

III. RESULTS AND DISCUSSION

A. Materials characterization

The deposited films' morphology was observed using SEM (SCIOS DualBeam from FEI). Fig. 2 shows that the ALD films are very uniform when deposited directly over the Si wafer. On the other hand, the Al films present notable defects like Al accumulations in the form of grains as well as holes in the film.

EDS analysis was performed on the deposited films to confirm their elemental composition. The EDS spectra for the Al_2O_3/Si sample (Fig. 2d) shows an atomic composition with excess oxygen resulting in an out of stoichiometry aluminum oxide: $Al_2O_{3.2}$.

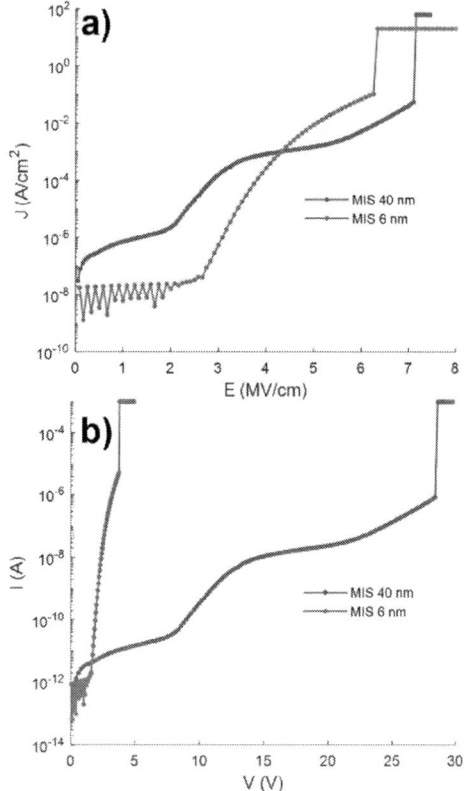

Fig. 3. a) I-V curves, b) E-J curves for Al/ Al_2O_3/Si MIS capacitors.

The average surface roughness of the films was obtained after AFM analysis (Fig. 2e-g). The surface roughness value for the Al films is 3 nm, which is very high considering JJs require an insulator with a thickness of 1 or 2 nm. In that case, the roughness of the Al will be greater than the insulator thickness. Although the fabricated devices have demonstrated an adequate electrical behavior, it is necessary to minimize the Al roughness to further optimize the fabrication process and increase reproducibility. For the Al_2O_3/Si sample, the Al_2O_3 surface roughness is 0.7 nm, which shows the capacity of ALD to deposit ultra-thin films with a very uniform thickness.

4-probe measurements and profilometry measurements were used to obtain the resistivity of the Al deposits in all the different samples. The calculated resistivity for the evaporated Al in the different samples was 2.8-3.1 μΩ·cm, close to the resistivity value for bulk Al, which is 2.65 μΩ·cm.

B. Ambient temperature electrical characterization

A semiconductor parameter analyzer (Keithley 4200 SCS) was used to obtain the current-voltage (I-V) and capacitance-voltage (C-V) characteristics of the capacitors at ambient temperature to electrically characterize the dielectric and to analyze the charge transport through the capacitor.

Fig. 3a shows the results of I-V measurements of the MIS capacitors. In Fig. 3a it is evident that the leakage current is greater and dielectric breakdown voltage is lower for the thinner insulator. When these curves are compared in the form of current density vs electric field (Fig. 3b), the electric field for breakdown is very similar for both thicknesses ($E_{bkd} = 6 - 7$ MV/cm). This value for E_{bkd} corresponds to aluminum oxide with high electrical quality.

In both curves, current starts increasing dramatically when the applied electric field surpasses 2 MV/cm. The shapes of the curves after this value correspond to different conduction mechanisms.

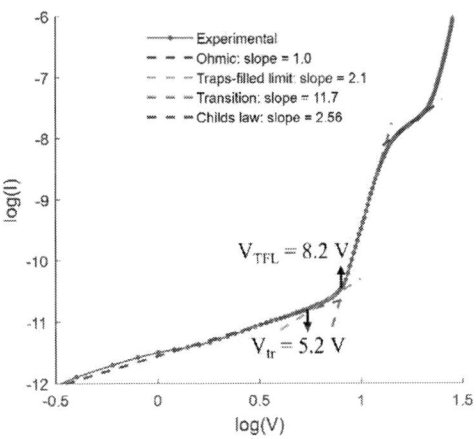

Fig. 4. SCLC plot for 40 nm MIS capacitors.

In the 40 nm MIS capacitors, space charge limited conduction (SCLC) can be identified. The typical log(I)-log(V) SCLC plot for this sample is shown in Fig. 4. For low voltages, there is a region which follows Ohm's law (I ∝ V) according to (1). As the applied voltage increases, a trap-filled limit current appears, which follows a relation of I ∝ V^2 according to (2). The transition voltage is $V_{tr} = 5.2$ V, at this voltage the density of injected carriers becomes greater than the thermally generated free carriers, so they can travel across the dielectric, charge redistribution occurs, and a space charge is generated. The traps-filled limit voltage $V_{TRL} = 8.2$ V is the point at which the traps in the dielectric become filled and the injected carriers coming from strong injection are free to move through the dielectric.

After a strong current increase, a region dominated by Child's law (I ∝ V^2) appears (3). [7]

$$J = qn_0\mu\frac{V}{d} \qquad (1)$$

$$J = 9\mu\varepsilon\theta\frac{V^2}{8d^3} \qquad (2)$$

$$J = 9\mu\varepsilon\frac{V^2}{8d^3} \qquad (3)$$

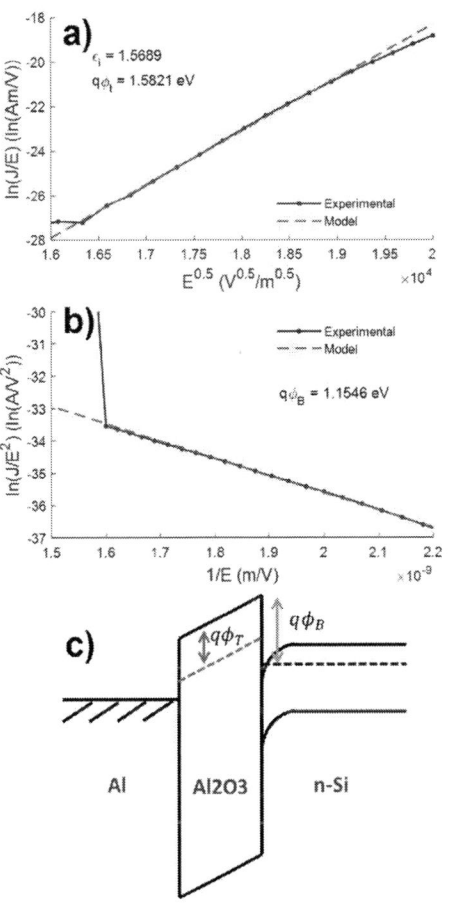

Fig. 5. Modeling of conduction mechanisms for 6 nm Al/Al₂O₃/Si MIS capacitor: a) Poole-Frenkel plot, b) Fowler-Nordheim plot, c) simplified energy band diagram.

For the 6 nm MIS capacitors, semi empirical equations for Poole-Frenkel (PF) emission (4) and Fowler-Nordheim (FN) tunneling (5) were used to model the current in the regions from 3 to 4 MV/cm and 5 to 6 MV/cm, respectively. [7]

$$J = q\mu N_C E e^{-q(\phi_T - \sqrt{qE/(\pi\varepsilon_i\varepsilon_0)})/k_BT} \qquad (4)$$

$$J = \frac{q^3E^2}{8\pi hq\phi_B} e^{-8\pi\sqrt{2m^*}(q\phi_B)^{3/2}/(3qhE)} \qquad (5)$$

PF and FN plots for the corresponding regions are shown in Fig. 5. The energy for the trap level associated with PF as well as the triangular barrier height for the FN mechanism were extracted, with values of $\phi_T = 1.58$ eV and $\phi_B = 1.15$ eV, in agreement with previous experiments. [8-9]

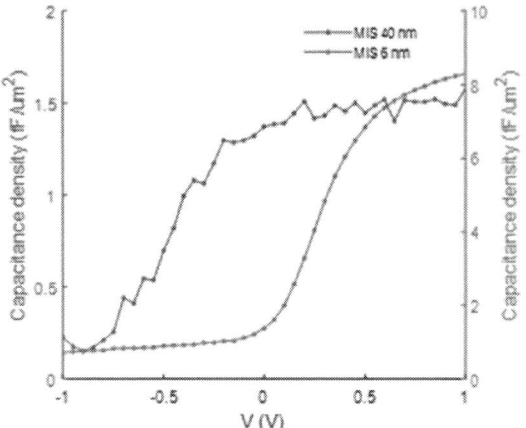

Fig. 6. C-V curves at 1 MHz for MIS capacitors.

C-V curves were also obtained, results are shown in Fig. 6. Using the capacitance value in the accumulation region, the dielectric constant of the Al_2O_3 was calculated, obtaining a value of k = 7.5.

C. Cryogenic electrical characterization

Electrical characterization at cryogenic temperatures was performed on the MIS capacitors with thick SiO_2 oxide. A closed-cycle cryostat system from ColdEdge Technologies equipped with a He3-He4 fridge and a Lakeshore 218 temperature monitor was used to cool the devices from ambient temperature to a base cryogenic temperature of 3.6 K. A single channel SMU (Keithley 2450) was used to perform I-V measurements on capacitors packaged and wired to a CDIP-24 placed inside the cryostat.

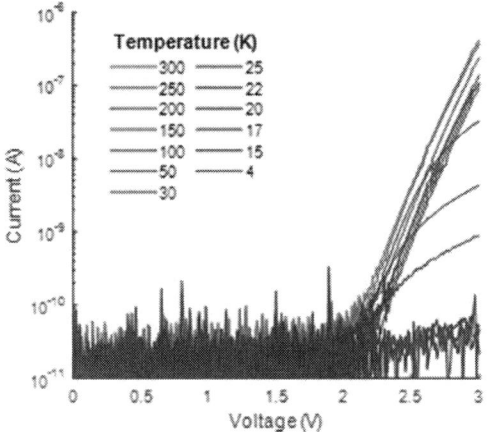

Fig. 7. I-V curves for 6 nm $Al/Al_2O_3/Si$ MIS capacitor from 300 to 4 K.

The graph in Fig. 7 contains I-V measurements taken at different temperatures. It can be noted that the current decreases with decreasing temperature. This result is expected as the current injection is done from the Si substrate, in which the density of free charge carriers decreases notably at low temperatures. Also, the temperature dependance of the PF emission and the increase of the energy barrier of the FN mechanism due to the widening of the insulator's band gap with lower temperatures contribute to this trend. An abrupt decrease of current from 25 to 15 K can be attributed to the Si entering a freeze-out regime in which there isn't enough energy to ionize the carriers provided by dopant atoms.

IV. CONCLUSIONS

MIS capacitors with a fabrication process compatible with standard IC fabrication technology were successfully processed using thermal ALD to deposit the Al_2O_3 insulator. Materials characterization showed that there is still work to be done in the optimization of the Al deposit to improve the uniformity and decrease the surface roughness of the films to deposit the insulator on top of the Al film. The Al films have a very low resistivity, which is important for devices working in a superconducting regime. Electrical characterization showed that high-quality Al_2O_3 was deposited over the Si substrate and some of the mechanisms that allow charge transport through the oxide were identified for different thicknesses. This Al_2O_3 has adequate characteristics to function as the central element of JJs. Finally, an experimental setup for cryogenic characterization of these devices was implemented and cryogenic measurements were carried out with the expected results. This system will be of great utility for the characterization of superconducting MIM capacitors.

V. REFERENCES

[1] M. A. Nielsen and I. L. Chuang, "Introduction and overview," in Quantum Computation and Quantum Information, Cambridge: Cambridge University Press, 2022, pp. 1–42.J. Clerk Maxwell, A Treatise on Electricity and Magnetism, 3rd ed., vol. 2. Oxford: Clarendon, 1892, pp.68–73.

[2] P. W. Shor, "Polynomial-time algorithms for prime factorization and discrete logarithms on a quantum computer," SIAM Review, vol. 41, no. 2, pp. 303–332, 1999.

[3] L. K. Grover, "A fast quantum mechanical algorithm for database search," Proceedings of the twenty-eighth annual ACM symposium on Theory of computing - STOC '96, 1996.

[4] A. Barone and G. Paternò, Physics and applications of the Josephson effect. Ann Arbor, Michigan: UMI Books on Demand, 1997.

[5] F. Lecocq, I. M. Pop, Z. Peng, I. Matei, T. Crozes, T. Fournier, C. Naud, W. Guichard, and O. Buisson, "Junction fabrication by shadow evaporation without a suspended bridge," Nanotechnology, vol. 22, no. 31, p. 315302, July 2011.

[6] R. Simmonds, K. Lang, D. Hite, S. Nam, D. Pappas, and J. Martinis, "Decoherence in Josephson phase qubits from junction resonators," Physical Review Letters, vol. 93, no. 7, p. 077003, August 2004.

[7] F.-C. Chiu, "A review on conduction mechanisms in dielectric films," Advances in Materials Science and Engineering, vol. 2014, pp. 1–18, Feb. 2014.

[8] S. Salas-Rodríguez, J. Molina-Reyes, J. Martínez-Castillo, R. M. Woo-Garcia, A. L. Herrera-May, and F. López-Huerta, "Modeling of conduction mechanisms in ultrathin films of Al2O3 deposited by ALD," Electronics, vol. 12, no. 4, p. 903, Feb. 2023.

[9] J. Molina, H. Uribe-Vargas, R. Torres, P.G. Mani-Gonzalez and A. Herrera-Gomez, "Accurate modeling of gate tunneling currents in Metal-Insulator-Semiconductor capacitors based on ultra-thin atomiclayer deposited Al2O3 and post-metallization annealing", Thin Solid Films, Vol. 638 pp. 48-56 (2017).

Implementation of Step-Hetero-Oxide & Dual Material Gate Designs on Lateral β-Ga₂O₃ MOSFET for Terahertz Applications

Priyanshi Goyal
Department of Electronic Science
University of Delhi South Campus
New Delhi-110021, India
priyanshigoel.31@gmail.com

Harsupreet Kaur
Department of Electronic Science
University of Delhi South Campus
New Delhi-110021, India
harsupreetkaur@gmail.com

Abstract— **The work presents extensive simulations to study the impact of Step-Hetero-Oxide and Dual Material Gate designs on lateral β - Ga₂O₃ MOSFET. Various device attributes of the proposed device are examined and compared with those obtained for conventional device. Furthermore, key metrics such as cut-off frequency, transconductance gain product and gain bandwidth product etc., that are critical for assessing high frequency performance of the device are also evaluated. It is noteworthy to mention that the proposed device demonstrates drastic improvement in high frequency metrics and it is apparent that the proposed device is a promising design for high frequency switching applications.**

Keywords—Gallium Oxide, Step-Hetero-Oxide, Dual Material Gate, Radio Frequency.

I. Introduction

Presently, ultra high frequency spectrum is being utilized extensively for varied applications such as military, aviation, broadcasting, marine and air navigation systems [1-2] etc. Due to various advantages such as compactness, light weight, high speed and low power consumption of semiconductor devices, these devices are being explored for ultra-high frequency applications [3-4]. Moreover, although many devices have been demonstrated for high frequency applications, however, constraints of size, mobility, and saturation velocity pose a challenge to achieve higher cut-off frequencies [5-6]. Therefore, materials with high scaling feasibility, high mobility and saturation velocity are being investigated to achieve ultra-high frequencies. In this regard, wide bandgap semiconductors such as Gallium Nitride (GaN) and Silicon Carbide (SiC) etc., are substituting the conventional material i.e., silicon for high frequency applications. These materials exhibit high critical electric field which enables aggressive scaling, high saturation velocity which translates into high frequency operation [7-8]. However, factors such as expensive substrates and low crystal quality are the major roadblocks due to which research is now being focussed on emerging ultra- wide bandgap semiconductors such as gallium oxide, diamond etc. Ga₂O₃ exhibits superior intrinsic properties such as very high critical electric field of 8 MV/cm, reasonable electron mobility of 300 cm²/V.s and high saturation velocity of 2x10⁷ V/cm [9-10]. Furthermore, the conventional melt-grown techniques such as Czochralski (Cz) method, Float Zone (Fz) method and edge defined film deposition etc., can be used to produce large diameter substrates of Ga₂O₃. Many studies have been reported on exploring Ga₂O₃ for numerous applications which include power and RF electronics, optoelectronics and sensors etc [11-12]. However, the work on gallium oxide is still at nascent stage, and in order to fully exploit the potential of gallium oxide-based devices, innovation at each level from material processing to device designing and packing is required.

Furthermore, with the objective of achieving improvement in the key metrics for high frequency applications, Step-Hetero-Oxide [13] and Dual Material Gate (DMG) architectures [14] have been incorporated on gallium oxide MOSFET in the present work. Step-Oxide comprises two steps in the oxide such that the oxide thickness is lowest at the source side and highest at the drain side whereas, oxide with moderate thickness is kept in between the two. Furthermore, different dielectric oxides are chosen in the steps such that lowest dielectric constant value (SiO₂ @ κ=3.9) is taken at the source side whereas dielectric with highest dielectric constant value (HfO₂ @ κ = 25) is taken at the drain side whereas, the dielectric with moderate dielectric constant (Al₂O₃ @ κ = 9) is considered between the two. This particular design leads to enhancement in the device performance by increasing the current drivability owing to thin oxide at source side and, it also decreases the effective capacitance due to thick oxide at the drain side. Additionally, the different dielectrics have been chosen such that the effective gate capacitance reduces while increasing the peak transconductance. Moreover, these oxides are compatible with gallium oxide [15-16]. Furthermore, the dual material gate design is considered in such a way that the gate with higher work function is taken at the source side whereas, the gate with lower work function is chosen at the drain side. It is anticipated that the amalgamation of Step-Hetero-Oxide [13] and Dual Material Gate [14] designs would result in superior RF performance as compared to its counterparts.

The paper is organized as follows. The structure of the proposed device and the dimensions and doping profile under consideration are explained in section II. The results and discussion is presented in section III which is followed by conclusion in section IV.

II. Device Structure and Simulation

The schematic cross section of the proposed device i.e., Step-Hetero-Oxide – Dual Material Gate (SHO-DMG) β – Ga₂O₃ MOSFET is shown in fig. 1. Here, three regions are considered such that oxide of thickness t_{ox1} spans from 0 to L_{g1} (30 nm), oxide with thickness t_{ox2} spans from L_{g1} to L_{g2} (30 nm) whereas, the oxide with thickness t_{ox3} is present under gate length (L_{g3}) of 60 nm respectively as shown in fig. 1. The thicknesses of oxides are taken as (t_{ox1} , t_{ox2} and t_{ox3} = 10 nm, 15 nm and 20 nm) respectively. The total length of the channel is 120 nm and the channel thickness is 60 nm. A

979-8-3503-1191-4/23 $31.00 © 2023 IEEE

uniform doping of 1×10^{18} cm^{-3} is taken in the channel. A metal with higher work function of 5.93 eV is considered near the source while a lower work function of 4.4 eV is considered near the drain. Henceforth, the conventional device will be denoted as device 1 and the proposed device will be denoted as device 2.

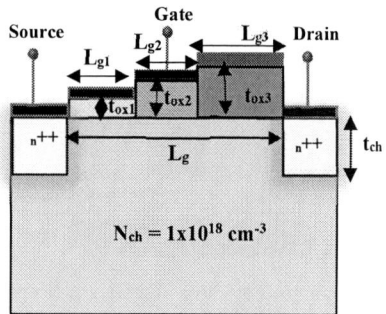

Fig. 1 Schematic cross section of proposed Step-Hetero-Oxide – Dual Material Gate -β - Ga$_2$O$_3$ MOSFET.

III. RESULTS AND DISCUSSION

In this section, various analog and RF characteristics of the proposed device are discussed. Exhaustive simulations have been performed on TCAD silvaco [17] and the relevant models such as drift-diffusion models, fermi-dirac statistics etc., have been invoked. The obtained characteristics have also been compared with the conventional device of similar dimensions. The conventional device has a uniform oxide (Al$_2$O$_3$) with thickness of 15 nm and a uniform gate with work function of 4.4 eV.

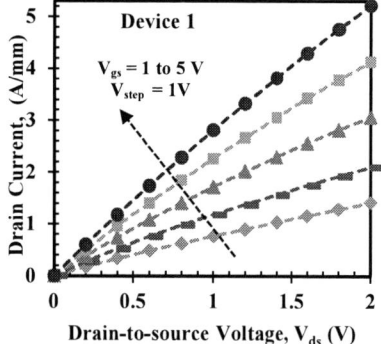

Fig. 2 (a) Output characteristics of conventional device for different V_{gs}

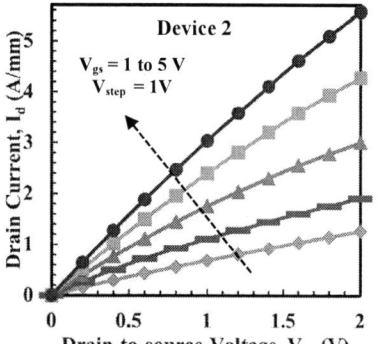

Fig. 2 (b) Output characteristics of the proposed device for different V_{gs}

Further, fig. 2(a-b) elucidates the output characteristics of the conventional and the proposed device respectively for different values of V_{gs}. On comparing these output characteristics, it can be seen that value of peak drain current corresponding to each V_{gs} is slightly higher for the proposed device. This is due to the combined effects of the Step-Hetero-Oxide and Dual Material Gate designs.

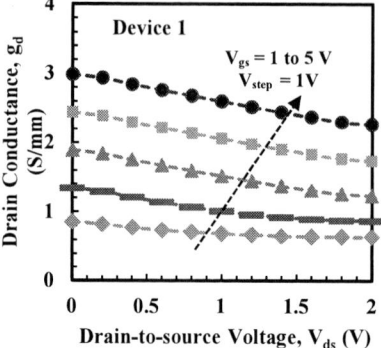

Fig. 3(a) g_d – V_{ds} characteristics of conventional device for different V_{gs}

The variation of drain conductance with V_{ds} is shown for different V_{gs} in fig. 3(a-b) for both the devices respectively. As expected, the drain conductance decreases with V_{ds} and increases with V_{gs}. It can be further noted that both the devices exhibit almost similar values of peak drain conductance for entire V_{gs} range.

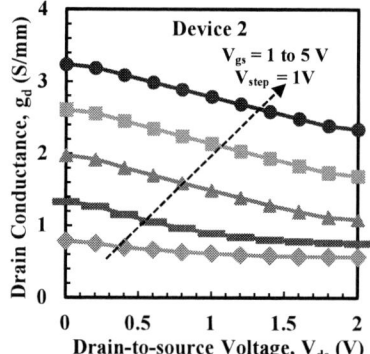

Fig. 3(b) g_d – V_{ds} characteristics of proposed device for different V_{gs}

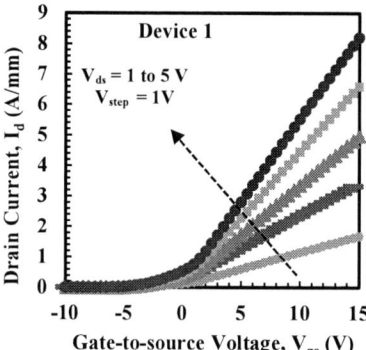

Fig. 4(a) Transfer characteristics of conventional device for different V_{ds}

Fig. 4 (a-b) demonstrates the transfer characteristics of both the devices respectively for different V_{ds} values. The value of drain current increases with V_{gs} can be seen for both the devices. A slightly higher drain current is seen for the proposed device as compared to that of the conventional device for the same reasons mentioned earlier.

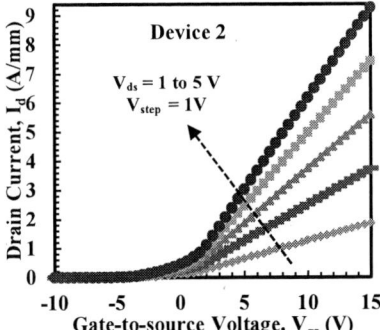

Fig. 4(b) Transfer characteristics of proposed device for different V_{ds}

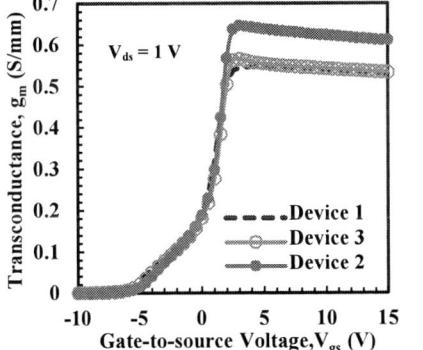

Fig. 5 Transconductance (g_m) – Gate-to-source Voltage (V_{gs}) for all three devices.

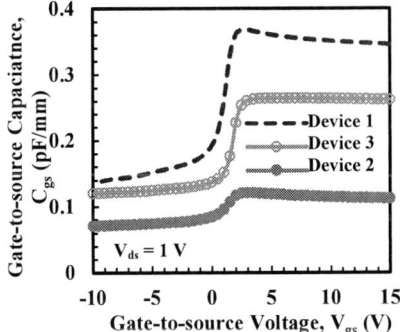

Fig. 6 (a) Gate-to-source Capacitance (C_{gs}) with variation in V_{gs} for all three devices.

Further, with the aim to explore the performance of these devices for high frequency applications, RF metrics of both the devices have been studied. In order to further gain insights about the impact of step-oxide and the impact of choice of different dielectric constants of the oxide on RF performance, the comparison of both device 1 and device 2 will also be done with the device with the step-oxide alongwith dual material gate keeping the dielectric constant same for each step and this is further indicated as device 3. Furthermore, variation of transconductance with change in V_{gs} has been plotted for fixed value of V_{ds} of 1 V for all three devices and is shown in fig. 5. It is evident from figure that value of g_m is same for device 1 and device 3 whereas, it is highest for device 2. This signifies that although incorporation of step-oxide and dual material gate designs do not have pronounced effect on transconductance, however, the hetero- oxide results in increase in transconductance due to high value of the effective dielectric constant.

Parasitic capacitances play a significant role in determining RF performance of a device. These capacitances are inevitable and are also the function of applied biasing voltages. High parasitic capacitances cause delay in the device and impede high frequency performance therefore, these need to be reduced. The parasitic capacitances have been extracted and shown in fig. 6(a-c).

Fig 6 (a) shows gate-to-source capacitance for all devices with variation in V_{gs} for V_{ds} = 1 V. It can be seen that the value of C_{gs} is highest for the conventional device, it decreases for device 3 due to higher work function at the source side while, C_{gs} is lowest for device 2. It may be attributed to the presence of low dielectric constant at the source side in device 2. Furthermore, the gate-to-drain capacitance with V_{gs} is elucidated in fig. 6(b). It can be seen that device 1 exhibits highest C_{gd} which decreases with the incorporation of thick oxide near the drain in device 2 and device 3 whereas, the hetero oxide does not have pronounced effect on C_{gd}.

The total gate capacitance (C_{gg}) with V_{gs} is shown in fig. 6(c) for all three devices. It is apparent that C_{gg} is highest for device 1 which is followed by device 3 whereas device 2 shows lowest value of C_{gg} among all three devices.

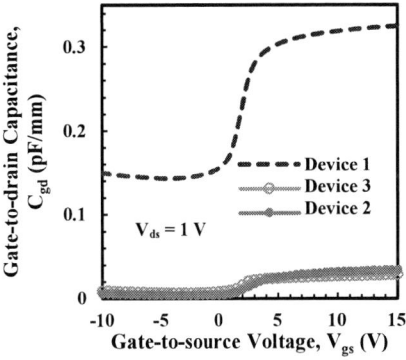

Fig. 6 (b) Gate-to-drain Capacitance (C_{gd}) with variation in (V_{gs}) for all three devices

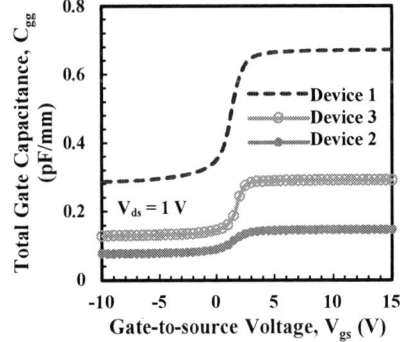

Fig. 6 (c) Total Gate Capacitance (C_{gg}) with variation in (V_{gs}) for all three devices

The variation in cut-off frequency with V_{gs} for all three devices is shown in fig. 7 at fixed V_{ds} of 1 V. The cut-off frequency is a critical parameter which gives insights about the operating frequency range of a device. Higher cut-off frequency translates to high operating frequency. It is a direct function of g_m and it varies inversely with C_{gg} as also given by equation (1). It can be observed that the value of f_t is highest for the proposed device owing to lowest C_{gg} and highest g_m. This is followed by device 3 whereas, device 1 exhibits lowest

979-8-3503-1191-4/23 $31.00 © 2023 IEEE

f_t. This suggests that proposed device is the most suitable for high frequencies applications as compared to its counterparts.

$$f_t = \frac{g_m}{2\Pi(c_{gd} + c_{gs})} \qquad (1)$$

where, g_m, C_{gd} and C_{gs} are transconductance, gate-to-drain capacitance and gate-to-source capacitance respectively.

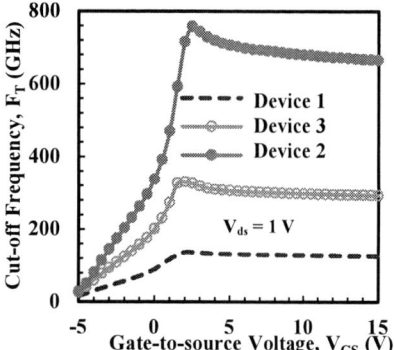

Fig. 7 Cut-off Frequency (f_t) – Gate-to-source Voltage (V_{gs}) for all three devices.

The Transconductance Frequency Product (TFP) with V_{gs} is shown in fig. 8 for all three devices. It is another key metric for assessing the trade-off between power and bandwidth and is expressed as eq. 2. It is evident that device 2 demonstrates highest TFP due to the highest transconductance and cut-off frequency for device 2 as observed in fig. 7 and 8. Device 3 shows slightly lower value of TFP whereas, lowest value of TFP is observed for device 1.

$$TFP = \frac{g_m f_t}{I_d} \qquad (2)$$

where, g_m, f_t and I_d are transconductance, cut-off frequency and drain current respectively.

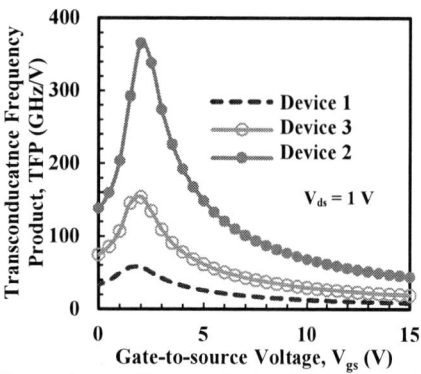

Fig. 8 Transconductance Frequency Product (TFP) with V_{gs} for all three devices.

Furthermore, Gain Bandwidth product (GBP) with variation in V_{gs} is shown in fig. 9. It determines the trade-off between the gain and bandwidth and is evaluated as (3). Highest value of GBW is obtained for device 2. This is followed by device 3 whereas, device 1 exhibits lowest GBP among all three devices.

$$GBP = \frac{g_m}{2\Pi c_{gd}} \qquad (3)$$

where, g_m and C_{gd} are transconductance and gate-to-drain capacitance respectively. It can be noted that the proposed device with the incorporation of Step-Hetero-Oxide and dual material gate designs leads to significant improvement in high power switching applications.

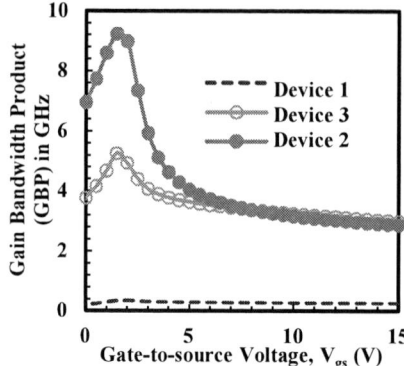

Fig. 9 Gain Bandwidth Product (GBP) with V_{gs} for all three devices.

IV. CONCLUSION

In conclusion, the Step-Hetero-Oxide along with Dual Material Gate design has been implemented on $\beta - Ga_2O_3$ MOSFET with the objective of improving high frequency performance. The critical parameters of the proposed device design were optimized in order to obtain superior high frequency figure of merits. The characteristics of the proposed device were extensivley analyzed and several RF figure of merits were evaluated and compared with the characteristics of the conventional device. It has been demonstrated that the proposed device design is a suitable stratagey particularly for high frequency switching performance.

ACKNOWLEDGMENT

The work of P. Goyal was supported by University Grants Commission, Government of India under Grant (200510172483). The authors also acknowledge Faculty Programme Grant – IoE – 2022.

REFERENCES

[1] M. Tao, J. Su, Y. Huang, L. Wang, " Mitigation of radio frequency interference in synthetic aperture radar data: Current status and future trends, " Remote Sensing, 2019 Oct 21;11(20):2438.

[2] S. Zeng, M. Li, G. Li, W. Lv, X. Liao, L. Wang, "Innovative applications, limitations and prospects of energy-carrying infrared radiation, microwave and radio frequency in agricultural products processing," Trends in Food Science & Technology, 2022 Jan 31.

[3] P K. Lu, AD. Fernandez Olvera, D. Turan, TS. Seifert, NT. Yardimci, T. Kampfrath, S. Preu, M.Jarrahi, " Ultrafast carrier dynamics in terahertz photoconductors and photomixers: beyond short-carrier-lifetime semiconductors," Nanophotonics, 2022 Mar 10;11(11):2661-91.

[4] M. Lavanya, MJ. Priya, P. Janet, KP. Kalyan, V. Vallabhuni, " Advanced 18 nm FinFET Node-Based Energy Efficient and High-Speed Data Comparator Using SR Latch," Advances in Signal Processing and Communication Engineering: Select Proceedings of ICASPACE 2021 2022 Dec 2 (pp. 327-334). Singapore: Springer Nature Singapore.

[5] J. Jeong, SK. Kim, J. Kim, DM. Geum, D. Kim, E. Jo, H. Jeong, J. Park, JH. Jang, S. Choi, I. Kwon, "Heterogeneous and monolithic 3D integration of III–V-based radio frequency devices on Si CMOS circuits," ACS nano. 2022 Apr 19;16(6):9031-40.

[6] R. Maram, S. Kaushal, J. Azaña, LR. Chen, "Recent trends and advances of silicon-based integrated microwave photonics," Photonics 2019 Jan 30 (Vol. 6, No. 1, p. 13). MDPI.

979-8-3503-1191-4/23 $31.00 © 2023 IEEE

[7]   Y. Zhang, A. Zubair, Z. Liu, M. Xiao, J. Perozek, Y. Ma, T. Palacios " GaN FinFETs and trigate devices for power and RF applications: Review and perspective", Semiconductor Science and Technology. 2021 Mar 31;36(5):054001.

[8]   Y. Cui, Y. Cao, M. Pilla, E. Beam, A. Xie, C. Lee, A. Ketterson, M. Roach, A. Geiler, M. Geiler, L. Burns, "Integration of self-biased circulators on GaN/SiC for Ka-band RF application," 2019 Device Research Conference (DRC) 2019 Jun 23 (pp. 41-42). IEEE.

[9]   SJ. Pearton, J. Yang, PH. Cary IV, F. Ren, J. Kim, MJ. Tadjer, MA. Mastro, "A review of Ga_2O_3 materials, processing, and devices. Applied Physics Reviews. 2018 Mar 11;5(1):011301.

[10]   C.Wang, J. Zhang, S. Xu, C. Zhang, Q. Feng, Y. Zhang, J. Ning, S. Zhao, H. Zhou, Y. Hao, "Progress in state-of-the-art technologies of Ga_2O_3 devices," Journal of Physics D: Applied Physics. 2021 Mar 31;54(24):243001.

[11]   R.Singh, TR Lenka, DK. Panda, RT. Velpula, B. Jain, HQ. Bui, HP. Nguyen,"The dawn of Ga_2O_3 HEMTs for high power electronics- A review," Materials Science in Semiconductor Processing. 2020 Nov 15;119:105216.

[12]   D. Guo, Q. Guo, Z. Chen, Z. Wu, P. Li, WJ. Tang, "Review of Ga_2O_3 -based optoelectronic devices," Materials Today Physics. 2019 Dec 1;11:100157.

[13]   R. Sithanandam, MJ. Kumar, "A new hetero-material stepped gate (HSG) SOI LDMOS for RF power amplifier applications," In2010 23rd International Conference on VLSI Design 2010 Jan 3 (pp. 230-234). IEEE.

[14]   W. Long, H. Ou, JM. Kuo, KK. Chin, "Dual-material gate (DMG) field effect transistor," IEEE Transactions on Electron Devices. 1999 May;46(5):865-70.

[15]   CV. Prasad, YS.Rim, "Review on interface engineering of low leakage current and on-resistance for high-efficiency Ga_2O_3-based power devices," Materials Today Physics. 2022 Jul 14:100777.

[16]   PH. Carey IV, F. Ren, D. Hays, BP. Gila, S. Pearton, "Band alignments of dielectrics on (-201) β-Ga_2O_3," Gallium Oxide 2019 Jan 1 (pp. 287-311). Elsevier.

[17]   ATLAS : 2-D Device Simulator, Version 5.19.20.R, Silvaco, Santa clara, CA, USA (2014).

2023 IEEE Latin American Electron Devices Conference (LAEDC) Puebla, México, July 3-5, 2023

TCAD Investigation of Step-Oxide and Asymmetric Doping Design with Electrode Engineering on Lateral β-Ga₂O₃ MOSFET for Terahertz Applications

Priyanshi Goyal
Department of Electronic Science
University of Delhi South Campus
New Delhi-110021, India
priyanshigoel.31@gmail.com

Harsupreet Kaur
Department of Electronic Science
University of Delhi South Campus
New Delhi-110021, India
harsupreetkaur@gmail.com

Abstract— **This work presents exhaustive TCAD investigation of Step-Oxide- Dual Material Gate with Asymmetric Doping engineering on lateral β – Ga₂O₃ MOSFET. The objective of this work is to achieve high frequency performance. Several device attributes such as output and transfer characteristics, transconductance, parasitic capacitances and cut-off frequency etc., have been studied. Further, various critical figure of merits such as transconductance frequency product, gain bandwidth product etc., have also been obtained. A comparison of the proposed device is drawn with the conventional device. The results highlight that the proposed device is a promising design strategy for RF applications.**

Keywords—Gallium Oxide, Step-Oxide, Dual Material Gate, Asymmetric Doping, Radio Frequency.

I. INTRODUCTION

In the last few decades, significant progress has been made in the field of RF electronics for a wide range of applications. Numerous semiconducting materials such as Silicon (Si), Gallium Arsenide (GaAs) and Indium Phosphide (InP) etc., have already been explored for RF applications and demonstrated high performance [1-2]. However, these materials have reached their intrinsic limits, therefore, advanced materials such as Gallium Nitride (GaN) and Gallium Oxide (Ga₂O₃) etc., are now being investigated to serve the purpose of achieving high frequency performance [3-4]. Although, GaN is more mature technology in terms of RF key metrics as compared to Ga₂O₃, nonetheless, the lack of large-diameter substrates and low-quality crystals pose a limit on fully adoption of GaN [5-6]. On the other hand, Ga₂O₃ does not possess such fabrication constraints and can be processed at much lower cost with ease of mass production as compared to its counterparts [7-8]. Further, the intrinsic properties of gallium oxide such as high critical electric field, high saturation velocity and high radiation hardness aid to the adoption of Ga₂O₃ for various applications [9-10]. Several experimental studies have been reported on gallium oxide and various design strategies such as recessed gate MOSFET [11], trench gate MOSFET [12], field plated MOSFET [13], Vertical MOSFET [14], FinFET [15] etc., have also been demonstrated to unfold its potential for diversified applications. Although key metrics particularly for power and RF applications have been significantly improved, however, these are still far below than the predicted theoretical values. In view of this, in the present work, step-oxide design along with dual material gate design and asymmetric doping profile in the channel has been proposed on lateral Ga₂O₃ MOSFET.

The objective is to improve the key metrics such as intrinsic gain, and bandwidth etc., for achieving ultra-high frequencies.

The proposed device has various advantages such as reduction in parasitic capacitances owing to Step-Oxide[16], enhancement in transport efficiency due to dual material gate design [17] and improvement in transconductance and bandwidth on account of the asymmetric doping profile in the channel [18].

The paper is organized as follows. The device structure is presented in section II. Section III presents the results and discussions. Finally, the conclusion in presented in section IV.

II. DEVICE STRUCTURE AND SIMULATION

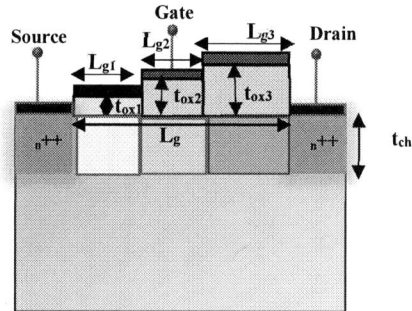

Fig. 1 Schematic cross section of proposed Step-Oxide – Dual Material Gate Engineered and Asymmetric Doped -β - Ga₂O₃ MOSFET.

Fig. 1 shows the schematic cross section of the proposed, Step-Oxide - Dual Material Gate Engineered and Asymmetric Doped β – Ga₂O₃ MOSFET. The non-uniform oxide is considered such that the thickness of the oxide from 0 o L_{g1} is t_{ox1} (10 nm), from L_{g1} to L_{g2} is t_{ox2} (15 nm) whereas, from L_{g2} to L_{g3} is t_{ox3} (20 nm) respectively. The total channel length ($L_g = L_{g1} + L_{g2} + L_{g3}$) is (120 nm = 30 nm + 30 nm+ 60 nm). The channel thickness is 240 nm. A non-uniform doping is considered in the channel such that 0.5×10^{18} cm⁻³ in region L_{g1}, 1×10^{18} cm⁻³ in region L_{g2} whereas, 1.5×10^{18} cm⁻³ is taken in region L_{g3}. The work function of the gate at the source side (L_{g1} and L_{g2}) is taken as 5.93 eV while, 4.4 eV is taken for the gate near drain.

III. RESULTS AND DISCUSSION

This section deals with the simulated results of the proposed device and the conventional device. Exhaustive simulations have been carried out on TCAD silvaco [19] and

979-8-3503-1191-4/23 $31.00 © 2023 IEEE

the suitable models such as drift-diffusion models etc., have been invoked throughout the analysis. The dimensions of the conventional device are kept similar to the proposed device for fair assessment. A uniform channel doping of 1×10^{18} cm^{-3} is considered in conventional device. Further, a fixed oxide (Al$_2$O$_3$) with thickness of 15 nm is taken in the conventional device. The work function of the metal gate is chosen as 4.4 eV. Rest of the dimensions are kept same as mentioned in section II. Furthermore, the conventional device is denoted as device 1 whereas, the proposed device is denoted as device 2 henceforth.

The output characteristics of both the devices under consideration are shown in fig. 2(a-b) respectively for wide range of V_{ds}. It can be clearly seen that the value of drain current is higher for the proposed device corresponding to each V_{gs}. This is due to the incorporation of dual material gate design and asymmetric doping in the channel since dual material gate design leads to increase in transport efficiency whereas, high doping at the drain side further increases the drain current of the proposed device.

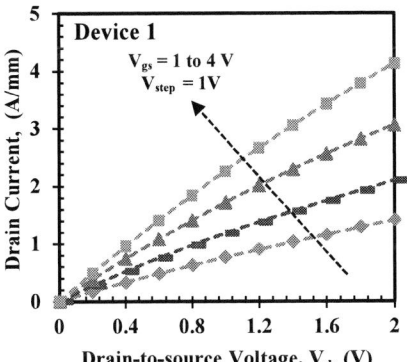

Fig. 2(a) $I_d - V_{ds}$ characteristics of the conventional device for different V_{gs}

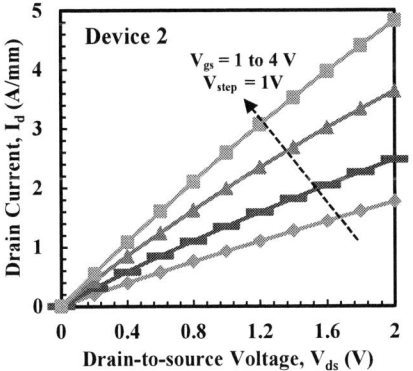

Fig. 2(b) $I_d - V_{ds}$ characteristics of the proposed device for different V_{gs}

The variation in drain conductance with change in V_{ds} is elucidated in fig. 3(a-b) for both the devices for different V_{gs} values. A lower value of g_d signifies higher intrinsic gain, therefore, a low value of g_d is desirable. It is apparent from fig. 3 that the proposed device exhibits slightly lower value of g_d as compared to the conventional device for entire range of V_{gs}. This implies that the proposed device is suitable design for high gain applications as well. Furthermore, the transfer characteristics of both the devices are demonstrated in fig. 4(a-b) for different V_{ds}. Similar pattern as was seen in fig, 2(a-b) can be observed here as well for the same reasons mentioned earlier. The device characteristics shown in fig. 2 to 4 indicate that the proposed device exhibits good dc characteristics.

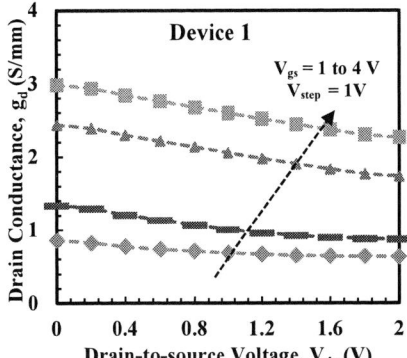

Fig. 3(a) $g_d - V_{ds}$ characteristics of conventional device for different V_{gs}

Furthermore, in order to analyze RF performance of both the devices, the RF metrics have further been evaluated. Moreover, to obtain insights about the impact of asymmetric doping in the proposed device, the proposed device with uniform doping in the channel is also considered and will be denoted as device 3.

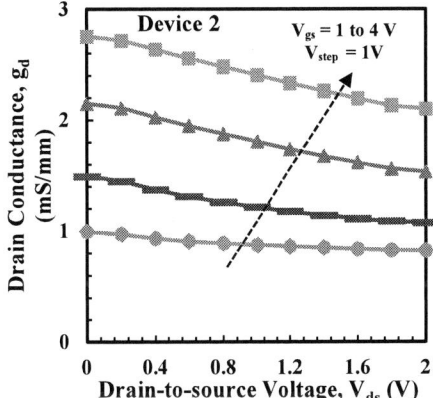

Fig. 3(b) $g_d - V_{ds}$ characteristics of conventional device for different V_{gs}

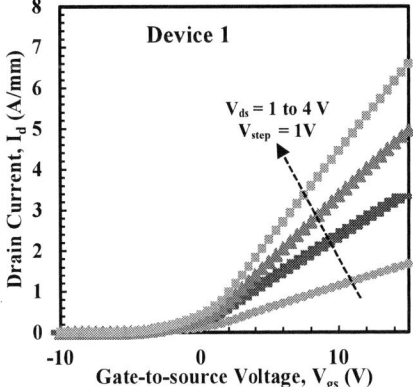

Fig. 4(a) $I_d - V_{gs}$ characteristics of conventional device for different V_{ds}

The variation in transconductance with change in V_{gs} is shown in fig. 5 for all three devices for $V_{ds} = 1$V. It is evident from the figure that device 1 exhibits lowest value of peak g_m which is followed by device 3 which incorporates step oxide and dual material gate designs whereas, the proposed device with additional asymmetric doping in the channel further increases the peak g_m. This is due to the asymmetric doping profile

979-8-3503-1191-4/23 $31.00 © 2023 IEEE

wherein, high doping at the drain side is considered which increases the transport efficiency and further enhances transconductance. Transconductance can be evaluated as:

$$g_m = \frac{\partial I_d}{\partial v_{gs}} \qquad (1)$$

where, I_d and V_{gs} are drain current and gate-to-source voltage respectively.

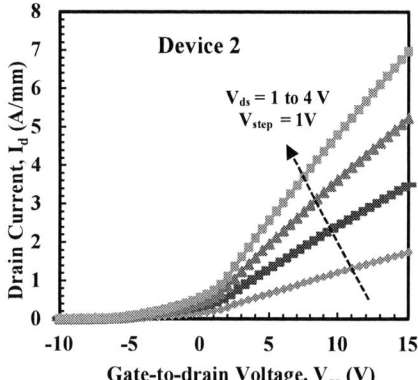

Fig. 4(b) $I_d - V_{gs}$ characteristics of proposed device for different $V_{ds.}$

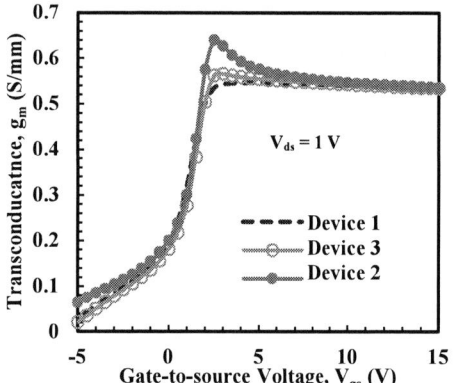

Fig. 5 $g_m - V_{gs}$ characteristics of all three devices for $V_{ds} = 1V$

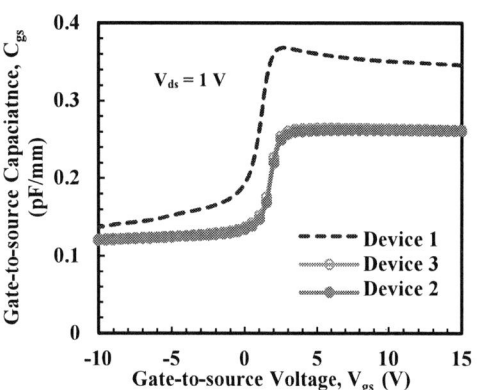

Fig. 6(a) $C_{gs} - V_{gs}$ characteristics of all three devices for $V_{ds} = 1V$

The parasitic capacitances for all three devices are studied and are shown in fig. 6 (a-c). Fig 6(a) shows gate-to-source capacitance with variation in V_{gs} for all devices under consideration. It can be noted that device 2 and device 3 exhibit similar values of capacitances corresponding to each V_{gs} however, device 1 shows highest C_{gs}. It can be inferred that the incorporation of step oxide and dual material gate

designs have pronounced effects on capacitance since the thick oxide at the drain significantly decreases the capacitance of the device. Additionally, high work function at the source side further reduces the gate charge under the source and reduces gate-to-source capacitance. Furthermore, it can be noted that since there is slight change in doping concentration in the three regions therefore, significant impact of asymmetric doping is not seen on capacitances.

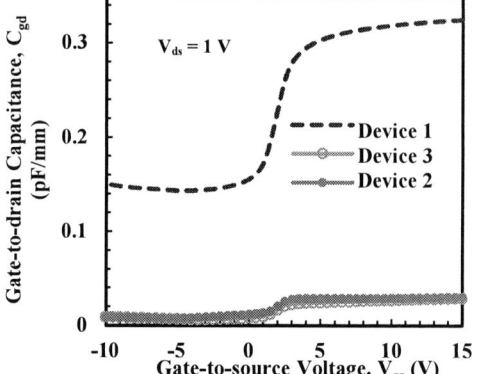

Fig. 6(b) $C_{gd} - V_{gs}$ characteristics of all three devices for $V_{ds} = 1V$

Furthermore, gate-to-drain and total gate capacitances are shown in fig. 6(b-c) respectively. The same trend as was observed in fig. 6(a) is evident here as well. Both C_{gd} and C_{gg} are highest for conventional device i.e., device 1 and show significant reduction in C_{gd} and C_{gg} for the proposed designs.

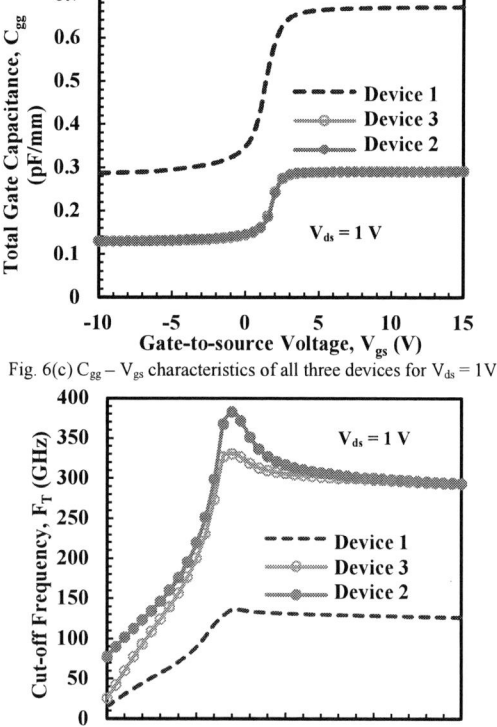

Fig. 6(c) $C_{gg} - V_{gs}$ characteristics of all three devices for $V_{ds} = 1V$

Fig. 7 $f_t - V_{gs}$ characteristics of all three devices for $V_{ds} = 1V$

Fig. 7 elucidates the cut-off frequency with V_{gs} for all three devices. It is apparent from figure that device 1 exhibits lowest

979-8-3503-1191-4/23 $31.00 © 2023 IEEE

value of f_t. The incorporation of step-oxide reduces the parasitic capacitances whereas, the integration of dual material gate designs improves transconductance by enhancing transport efficiency. Due to these reasons, higher f_t is observed in device 2 and device 3 which incorporate step-oxide and dual material gate designs. Furthermore, the doping engineering in the channel in device 2 leads to further increase in transconductance which translates to maximum cut-off frequency of device 2. The expression for evaluating f_t is expressed as (2).

$$f_t = \frac{g_m}{2\,\Pi\,(C_{gd} + C_{gs})} \qquad (2)$$

where, g_m, C_{gd} and C_{gs} are transconductance, gate-to-drain capacitance and gate-to-source capacitance respectively.

Fig. 8 demonstrates g_m/I_d with V_{gs} for all three devices. It is an important figure of merit which indicates efficiency of a device. It is a measure of how effectively the drain current translates into transconductance. Higher value of g_m/I_d is required for potential current modulation. It can be observed that highest value of g_m/I_d is exhibited by the proposed device i.e., device 2 while, slightly lower value is exhibited by device 3 whereas, device 1 shows lowest value of g_m/I_d among all three devices.

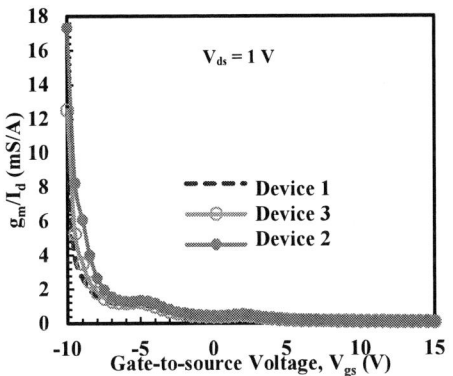

Fig. 8 g_m/I_d – V_{gs} characteristics of all three devices for $V_{ds} = 1$ V

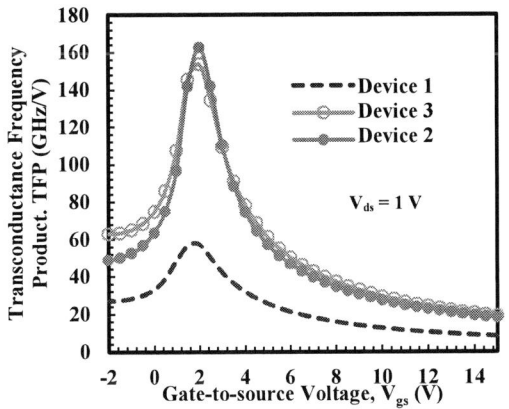

Fig. 9 TFP with V_{gs} of all three devices for $V_{ds} = 1$ V

Transconductance Frequency Product (TFP) with V_{gs} is shown in fig. 9. It accounts for the trade-off between the power and bandwidth and is given as equation (3). It can be seen that device 2 exhibits the highest value of TFP among all three devices which is followed by device 3 whereas, device 1

shows lowest value of TFP.

$$TFP = \frac{g_m f_t}{I_d} \qquad (3)$$

where, g_m, f_t and I_d denote transconductance, cut-off frequency and drain current respectively.

Further, the Gain Bandwidth Product (GBP) with variation in V_{gs} for all three devices is illustrated in fig. 10. It depends on the intrinsic gain and cut-off frequency as shown in equation (4). It can be noted that highest intrinsic gain (from fig. 8) and highest cut-off frequency (from fig. 7) translates to high GBP of device 2 as compared to other two devices. As expected, device 3 follows the proposed device which is further followed by device 1.

$$GBP = \frac{g_m}{2\,\Pi\,C_{gd}} \qquad (4)$$

where, g_m and C_{gd} denote transconductance and gate-to-drain capacitance respectively.

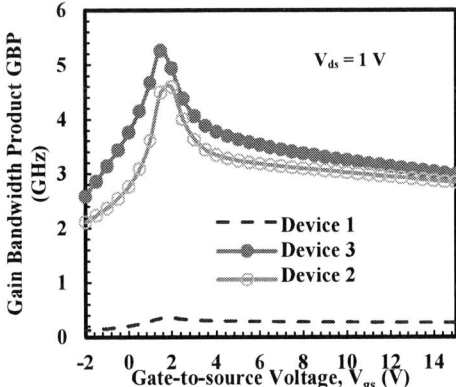

Fig. 10 GBP with V_{gs} of all three devices for $V_{ds} = 1$V

IV. CONCLUSION

In conlcusion, the impact of asymmetric doping profile, step- oxide and work function engineering have been studied on lateral $\beta - Ga_2O_3$ MOSFET using TCAD Silvaco. It can be seen that step-oxide design drastically reduces the parastic capacitances whereas the amalgamation of dual material gate design improves the transport efficiency. Further, the asymmetric doping in the channel leads to improvement in transconductance. It has been demonstrated that the proposed device is an effective device with remarkable high frequency figure of merits.

ACKNOWLEDGMENT

The work of P. Goyal was supported by University Grants Commission, Government of India under Grant (200510172483). The authors also acknowledge Faculty Programme Grant – IoE – 2022.

REFERENCES

[1]   Y.H. Jung, H. Zhang, IK. Lee, JH. Shin, TI. Kim, Z. Ma," Releasable high-performance GaAs Schottky diodes for gigahertz operation of flexible bridge rectifier," Advanced Electronic Materials, 2019 Feb;5(2):1800772.

[2]   N. Gowthaman, V. Srivastava, "Design of Hafnium Oxide (HfO₂) sidewall in InGaAs/InP for high-speed electronic devices," InKey Engineering Materials 2022 (Vol. 907, pp. 10-16). Trans Tech

Publications Ltd.

[3] J. Zhang, J. Shi, DC. Qi, L. Chen, KH. Zhang, "Recent progress on the electronic structure, defect, and doping properties of Ga_2O_3," APL Materials. 2020 Feb 1;8(2):020906.

[4] C. Wang, J. Zhang, S. Xu, C. Zhang, Q. Feng, Y. Zhang, J. Ning, S. Zhao, H. Zhou, Y. Hao, "Progress in state-of-the-art technologies of Ga_2O_3 devices," Journal of Physics D: Applied Physics. 2021 Mar 31;54(24):243001.

[5] T. Narita, T. Kachi, "Future challenges: Defects in GaN power Device due to fabrication processes", In Characterization of Defects and Deep Levels for GaN Power Devices 2020 Dec 16 (pp. 8-1). Melville, New York: AIP Publishing LLC.

[6] D. Marcon, B. De Jaeger, S. Halder, N. Vranckx, G. Mannaert, M. Van Hove, S. Decoutere," Manufacturing challenges of GaN-on-Si HEMTs in a 200 mm CMOS fab," IEEE transactions on semiconductor manufacturing. 2013 Mar 29;26(3):361-7.

[7] M. Higashiwaki, " β-Ga_2O_3 material properties, growth Technologies, and devices: a review," AAPPS Bulletin. 2022 Jan 17;32(1):3.

[8] S. Stepanov, V. Nikolaev, V. Bougrov, A. Romanov, " Gallium oxide: Properties and applications", 498 a review. Rev. Adv. Mater. Sci. 2016;44:63-86.

[9] N. Moser, K. Liddy, A. Islam, N. Miller, K. Leedy, T. Asel, S. Mou, A. Green, K. Chabak, "Toward high voltage radio frequency devices in β-Ga_2O_3," Applied Physics Letters. 2020 Dec 14;117(24):242101.

[10] D. Guo, Q. Guo, Z. Chen, Z. Wu, P. Li, WJ. Tang, :Review of Ga_2O_3 -based optoelectronic devices," Materials Today Physics. 2019 Dec 1;11:100157.

[11] KD. Chabak, JP. McCandless, NA. Moser, AJ. Green, K. Mahalingam, A. Crespo, N. Hendricks, BM. Howe, SE. Tetlak, K. Leedy, RC.Fitch, "Recessed-Gate Enhancement-Mode β-Ga_2O_3 MOSFETs," IEEE Electron device letters. 2017 Dec 4;39(1):67-70.

[12] X. Chen, F. Li, H. Hess, "Trench Gate β-Ga_2O_3 MOSFETs: A Review," Engineering Research Express. 2023 Feb 28.

[13] S. Kumar, H. Murakami, Y. Kumagai, M. Higashiwaki, "Vertical Ga_2O_3 Schottky barrier diodes with trench staircase field plate," Applied Physics Express. 2022 Apr 8;15(5):054001.

[14] X. Zhou, Y. Ma, G. Xu, Q. Liu, J. Liu, Q. He, X. Zhao, S. Long, "Enhancement-mode β-Ga_2O_3 U-shaped gate trench Vertical MOSFET realized by oxygen annealing," Applied Physics Letters. 2022 Nov 28;121(22):223501.

[15] RM. Kotecha, A. Zakutayev, WK. Metzger,P. Paret, G. Moreno, B. Kekelia, K. Bennion, B. Mather, S. Narumanchi, S. Kim, S. Graham, "Electrothermal Modeling and Analysis of Gallium Oxide Power Switching Devices," In International Electronic Packaging Technical Conference and Exhibition 2019 Oct 7 (Vol. 59322, p. V001T06A017). American Society of Mechanical Engineers.

[16] R. Sithanandam, MJ. Kumar, "A new hetero-material stepped gate (HSG) SOI LDMOS for RF power amplifier applications," In2010 23rd International Conference on VLSI Design 2010 Jan 3 (pp. 230-234). IEEE.

[17] W. Long, H. Ou, JM. Kuo, KK. Chin, "Dual-material gate (DMG) field effect transistor," IEEE Transactions on Electron Devices. 1999 May;46(5):865-70.

[18] J. Saltin, S. Tian,F. Ding, HY. Wong, "Novel doping engineering techniques for gallium oxide MOSFET to achieve high drive current and breakdown voltage," In2019 IEEE 7th Workshop on Wide Bandgap Power Devices and Applications (WiPDA) 2019 Oct 29 (pp. 261-264). IEEE.

[19] ATLAS : 2-D Device Simulator, Version 5.19.20.R, Silvaco, Santa clara, CA, USA (2014).

Effect of the thermal annealing temperature on the luminescent and morphological properties of silicon rich oxide bilayer structures.

J. Juan Avilés Bravo
National Institute of Astrophysics, Optics and Electronics
Puebla, Mexico
juan.aviles@inaoep.mx.

A. Morales Sánchez
National Institute of Astrophysics, Optics and Electronics
Puebla, Mexico
alfredom@inaoep.mx.

L. Palacios Huerta
Instituto Politécnico Nacional, Unidad Profesional Interdisciplinaria de Ingeniería Campus Tlaxcala,
Tlaxcala, Mexico
lpalaciosh@ipn.mx.

J. Federico Ramirez Rios
National Institute of Astrophysics, Optics and Electronics
Puebla, Mexico
juan.ramirez@inaoep.mx

M. Moreno Moreno
National Institute of Astrophysics, Optics and Electronics
Puebla, Mexico
mmoreno@inaoep.mx

Abstract—This work studies the effect of the thermal annealing temperature on the optical and morphological properties of SRO_x/SRO_y bilayer structures when the SRO_y contains a high silicon excess. SRO_x/SRO_y bilayer structures were deposited by low pressure chemical vapor deposition and then thermally annealed at temperatures ranging from 900-1100 °C. In addition, SRO_x and SRO_y monolayers were deposited for comparison. The refraction index and the average roughness decrease when the thermal annealing temperature increases as a result of structural changes within the SRO_x/SRO_y bilayers. A two-fold improvement in the photoluminescence (PL) intensity (65 to 136 a.u/nm) was obtained as the annealing temperature is reduced from 1100 to 1050 °C. A relationship between PL emission and surface morphology was analyzed.

Keywords—Silicon rich oxide, bilayer structure, thermal annealing, surface morphology, photoluminescence.

I. INTRODUCTION

Over the last several decades, silicon nanoparticles (Si-nps) embedded in a dielectric matrix have become an excellent alternative for the development of light-emitting devices compatible with the Si-based technology [1-2]. These devices have been obtained using different dielectric matrices such as silicon-rich silicon nitride (SRN) [3-4], silicon-rich silicon oxide (SRO) [5-6], and silicon carbide [7-8]. The Si-excess, the time and the temperature of thermal annealing determine the Si-nps size and their amorphous/crystalline nature [9]. SRO films have been obtained by different techniques including low pressure chemical vapor deposition (LPCVD) and plasma-enhanced chemical vapor deposition (PECVD) where the highest photoluminescence (PL) is obtained in those deposited by LPCVD [10]. The Si-excess in SRO-LPCVD films is controlled by the ratio (Ro) of the precursor gasses, as defined by Eq. (1):

$$Ro = P(N_2O)/P(SiH_4) \qquad (1)$$

where $P(N_2O)$ and $P(SiH_4)$ are the partial pressures of nitrous oxide and silane, respectively. It has been reported that Ro=10, Ro=20 and Ro=30 produce SRO films with 12, 6, 4 at. % of Si-excess, respectively [11]. Also, it was observed that SRO-LPCVD films with a Si-excess about 4-6 at. % thermally annealed at 1100 °C for 180 min emit the strongest

PL [10]. As mentioned, the Si-excess is extremely important since if it is increased, under the same thermal annealing conditions, the PL drastically decreases. This effect is related to the formation of Si-nps larger than 5 nm, which are not luminescent [12-13]. High Si-excess ($\geq$12 at. %) allows more silicon atoms available within the SRO matrix to form Si-nps. Therefore, less thermal energy could be required to form Si-nps (< 5nm) that contribute to PL emission. Nevertheless, the surface of the SRO films with high Si-excess tends to be oxidized after thermal annealing [14]. This effect can be reduced if we use a SRO film with low Si-excess acting as passivating layer on the surface of that with high Si-excess in the form of SRO_x/SRO_y bilayer. Therefore, in this work we study the effect of the thermal annealing temperature on the optical and morphological properties of SRO_x/SRO_y bilayer structures.

II. EXPERIMENTAL DETAILS

A. Mono and bilayer structure deposition

In the present study, mono- and bi-layer of SRO films were deposited on n-type (100)-Si substrate (resistivity 1-5 Ω-cm) by LPCVD at 732 °C. Ro ($P(N_2O)/P(SiH_4)$) values of 20 (0.8 Torr/1 Torr) and 5 (0.2 Torr/1 Torr) were used for this experiment, which allows to obtain SRO films with 6 and 14 at. % of Si-excess, respectively [11]. SRO_{20} (~5 nm)/SRO_5 (~45 nm) bilayers were deposited and subsequently thermally annealed at different temperature (900 °C–1100 °C) for 180 min in nitrogen atmosphere, as specified in Table I. The subscript in SRO_x indicates the Ro value. SRO_{20} (~48.5 nm) and SRO_5 (~58 nm) films (monolayer) were also deposited and thermally annealed at 1100 °C for 180 min in nitrogen atmosphere for reference.

B. Measurements details

The thickness and refractive index of the SRO and SRO_{20}/SRO_5 bilayer structures were measured with a Gaertner

TABLE I. SAMPLE LABEL, Ro VALUE, AND THERMAL ANNEALING TEMPERATURE FOR EACH STRUCTURE.

Sample label	Ro	Temperature (°C)
SRO$_{20}$	20	
SRO$_5$	5	1100
B-1100		
B-1050		1050
B-1000	20/5	1000
B-950		950
B-900		900

L117 ellipsometer (632.8 nm). The surface morphology was measured with a Nanosurf EasyScan atomic force microscopy (AFM) operated in no contact mode. A 4×4 μm^2 scanned area was used for each topographic image. AFM images were analyzed using Gwyddion 2.59 software. The PL spectra were measured with a Duetta spectrometer of Horiba. The samples were excited with a wavelength (λ) of 300 nm and the luminescence was measured from 450 to 1100 nm.

III. RESULTS AND DISCUSSION

Table II shows the thickness and refractive index of all samples before and after thermal annealing. The thickness of reference samples SRO$_{20}$ and SRO$_5$ are 48.53 ± 0.17 and 58.76 ± 0.20 nm, respectively. Whereas the thickness of all bilayers is ~50 nm, as planned. The refractive index of the reference samples SRO$_{20}$ and SRO$_5$ are 1.625 ± 0.017 and 1.790 ± 0.003, respectively. The refractive index obtained for the SRO$_{20}$ agrees with Ro=20 reported before, but the value for sample SRO$_5$ is lower than expected for Ro=5 [15]. While the refractive index all as deposited SRO$_{20}$/SRO$_5$ bilayers is ~1.7 indicating a mixture between both SRO layers. After thermal annealing, a decrease in refractive index is observed in all samples.

Figure 1 shows more clearly the difference in refractive index before and after thermal annealing of all samples. The decrease in refractive index after thermal annealing is an unusual behavior because it is known that the refractive index of SRO samples increases, tending to the value of silicon (4.01), due to the phase separation between Si and SiO$_2$ that allows the formation of amorphous or crystalline Si-nps [16]. In our case, the refractive index decreases after thermal

TABLE II. THICKNESS AND REFRACTIVE INDEX OF ALL SAMPLE BEFORE AND AFTER THERMAL ANNEALING.

Sample label	Thickness (nm)		Refractive index	
	Before	*After*	*Before*	*After*
SRO$_{20}$	48.53 ± 0.17	48.23 ± 0.55	1.625 ± 0.017	1.529 ± 0.024
SRO$_5$	58.76 ± 0.20	57.36 ± 1.53	1.790 ± 0.003	1.631 ± 0.017
B-1100	51.70 ± 1.01	51.86 ± 0.73	1.695 ± 0.026	1.553 ± 0.006
B-1050	49.12 ± 0.83	49.06 ± 0.63	1.685 ± 0.016	1.576 ± 0.017
B-1000	51.03 ± 0.97	50.93 ± 0.92	1.689 ± 0.020	1.592 ± 0.010
B-950	50.40 ± 0.87	50.43 ± 1.53	1.682 ± 0.022	1.595 ± 0.030
B-900	51.23 ± 1.26	50.06 ± 1.50	1.719 ± 0.018	1.640 ± 0.006

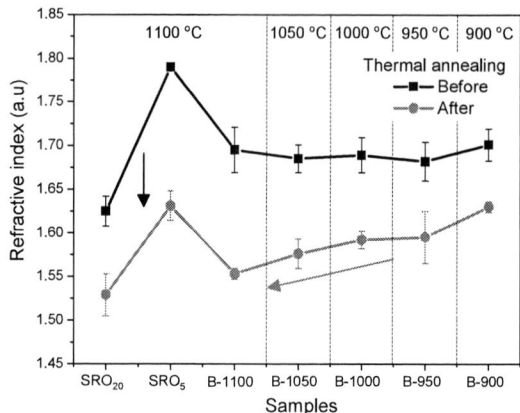

Fig. 1: Refractive index before and after thermal annealing of SRO and SRO$_{20}$/SRO$_5$ bilayer.

annealing and it tends towards the value of a SiO$_2$ (1.46) (black arrow) as the temperature increases, which indicates that oxygen atoms are bonded into the matrix. This can be explained if we consider a surface oxidation after the thermal annealing process [14]. Due to the high temperature of the sample, the surface reacts with the oxygen present in the environment, and it is incorporated into the matrix.

Surface oxidation can be related to: (1) silicon excess and (2) thermal annealing temperature. The high excess of silicon atoms available to react with the environment oxygen increases the oxidation. Similarly, the high surface temperature of the sample when exposed to ambient increases this reaction. The trend (1) can be observed by comparing the change in refractive index between sample SRO$_{20}$ and SRO$_5$ which decreases from 1.62 to 1.52 and 1.79 to 1.63, respectively (black arrow in Fig. 1). While trend (2) can be observed in the change of the refractive index of the SRO$_{20}$/SRO$_5$ bilayer samples where it decreases as the thermal annealing temperature increases (red arrow in Fig. 1).

Figure 2 shows the AFM images of the SRO and SRO$_{20}$/SRO$_5$ bilayers. All samples exhibit a rough surface. The average roughness (S_a) of sample SRO$_{20}$ is lower than SRO$_5$ (Fig. 3), as observed in the 3D images shown in Fig. 2 (a) and 2 (b).

In the literature, S_a has been related to the Si-excess and the size of the Si-nps formed after thermal annealing. This relationship indicates that high Si-excess within a SRO layer, largest the size of the Si-nps and high surface roughness [17]. On the other hand, for the SRO$_{20}$/SRO$_5$ bilayers, it was expected a lower S_a than SRO$_5$, but this is not the case, as shown in Figure 3. The lower S_a obtained in the SRO$_5$ sample with respect to B-1100 could be related to the surface oxidation at the end of the thermal annealing, as mentioned above. As more oxygen atoms are introduced on the surface of the SRO$_5$ sample, the surface of the sample becomes more ordered and homogeneous [14]. Moreover, a SRO$_{20}$ (low Si-excess) layer is at the surface in the SiO$_{20}$/SiO$_5$ bilayer reducing the oxidation as compared to the single SRO$_5$ layer. A decrease in S_a is also observed when the annealing temperature decreases (except for the sample annealed at 950 °C). This can be ascribed to the change in size of the Si-nps within the SiO$_2$ matrix caused by the decrease in temperature [18].

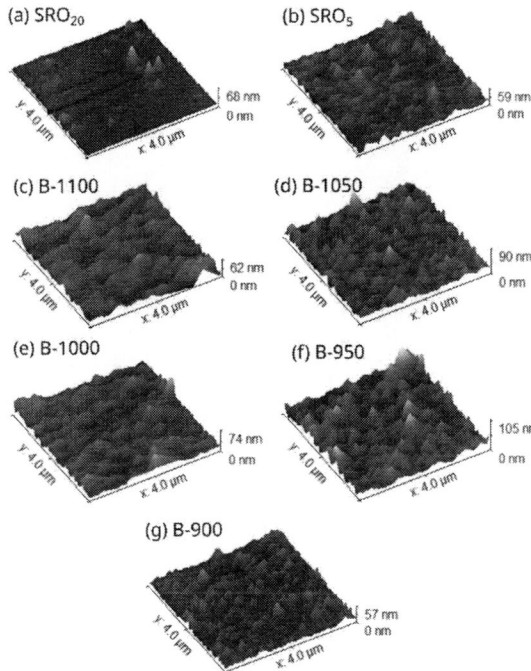

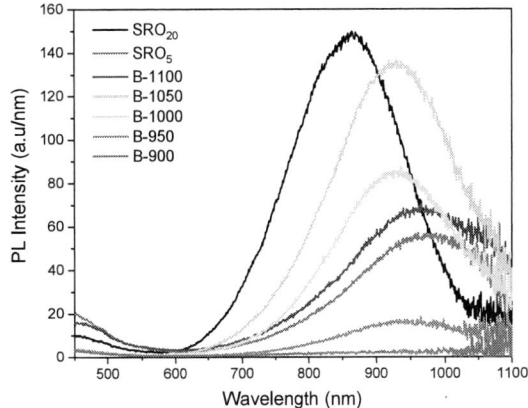

Fig. 4: PL spectra of thermally annealed SRO and SRO_{20}/SRO_5 bilayers.

Fig. 2: 3D AFM images of thermally annealed (a) SRO_{20}, (b) SRO_5, (c) B-1100, (d) B-1050, (e) B-1000, (f) BL-950 and (g) B-900 samples. Scanned area: 4×4 μm^2.

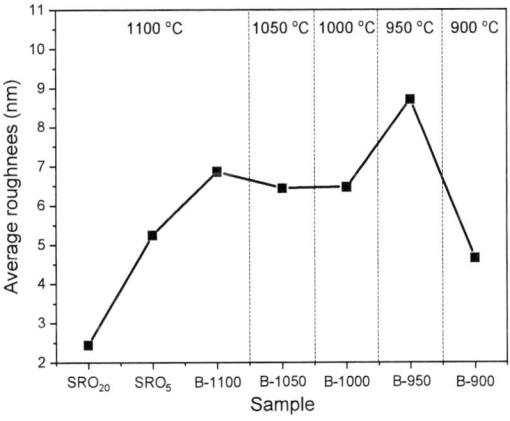

Fig. 3: Average roughness of thermally annealed SRO and SRO_{20}/SRO_5 bilayers.

Figure 4 shows the PL emission spectra for all samples after thermal annealing. It was observed that all samples presented two broad emission bands: from 450 to 550 nm and from 650 to 1100 nm. The first PL band (450 to 550 nm) has been related to the combination of oxygen defect center (ODC) and the E'_δ (Si↑Si≡Si) defects [14-19]. While the second PL band (650 to 1100 nm) is related to band-to-band transitions in Si-nanocrystals (Si-ncs) due to the quantum confinement effects (QCE) [20], where their emission energy depended on the size of Si-ncs.

Figure 5 shows the relationship between the intensity and the maximum emission peak (of the second PL band) for each sample. It is observed that the maximum PL intensity (149 a.u./nm) is obtained in the SRO_{20} sample with its main peak

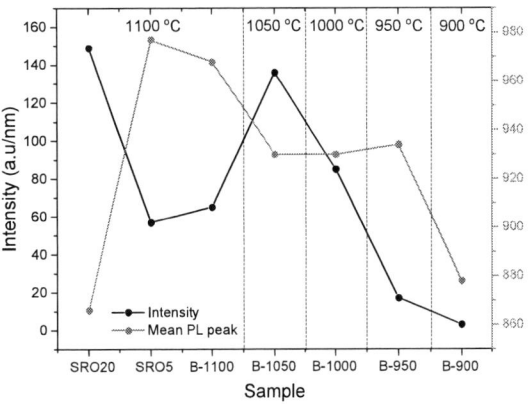

Fig. 5: Intensity and wavelength of main PL peak of each thermally annealed samples.

located at 866 nm. For the SRO_5 sample, the PL intensity strongly decreases (57 a.u./nm) and it exhibits a redshift to 977 nm. It has been reported that SRO films with a Si-excess > 12 at. %, thermally annealed at 1100 °C, contain Si-ncs with sizes larger than 5 nm, which do not contribute to the radiative processes [12-13]. However, the B-1100 bilayer shows an increase in intensity (57 to 65 a.u./nm) and a blueshift (977 to 968 nm) with respect to the SRO_5 sample. This is indicative of a smaller growth in the size of Si-ncs due to a diffusion of silicon atoms between the two SRO layers [18]. In addition, as the annealing temperature decreases to 1050 °C, 1000 °C and 950 °C, a higher blueshift (932 ± 2 nm) is observed in the SiO_{20}/SiO_5 bilayers with respect to the bilayer annealed at 1100 °C. This effect indicates that the size of Si-ncs is reduced by decreasing the annealing temperature. The reduction of the Si-ncs size agrees with the decrease in S_a, as mentioned above. An improvement in the PL intensity (65 to 136 a.u./nm) is observed for sample B-1050, while it reduces again for samples B-1000 and B-950. It is possible that the 1000 °C and 950 °C annealing temperatures produce a lower number of luminescent centers (Si-ncs) than 1100 °C. Finally, sample B-900 presents the lowest PL intensity (3 a.u./nm) with its main peak located at 878 nm. However, in order to ensure the decrease in the size of Si-ncs it is necessary to perform TEM characterization in the future. The improvement in PL intensity for near-infrared wavelengths in bilayers opens the opportunity to design luminescent devices for biomedical applications [21].

IV. CONCLUSION

Bilayer structures composed of a SRO_5 (45 nm) film followed by a SRO_{20} (5 nm) film as a passivating layer were studied. A minor decrease in refractive index was observed after thermal annealing by reducing the temperature. This indicates less surface oxidation after thermal annealing due to the presence of the passivating layer. In addition, the average surface roughness reduces from 6.86 to 6.40 nm as the annealing temperature decreases from 1100 °C to 1000 °C indicating a decreasing of the Si-nps size within the SRO_5 film. This effect is corroborated by a blueshift of the main PL peak from 968 to 932 nm. Finally, there is a two-fold improvement in the PL intensity (65 to 136 a.u/nm) as the annealing temperature is reduced from 1100 to 1050 °C. Although this improvement decreases (136 to 85 a.u/nm) for the 1000 °C sample, it is still more intense than the thermally annealed at 1100 °C.

ACKNOWLEDGMENT

J. J. Avilés Bravo and J. F. Ramirez Rios acknowledge to the Consejo Nacional de Humanidades, Ciencias y Tecnologías (CONAHCYT) of Mexico for the PhD scholarship grant No. CVU 852431 and CVU 869500, respectively. Authors would like to thank CONAHCYT for providing support through the project CONAHCYT-CB A1-S-8205. The help of technicians Victor Aca, Armando Hernández and Leticia Tecuapetla from INAOE is also appreciated.

REFERENCES

[1] B. H. Lai, C. H. Cheng, and G. R. Lin, "Multicolor ITO/SiOx/p-Si/Al light emitting diodes with improved emission efficiency by small Si quantum dots," *IEEE J. Quantum Electron.*, vol. 47, no. 5, pp. 698–704, 2011, doi: 10.1109/JQE.2011.2109699.

[2] Y. Xu, S. Terada, Y. Xin, H. Ueda, and K. I. Saitow, "Ligand Effects on Photoluminescence and Electroluminescence of Silicon Quantum Dots for Light-Emitting Diodes," *ACS Appl. Nano Mater.*, vol. 5, no. 6, pp. 7787–7797, Jun. 2022, doi: 10.1021/acsanm.2c00811.

[3] M. Xie, D. Li, F. Wang, and D. Yang, "Luminescence Properties of Silicon-Rich Silicon Nitride Films and Light Emitting Devices," no. May 2014, pp. 3–19, 2011, doi: 10.1149/1.3647900.

[4] T. W. Kim, C. H. Cho, B. H. Kim, and S. J. Park, "Quantum confinement effect in crystalline silicon quantum dots in silicon nitride grown using SiH4 and NH3," *Appl. Phys. Lett.*, vol. 88, no. 12, pp. 7–10, 2006, doi: 10.1063/1.2187434.

[5] A. Sarkar, R. Bar, S. Singh, R. K. Chowdhury, S. Bhattacharya, A. K. Das, and S. K. Ray, "Size-tunable electroluminescence characteristics of quantum confined Si nanocrystals embedded in Si-rich oxide matrix," *Appl. Phys. Lett.*, vol. 116, no. 23, 2020, doi: 10.1063/5.0001840.

[6] Z. Y. Yu, Y. C. Zhang, S. Li, X. Y. Dai, X. Y. Xue, H. Shen, S. Y. Wang, and M. Lu, "A synergistic approach of interface engineering to improve the performance of silicon nanocrystal light-emitting diode," *Vacuum*, vol. 197, no. December 2021, p. 110822, 2022, doi: 10.1016/j.vacuum.2021.110822.

[7] M. Meneses-meneses, M. Moreno-moreno, A. Morales-s, A. Ponce-pedraza, J. Flores-m, J. C. Mendoza-cervantes, and L. Palacios-huerta, "Development of Heterojunction c-Si / a-Si 1 − x C x : H PIN Light-Emitting Diodes," vol. 13, 2022.

[8] H.-Y. Tai, C.-T. Lee, L.-H. Tsai, Y.-H. Lin, Y.-H. Pai, C.-I. Wu, and G.-R. Lin, "SiC and Si Quantum Dots Co-Precipitated Si-Rich SiC Film with n- and p-Type Dopants Grown by Hydrogen-Free PECVD," *ECS J. Solid State Sci. Technol.*, vol. 2, no. 9, pp. N159–N164, 2013, doi: 10.1149/2.002309jss.

[9] A. Morales-Sánchez, K. M. Leyva, M. Aceves, J. Barreto, C. Domínguez, J. A. Luna-López, J. Carrillo, and J. Pedraza, "Photoluminescence enhancement through silicon implantation on SRO-LPCVD films," *Mater. Sci. Eng. B Solid-State Mater. Adv. Technol.*, vol. 174, no. 1–3, pp. 119–122, 2010, doi: 10.1016/j.mseb.2010.03.031.

[10] A. Morales, J. Barreto, C. Domínguez, M. Riera, M. Aceves, and J. Carrillo, "Comparative study between silicon-rich oxide films obtained by LPCVD and PECVD," *Phys. E Low-Dimensional Syst. Nanostructures*, vol. 38, no. 1–2, pp. 54–58, 2007, doi: 10.1016/j.physe.2006.12.056.

[11] D. Dong, E. A. Irene, and D. R. Young, "Preparation and Some Properties of Chemically Vapor-Deposited Si-Rich SiO2 and Si3N4, Films," *J. Electrochem. Soc. SOLID-STATE Sci. Technol.*, vol. 125, no. 5, pp. 819–823, 1977.

[12] E. G. Barbagiovanni, D. J. Lockwood, P. J. Simpson, and L. V. Goncharova, "Quantum confinement in Si and Ge nanostructures," *J. Appl. Phys.*, vol. 111, no. 3, 2012, doi: 10.1063/1.3680884.

[13] S. Öğüt, J. R. Chelikowsky, and S. G. Louie, "Quantum confinement and optical gaps in si nanocrystals," *Phys. Rev. Lett.*, vol. 79, no. 9, pp. 1770–1773, 1997, doi: 10.1103/PhysRevLett.79.1770.

[14] R. Salh, "Defect Related Luminescence in Silicon Dioxide Network: A Review," *Cryst. Silicon - Prop. Uses*, p. 13, Jul. 2011, doi: 10.5772/22607.

[15] J. Alarcón-Salazar, I. E. Zaldívar-Huerta, and M. Aceves-Mijares, "Electrical and electroluminescent characterization of nanometric multilayers of SiOX/SiOY obtained by LPCVD including non-normal emission," *J. Appl. Phys.*, vol. 119, no. 21, 2016, doi: 10.1063/1.4952730.

[16] O. Sublemontier, C. Nicolas, D. Aureau, M. Patanen, H. Kintz, X. Liu, M. A. Gaveau, J. L. Le Garrec, E. Robert, F. A. Barreda, A. Etcheberry, C. Reynaud, J. B. Mitchell, and C. Miron, "X-ray photoelectron spectroscopy of isolated nanoparticles," *J. Phys. Chem. Lett.*, vol. 5, no. 19, pp. 3399–3403, 2014, doi: 10.1021/jz501532c.

[17] J. A. Luna-López, A. Morales-Sánchez, M. Aceves-Mijares, Z. Yu, and C. Domínguez, "Analysis of surface roughness and its relationship with photoluminescence properties of silicon-rich oxide films," *J. Vac. Sci. Technol. A Vacuum, Surfaces, Film.*, vol. 27, no. 1, pp. 57–62, 2009, doi: 10.1116/1.3032915.

[18] R. Limpens, A. Lesage, M. Fujii, and T. Gregorkiewicz, "Size confinement of Si nanocrystals in multinanolayer structures," *Sci. Rep.*, vol. 5, pp. 1–6, 2015, doi: 10.1038/srep17289.

[19] G. R. Lin, C. J. Lin, and C. T. Lin, "Low-plasma and high-temperature PECVD grown silicon-rich SiOx film with enhanced carrier tunneling and light emission," *Nanotechnology*, vol. 18, no. 39, pp. 2–7, 2007, doi: 10.1088/0957-4484/18/39/395202.

[20] X. J. Hao, A. P. Podhorodecki, Y. S. Shen, G. Zatryb, J. Misiewicz, and M. A. Green, "Effects of Si-rich oxide layer stoichiometry on the structural and optical properties of Si QD/SiO 2 multilayer films," *Nanotechnology*, vol. 20, no. 48, 2009, doi: 10.1088/0957-4484/20/48/485703.

[21] J. Watanabe, H. Yamada, H. T. Sun, T. Moronaga, Y. Ishii, and N. Shirahata, "Silicon Quantum Dots for Light-Emitting Diodes Extending to the NIR-II Window," *ACS Appl. Nano Mater.*, vol. 4, no. 11, pp. 11651–11660, 2021, doi: 10.1021/acsanm.1c02223.

Increasing the doping efficiency by post-deposition annealing in a-SiGe:H films synthesized by PECVD

Ernesto Franco
Electronics Department
National Institute for Astrophysics,
Optics and Electronics
Puebla, Mexico
ernestofp@inaoep.mx

Alfonso Torres
Electronics Department
National Institute for Astrophysics,
Optics and Electronics
Puebla, Mexico
atorres@inaoep.mx

Mario Moreno
Electronics Department
National Institute for Astrophysics,
Optics and Electronics
Puebla, Mexico
mmoreno@inaoep.mx

Abstract— **An acceptable thermoelectric efficiency is related to a figure of merit ZT ≥ 1, its value increases with a high value of Seebeck coefficient and electrical conductivity, and with a low thermal conductivity. The amorphous SiGe:H has a high Seebeck coefficient and low thermal conductivity, the only thing that they need to become a good thermoelectric material is a high electrical conductivity, for this reason the present work demonstrates the increase of the electrical conductivity of highly doped films of a-SiGe:H types P and N by annealing. The highest electrical conductivity value of the films was found in 4% of doping in gas phase. The P and N type samples were doped in the range from 4% to 16% with Boron and Phosphorus. After thermal treatment, both had an increase in their electrical conductivity value of three and two order of magnitude respectively. Also, the explanation of the activation mechanism of both impurities is given.**

Keywords—Hydrogenated amorphous silicon germanium alloys, doped amorphous silicon, conductivity, defects, increasing the doping efficiency, thermoelectric material.

I. INTRODUCTION

Despite efforts undertaken to only use clean energy sources, nowadays, all the countries have a dependence on oil and its products as a raw material to produce electrical energy [1]. For this reason, it is important to improve the efficiency of electronic devices and systems to reduce the damage caused by electricity production. One way to manage it, is through energy harvesting devices, named thermoelectric generators (TEGs), which can utilize a temperature gradient like a power source.

The thermoelectric materials must have a high figure of merit $ZT = \frac{\sigma \alpha}{k}$. Where: α is the Seebeck coefficient, σ is the electrical conductivity and k the thermal conductivity. The problem of improving the thermoelectric performance of the materials is that both, electrical and thermal conductivities, depend on the density of electronic and vibrational states respectively. Which are modified by the molecular arrangement.

In crystalline phase, it is necessary to add and control defects, a task that can be performed by: adding atoms with more mass to make alloys [2], the addition of nanoclusters in the molecular net [3], (3) generating grain border defects or changing the form and size of the grains [3], the addition of atoms to produce resonant states [4] and creating voids like

holey films [5]. On the other, hand complex molecular array can

reduce the thermal conductivity like amorphous materials, which are featured as thermal insulation materials [6,7], also the a-SiGe alloys have smaller thermal conductivity values, because the mass difference generates localization in the vibrational states with more energy [2]. In this phase, the materials also have a high value of Seebeck coefficient [7].

This work focuses on increasing the electrical conductivity through to activation of the dopants by annealing of highly doped B and P films of a-SiGe:H deposited by low frequency PECVD. The resulting films are proposed as a good thermoelectric materials.

II. EXPERIMENTAL

All the samples were deposited by PECVD capacitive discharge system (AMP-3300 of Applied Materials) at RF frequency of 110 kHz, its power density was $90 mW\backslash cm^2$ with a deposition temperature $T_s = 200$ °C (except in the samples of the series 1661, see table II). The reactive gasses were a mixture of Silane (SiH_4), Germane (GeH_4), their % in gas phase (Q ratio) was calculated with Eq. (1), the % in gas phase of dopants, Diborane (B_2H_6) and Phosphine (PH_3), was determined by the Eq. (2) and Eq. (3) respectively, additionally the dilution rate was defined by Eq. (3), where Q is the gas flux in sccm.

$$X_g = Q_{GeH_4} \times 100/(Q_{SiH_4} + Q_{GeH_4}) \quad (1)$$
$$B_{gas} = Q_{B_2H_6} \times 100/(Q_{SiH_4} + Q_{GeH_4}) \quad (2)$$
$$P_{gas} = Q_{PH_3} \times 100/(Q_{SiH_4} + Q_{GeH_4}) \quad (3)$$
$$R = Q_{H_2}/(Q_{SiH_4} + Q_{GeH_4}) \quad (4)$$

The deposition conditions of the films 1651-1 and 1651-2 were: deposition times were $t_d = 60$ min and $t_d = 20$ min respectively, chamber pressure $P_d = 0.6$ Torr, $B_{gas} = 4$, $X_g = 0.6$ and $R = 4$, . For the remainder of the samples, the conditions were: $P_d = 1.2$ Torr, $t_d = 60$ min, $X_g = 0.9$ the dopant percentages, B_{gas} and P_{gas}, were in a range from 4% to 16% (show in Table 1), and $R = 20$.

The annealing was done, at different temperatures (see Fig.1), inside an electric oven with a quartz chamber, through

979-8-3503-1191-4/23 $31.00 © 2023 IEEE

it the forming gas (N_2 60% and H_2 40%) was circulated. After annealing, mesas with circular shape appeared (see table III), their diameters were measured with the ruler of an optical microscope Orthoplan.

The room-temperature conductivity (σ_{RT}) was measured using source meter model 2401 of Keithley with 4-wire sensing and the thickness was estimated by using a profilometer model Dektak XT of Bruker. The a-SiGe:H films were deposited on substrates of Silicon wafer with 200nm of grown oxide and Corning glass 2947. For some wafers, 200nm of silicon nitride ($Si_{x-1}N_x$) was deposited by LPCVD on the oxide.

The thicknesses of two films with the same deposition conditions (one of them without annealing) were measured to find the thickness difference (mesas) that appeared after annealing (see table III), which are caused by Hydrogen effusion and the coalescence of the material. Thicknesses of both were compared to determine the thickness outside the mesas.

TABLE I. ELECTRICAL CONDUCTIVITY OF N AND P TYPE SAMPLES BEFORE (σ_{RT}) AND AFTER ANNEALING ($\sigma_{RT_{T_A}}$)

Sample	B_{gas} (%)	P_{gas} (%)	σ_{RT} ($\Omega^{-1}cm^{-1}$) on $Si_{x-1}O_x$	$\sigma_{RT_{Ta=500°C}}$ ($\Omega^{-1}cm^{-1}$) on $Si_{x-1}O_x$
1646-1	-	4	0.092	11.4225
1646-2	-	8	0.707	4.967
1646-3	-	16	0.533	0.749
1651-1	4	-	0.132	2.143
1651-2	4	-	0.0657	2.894
1654-1	4	-	3.25×10^{-3}	1.404
1654-2	8	-	5.63×10^{-3}	2.038
1654-3	16	-	7.16×10^{-4}	0.158

TABLE II. ELECTRICAL CONDUCTIVITY OF P TYPE SAMPLES OF THE SERIES 1661 BEFORE (σ_{RT}) AND AFTER ANNEALING ($\sigma_{RT_{T_A}}$) ON $SI_{x-1}O_x$ AND $SI_{x-1}0N_x$

Sample	T_s (°C)	B_{gas} (%)	σ_{RT} ($\Omega^{-1}cm^{-1}$) on $Si_{x-1}O_x$	$\sigma_{RT_{T_A=500°C}}$ on $Si_{x-1}O_x$ ($\Omega^{-1}cm^{-1}$)	$\sigma_{RT_{T_A=500°C}}$ on $Si_{x-1}N_x$ ($\Omega^{-1}cm^{-1}$)
1661-1	200	16	3.77×10^{-4}	0.433	No conductive
1661-2	350	16	3.48×10^{-4}	0.445	0.594
1661-3	320	16	1.70×10^{-4}	0.225	0.657
1661-4	300	16	1.49×10^{-4}	0.187	0.874

III. RESULTS AND DISCUSSION

After annealing at 500 °C most of the samples were damaged by the H effusion and circular mesas were formed because the coalescence of the material. The H effusion was so violent that pieces of the film were broken as can be seen in table III. Only in the samples 1651-1 and 1651-2 there were not broken pieces of film, this is because the dilution ratio is low (R=4). In the P type samples there is less damage than in the N type samples and it increases with the doping value rise.

In table III, it is observed that the effusion removes part of the film, but there is still continuity in the film that

allows to measure conductivity, and the effects of annealing in conductivity can be evaluated. The coalescence of the film in the mesa formation, results in a thin continuous film of around 260 nm thick. This is the characteristic result when the effusion has not broken the film.

In order to decrease the diameter of the mesas, the annealing was made in steps as follows: Step 1: Put the samples inside the furnace, the temperature was set to 355 °C, 370 °C and 400 °C for 40 minutes and then, Step 2: set temperature to 500°C and annealing for 30 minutes (see Fig.1). This method resulted in a gently H effusion which did not break the films but changed the ratio of areas of coalescence.

The change in the ratio of coalescence without breaking the film also produced an increase of electrical conductivity (compare the conductivity values $\sigma_{RT_{T_A=500°C}}$ of the samples on $Si_{x-1}O_x$ in the table II with the value of the sample 1654-3). Additionally, the diameter of coalesced regions increased with the temperature up until 400 °C as it is depicted in Fig. 2.

TABLE III. MESAS PRODUCED BY HYDROGEN EFFUSION

The change in the electrical conductivity is mainly produced by the dopant activation, and a trend is observed, the conductivity increases as the diameter of the closed mesa increases, see Fig. 3. The series samples 1661 (from 1661-1 to 1661-4) have an electrical conductivity on $Si_{x-1}N_x$ little larger than on $Si_{x-1}O_x$. Because these samples were deposited at 1.2 Torr, this pressure produced nano powders [7,8], which can become nano crystals, if there is enough H diffusion to produce the reaction Si-H+ Si-H → Si-Si + H-H [9], the grain size depends on the contact surface under the film [10]. For the N type samples, the nano crystal presence were confirmed [7].

The thermal expansion made cracks in the samples 1661-1 and 1661-4 after annealing at 500 °C, for the sample 1661-1 the damage was too strong, hence the film is not conductive (see Table III).

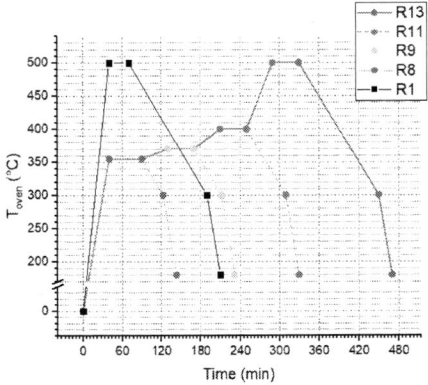

Fig. 1. Oven temperature T_{oven} vs time.

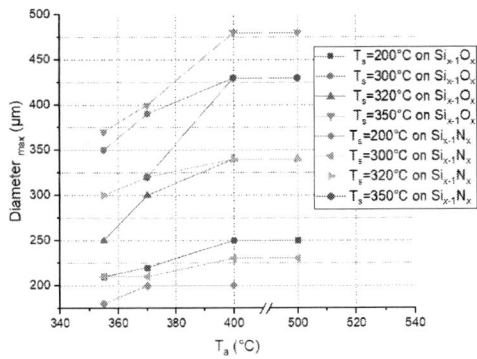

Fig. 2. Maximum diameter of the coalesced mesas produced by hydrogen effusion vs the annealing temperature T_a.

After annealing, the electrical conductivity at room temperature σ_{RT} increases two and three order of magnitude respectively, in the N and P type samples doped 4% (Fig. 4 and 5). The change was produced by the dopants activation ($B_4^- - T_3^+$), when passivated dopants ($H-B_4^0 + T_4^0 - H$) lose an hydrogen and get a dangling bond, they can interact with other dangling bonds (D^0) in the molecular network, while both dangling bonds have enough distance to make a bond (1.4 Å) [11]. For this reason, the thermal energy (for the diffusion) is necessary

to activate the dopants and increase conductivity to the values shown (Fig. 6).

But in N type samples when increasing the doping, the electrical conductivity decreases (Fig. 4), because the concentration of passivated dopants is large and they introduce new dangling bonds [12], therefore there are more electrons are trapped by dangling bonds than electrons added by active dopants. The reduction of conductivity is larger than the inverse square root rule of Street [12].

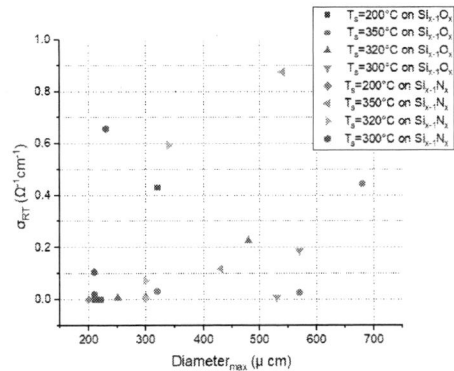

Fig. 3. Electrical conductivity of the samples of the series 1661 vs maximum diameter of the mesas.

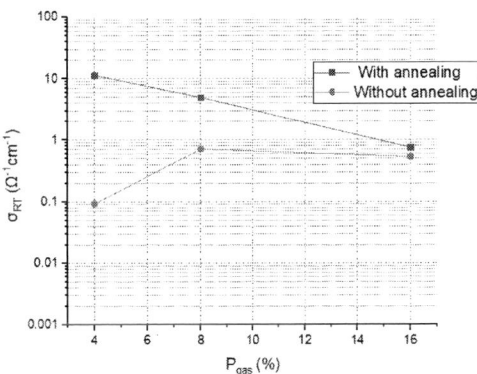

Fig. 4. Electrical conductivity vs the percentage of doping in gas phase of N type samples.

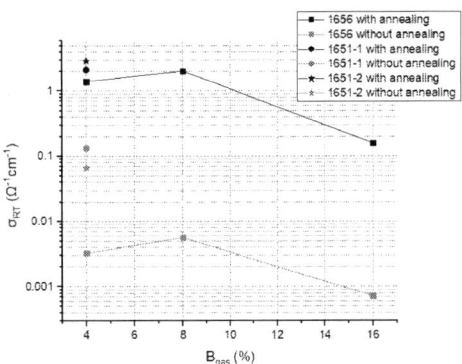

Fig. 5. Electrical conductivity vs the percentage of doping in gas phase of P type samples.

On the other hand, in the P type samples the increase of three order of magnitude persists for the samples with a dilution ratio R=20 (Fig. 5), this behavior can be related to results of measurements of neutron magnetic resonance spin-echo double-resonance (SEDOR) for a-Si:H doped samples. That study shows that 40% of B atoms in the amorphous film have hydrogen as a nearest neighbor (1.6 Å between two atoms) in P type samples. Whereas 50% of P atoms have H at a distance of 2.6Å [11], the hydrogen lies in a Si atom without tetrahedral configuration, which is a nearest neighbor of P atoms. Hence the acceptors in our P type samples may be activated by hydrogen effusion and their effect is reduced when the dilution ratio is small, like in the samples 1651-1 and 1651-2.

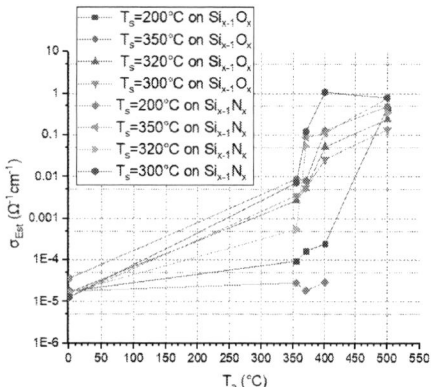

Fig. 6. a) Electrical conductivity of the samples of the series 1661 vs the maximum annealing temperature T_a, the value of the samples before annealing lie in Ta=0.

In contrast, based on the results in SEDOR measurements, the hydrogen is further from the donors (2.6 Å) in N type samples [11], hence less dopants may be activated by hydrogen effusion. Because not only the effusion is needed to produce dangling bonds, but also it is necessary that P atoms lie near to the dangling bonds in Si atoms.

Also, the tetrahedral configuration (hybridized sp$_3$) of P atoms need more energy than threefold Phosphorus coordination [13], while according to a computational study [14], the most stable configuration for B atoms is the tetrahedral configuration.

IV. CONCLUSION

An increase of electrical conductivity exists in two and three orders of magnitude for the N and P type samples at 4% in gas phase respectively.

The conductivity is reduced as much as the increase of the doping in the N type a-SiGe:H films, but in the P type samples the increase of three order of magnitude persists, this behavior may be related to the distance between Hydrogen and dopants. Also, the threefold Phosphorus coordination and fourfold Boron coordination in the amorphous N and P type films respectively may have influenced these results.

There is a small increase of the electrical conductivity in the samples deposited on $Si_{x-1}N_x$ ascribed to nanocrystals embedded in amorphous network, but the thermal expansion made cracks. For the sample 1661-1 the damage was too strong, and this sample is not conductive.

The reduction of the diameter of the mesas slightly increased the electrical conductivity value, but this change is negligible when it is compared with the rise of the conductivity produced by active dopants.

ACKNOWLEDGMENT

The authors want to thank all personnel in the laboratory of microelectronics at INAOE and to the CONACYT for the scholarship No. 794442.

REFERENCES

[1] L. Yang, Z.-G. Chen, M. S. Dargusch, and J. Zou, "High performance thermoelectric materials: progress and their applications," *Advanced Energy Materials*, vol. 8, no. 6, p. 1701797, 2018.

[2] T J. L. Feldman, M. D. Kluge, P. B. Allen, and F. Wooten, "Thermal conductivity and localization in glasses: Numerical study of a model of amorphous silicon," *Physical Review B*, vol. 48, no. 17, p. 12589, 1993.

[3] C. Fiedler, T. Kleinhanns, M. Garcia, S. Lee, M. Calcabrini, and M. Ibañez, "Solution-processed inorganic thermoelectric materials: Opportunities and challenges," *Chemistry of Materials*, vol. 34, no. 19, pp. 8471–8489, 2022.

[4] J. P. Heremans, B. Wiendlocha, and A. M. Chamoire, "Resonant levels in bulk thermoelectric semiconductors," *Energy & Environmental Science*, vol. 5, no. 2, pp. 5510–5530, 2012.

[5] J. Tang, H.-T. Wang, D. H. Lee, M. Fardy, Z. Huo, T. P. Russell, and P. Yang, "Holey silicon as an efficient thermoelectric material," *Nano letters*, vol. 10, no. 10, pp. 4279–4283, 2010.

[6] D. Banerjee, O. Vallin, K. M. Samani, S. Majee, S.-L. Zhang, J. Liu, and Z.-B. Zhang, "Elevated thermoelectric figure of merit of n-type amorphous silicon by efficient electrical doping process," *Nano Energy*, vol. 44, pp. 89–94, 2018.

[7] C. R. Ascencio-Hurtado, A. Torres, R. Ambrosio, M. Moreno, J. Álvarez-Quintana, and A. Hurtado-Macías, "N-type amorphous silicon-germanium thin films with embedded nanocrystals as a novel thermoelectric material of elevated zt," *Journal of Alloys and Compounds*, vol. 890, p. 161843, 2022.

[8] A. F. i Morral and P. R. i Cabarrocas, "Shedding light on the growth of amorphous, polymorphous, protocrystalline and microcrystalline silicon thin films," *Thin Solid Films*, vol. 383, no. 1-2, pp. 161–164, 2001.

[9] A. F. i Morral and P. R. i Cabarrocas, "Role of hydrogen diffusion on the growth of polymorphous and microcrystalline silicon thin films," *The European Physical Journal-Applied Physics*, vol. 35, no. 3, pp. 165–172, 2006.

[10] P. Roca i Cabarrocas, "New approaches for the production of nano-, micro-, and polycrystalline silicon thin films," *physica status solidi (c)*, vol. 1, no. 5, pp. 1115–1130, 2004.

[11] J. Boyce, S. Ready, and C. Tsai, "Local structure of dopants in compensated and singly-doped amorphous silicon," *Journal of Non-Crystalline Solids*, vol. 97, pp. 345–348, 1987.

[12] R. A. Street, "Doping and the fermi energy in amorphous silicon," Physical Review Letters, vol. 49, no. 16, p. 187–1190, 1982.

[13] R. A. Street, *Hydrogenated amorphous silicon.* Cambridge university press, 2005.

[14] P. Fedders and D. Drabold, "Theory of boron doping in a-si: H," *Physical Review B*, vol. 56, no. 4, p. 1864, 1997.

Long-Term Potentiation and Depression with Vertically Stacked Nanosheet FET

Nupur Navlakha
Microelectronics Research Center,
Electrical and Computer Engineering,
J.J. Pickle Research Campus,
The University of Texas at Austin,
Austin, TX-78758, United States
nupurnavlakha@utexas.edu

Md. Hasan Raza Ansari
SAMA Labs, Division of Computer, Electrical and
Mathematical Science and Engineering (CEMSE) Division,
Electrical and Computer Engineering, King Abdullah
University of Science and Technology (KAUST),
. Thuwal 23955-6900, Saudi Arabia
mdhasan.ansari@kaust.edu.sa

Abstract—**This work showcases the feasibility of vertically stacked nanosheet FET (NSFET) for charge trapping-based synapse for neuromorphic applications. The calibrated simulation models mimic the long-term potentiation (LTP) and depression (LTD) of biological synapses. Use of stacked nanosheet device facilitates a dense memory with high current. The work also evaluates the effect of number of pulses for LTP and LTD on the image classification accuracy of the MNIST dataset. The neural network results show high linearity, conductance, and symmetric behavior between LTP and LTD that aids achieves ~ 94.75 % accuracy in image classification.**

Keywords— *NSFET, Charge Trapping Memory, Synapse features, LTP, LTD.*

I. INTRODUCTION

An increase in the demands of non-volatile memories for neuromorphic computing requires new innovations in device design and/or new materials to improve the performance of artificial synapses and neurons [1], [2]. The essential building blocks for artificial hardware neural networks are synapses and neurons. Recently, researchers have been focusing on emerging non-volatile memories, such as Resistive random-access memories (RRAMs) [3], phase change memory (PCRAMs) [4], magnetic RAM (MRAMs) [5], Ferroelectric RAM (FeRAMs) [6], and three terminal charge-trapping memory as candidates for analog synaptic devices to achieve high accuracy in neural network and consume less energy in neuromorphic computing. Three-terminal artificial synapses, are structurally inspired by a transistor, have been under investigation for some time now for their potential application in a neuromorphic system [1]. However, for large-scale integration of neural networks, downscaling of the devices becomes necessary. Therefore, vertically stacked nanosheet FET (NSFET) for charge-trapping memory and its application as a synapse is essential [7]–[9].

Usually, higher nonlinearity in a synapse result in high energy and time costs in the training process [6]. Therefore, to build a low power and high-accuracy artificial neuromorphic network, it is necessary to improve the linearity and symmetry of an artificial synapse device.

Focusing on these factors, we evaluate the impact of states in a synapse using a vertically stacked nanosheet FET (NSFET) as a synapse for neuromorphic application for image classification. The NSFET devices have emerged as a promising technology for charge-trapping memory due to their ability to store charge in a small area, achieve high conductance and better linearity for long-term potentiation and depression, symmetric conductance between long-term potentiation and depression.

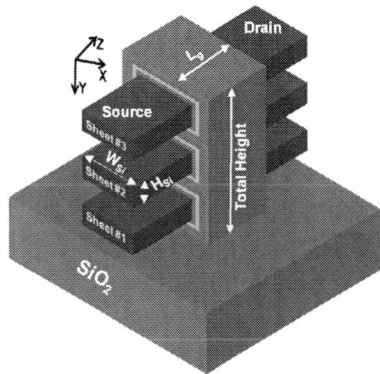

Fig. 1. Vertically stacked Nanosheet FET for charge trapping synapse. NSFET has three sheets in a vertical direction.

II. RESULTS AND DISCUSSION

Fig. 1 shows the schematic representation of a vertically stacked Nanosheet FET (NSFET). The device consists of three charge-trapping layers between the gate and silicon channel (Oxide-nitride-Oxide (ONO)). The device is simulated with a gate length of 50 nm, a silicon film thickness of 10 nm and a silicon width of 20 nm. For the charge trapping layer, the ONO layer is optimized to achieve high retention time and longer endurance. The ONO layer thickness is 6 nm of blocking oxide, 4 nm of the nitride layer, and 2 nm of tunnelling oxide.

The device is simulated through Synopsys Sentaurus with calibrated models [10]. Fig. 2 shows the comparison of simulated transfer characteristics with experimental transfer characteristics [11]. The synapse operation is based on charge

979-8-3503-1191-4/23 $31.00 © 2023 IEEE

trapping and de-trapping of holes from the nitride layer. Therefore, in this work, we have used Fowler-Nordheim, hot carrier injection models. The other models to generate carriers in the silicon are the dynamic non-local band-to-band tunnelling, impact ionization, SRH and Auger generation recombination models are incorporated.

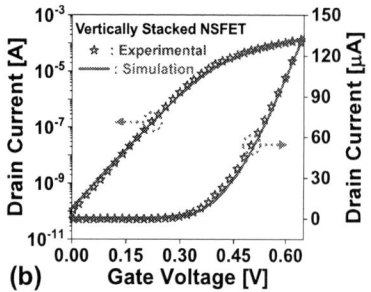

Fig. 2 Calibration of simulated transfer characteristics with experimental transfer characteristics [11].

Fig. 3 shows the schematic representation of a human brain and biological synapses which control the memorization and forgetting process of the brain. In the biological brain, the information is transferred from pre-synapse to post-synapse in terms of ions. Moreover, the charge-trapping in artificial electronic synapse shows the strengthening of the synapse through the trapping of positive charges in the nitride layer while the de-trapping of charges shows the forgetting processes of the brain.

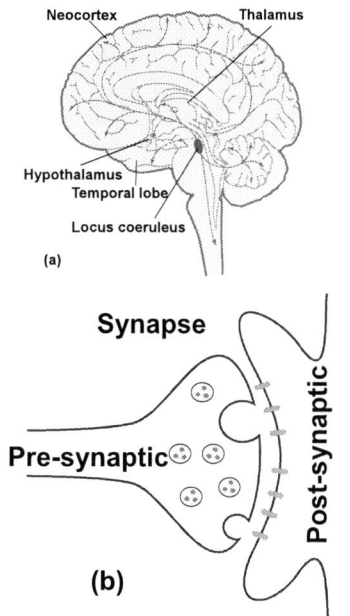

Fig. 3. Schematic representation of (a) human brain and (b) connections between two neurons. Synapse transforms the information between the two neurons.

Long-term potentiation (depression) is based on trapping (de-trapping) of holes in (from) the nitride layer, which contributes positive (negative) potential on the channel. Therefore, the conductance of the device during LTP increases while for LTD it decreases as shown in Fig. 4. The potentiation and depression operations are performed with an identical pulse (V_{GS} = -3.5 V, V_{DS} = 3.0 V with a pulse time of 500 ns) and incremental pulse (V_{GS} starts from 2 V with an increment of 50 mV and a pulse time of 100 ns), respectively. The simulation is performed for 32 and 64 pulses of LTP and LTD.

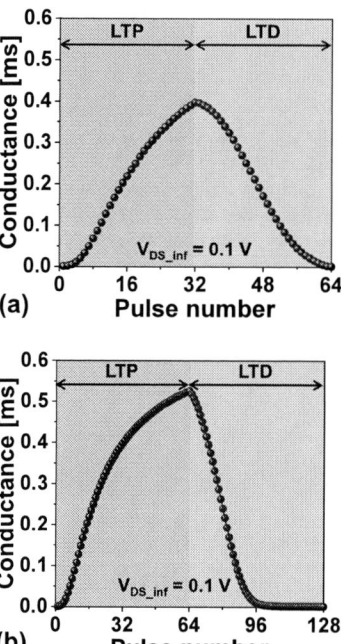

Fig. 4. Variation in long-term potentiation and depression conductance values during inference operation of the device with the total number of pulses, (a) 64 and (b) 128.

Fig. 4(a) and (b) show the variation of LTP and LTD conductance values during the inference operation at a lower drain voltage of 0.1 for the total pulse of 64 and 128, respectively. The results convey that an increase in number of pulses increases the conductivity of the device due to the trapping of holes in the nitride layer during potentiation operation. While increasing the pulse number during the depression, decreases the conductance due to the de-trapping of the holes from the nitride layer. The saturation behavior is due to the maximum number of trap charges. With the increase in number of pulses, we observe a higher I_{on}/I_{off} (dynamic range), however linearity is a trade-off as mentioned next.

After estimating the conductance values from the NSFET during potentiation and depression, their nonlinearity (*NL*) and asymmetric ratio (*AR*) between LTP and LTD are estimated with the following equations. These parameters are

2023 IEEE Latin American Electron Devices Conference (LAEDC)
Puebla, México, July 3-5, 2023

essential for estimating the accuracy of artificial neural networks [1].

$$G_{LTP} = \frac{G_{max} - G_{min}}{1 - exp\left(\frac{-P_{max}}{A}\right)}\left(1 - exp\left(\frac{-P}{A}\right)\right) + G_{min}$$

$$G_{LTD} = \frac{G_{min} - G_{max}}{1 - exp\left(\frac{-P_{max}}{A}\right)}\left(1 - exp\left(\frac{P - P_{max}}{A}\right)\right) + G_{max}$$

$$AR = \frac{max|G_{LTP}(n) - G_{LTD}(n)|}{G_{LTP}(n_{max}) - G_{LTD}(n_{max})}$$

where n =1-32 and 1-64 for 32 and 64 pulses of LTP and LTD, respectively, G_{max}, G_{min}, and P, P_{max} are maximum and minimum device conductance, number of pulses and maximum number of pulses applied for the device to transition from the lowest to highest conductance states, respectively. The A indicates the normalized value of potentiation and depression.

LTP and LTD achieves better performance in terms of NL_{LTP}, NL_{LTD}, and AR compared to the higher pulse. The reason for a better linearity with lower pulse number is the charges trapped in the nitride layer that increases linearly while for higher pulses, the trapped charges get saturated, and thus, the conductance also.

Finally, we investigate the learning and inference capabilities of NSFET devices for hardware-based neural networks (HNN). A multi-layer perceptron is simulated with one hidden layer as shown in Fig. 6(a). The network consists of 3 layers, the first layer consists of 784 neurons, which define the pixel of the image, and the output layer consists of 10 neurons. The Modified National Institute of Standards and Technology (MNIST) dataset is considered, in which 50000 images are used for training and 10000 for testing. In the simulation, non-linearity is activated through Rectified Linear Unit (ReLU) activation function and for weight update Adam is used as the optimizer.

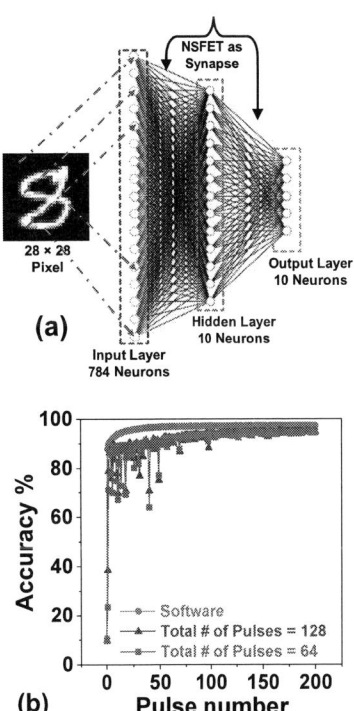

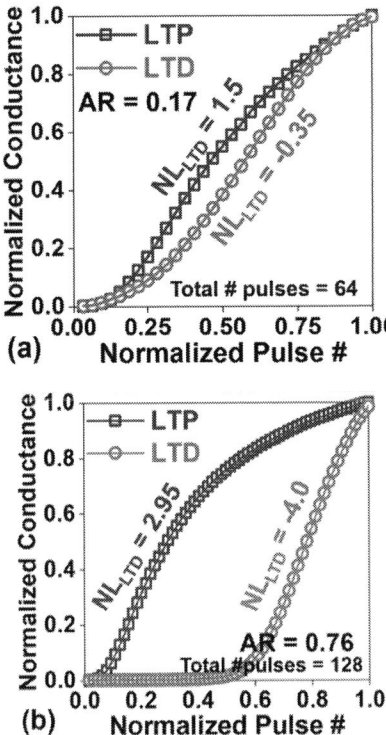

Fig. 5. Variation in normalized conductance value of the LTP and LTD with normalized pulse number for the total number of pulses is (a) 64 and (b) 128

Fig. 5(a) and (b) show the variation of normalized conductance values for the LTP and LTD with normalized pulse numbers for the total number of pulses of 64 and 128, respectively. The results convey that the lower pulse states of

Fig. 6 (a) Schematic representation of multilayer perceptron of neural network with one hidden layer with vertically stacked NSFET as a synapse for Modified National Institute of Standards and Technology (MNIST) digit recognition. (b) Digit recognition accuracy (%) as a function of the number of training epochs.

Fig. 6(b) shows the variation of the image classification accuracy with the number of epochs and compares the results between software-based and device weights extracted for the total number of pulses 64 and 128. The results convey that the digit recognition accuracy for the weight update from devices is 94.75% (total number of pulses is 64) and 94.33% (total number of pulses is 128), which is very close to the ideal software-based neural network accuracy of 97.33%.

979-8-3503-1191-4/23 $31.00 © 2023 IEEE   103

Thus, this reveals that the vertically stacked NSFET synaptic device is highly suited for neuromorphic inference applications.

III. CONCLUSION

In this work, the Vertically stacked Nanosheet Field Effect Transistor device with high density and current is utilized as an artificial synaptic transistor. The device works at lower drain bias at inference operation and achieves linear weight values for both long-term potentiation and depression, and also, a low asymmetric value of 0.17 which is suitable for low-power neuromorphic systems. The device shows a reliable and consistent digit recognition accuracy of 94.75% by a single-layer neural network with a single hidden layer on the MNIST dataset.

IV. REFERENCES

[1] D. Ielmini and S. Ambrogio, "Emerging neuromorphic devices," *Nanotechnology*, vol. 31, no. 9, p. 092001, Feb. 2020, doi: 10.1088/1361-6528/ab554b.

[2] A. Sebastian, M. Le Gallo, R. Khaddam-Aljameh, and E. Eleftheriou, "Memory devices and applications for in-memory computing," *Nat. Nanotechnol.*, vol. 15, no. 7, pp. 529–544, 2020, doi: 10.1038/s41565-020-0655-z.

[3] K. Moon *et al.*, "RRAM-based synapse devices for neuromorphic systems," *Faraday Discuss.*, vol. 213, pp. 421–451, Feb. 2019, doi: 10.1039/C8FD00127H.

[4] M. Bertuletti, I. M. Noz-Martín, S. Bianchi, A. G. Bonfanti, and D. Ielmini, "A Multilayer Neural Accelerator With Binary Activations Based on Phase-Change Memory," *IEEE Trans. Electron Devices*, vol. 70, no. 3, pp. 986–992, Mar. 2023, doi: 10.1109/TED.2022.3233292.

[5] M. Suri *et al.*, "Phase change memory as synapse for ultra-dense neuromorphic systems: Application to complex visual pattern extraction," in *2011 International Electron Devices Meeting*, Dec. 2011, pp. 4.4.1-4.4.4. doi: 10.1109/IEDM.2011.6131488.

[6] M. Jerry *et al.*, "Ferroelectric FET analog synapse for acceleration of deep neural network training," in *2017 IEEE International Electron Devices Meeting (IEDM)*, Dec. 2017, vol. 6, no. c, pp. 6.2.1-6.2.4. doi: 10.1109/IEDM.2017.8268338.

[7] M. H. R. Ansari, H. Li, and N. El-Atab, "Vertically Stacked Nanosheet FET: Charge- Trapping Memory and Synapse With Linear Weight Adjustability for Neuromorphic Computing Applications," *IEEE Trans. Electron Devices*, vol. 70, no. 3, pp. 1344–1350, Mar. 2023, doi: 10.1109/TED.2023.3234018.

[8] J. Lee, B. G. Park, and Y. Kim, "Implementation of Boolean Logic Functions in Charge Trap Flash for In-Memory Computing," *IEEE Electron Device Lett.*, vol. 40, no. 9, pp. 1358–1361, 2019, doi: 10.1109/LED.2019.2928335.

[9] J. Fu *et al.*, "Si-Nanowire Based Gate-All-Around Nonvolatile SONOS Memory Cell," *IEEE Electron Device Lett.*, vol. 29, no. 5, pp. 518–521, May 2008, doi: 10.1109/LED.2008.920267.

[10] Synopsys, "Sentaurus TCAD Datasheet," 2018.

[11] N. Loubet *et al.*, "Stacked nanosheet gate-all-around transistor to enable scaling beyond FinFET," in *2017 Symposium on VLSI Technology*, Jun. 2017, vol. 5, no. 1, pp. T230–T231. doi: 10.23919/VLSIT.2017.7998183.

2023 IEEE Latin American Electron Devices Conference (LAEDC)
Puebla, México, July 3-5, 2023

Experimental Analysis of HfO$_{2/X}$ ReRAM devices by the Capacitance Measurements

Fernando J. Costa
Student Member, IEEE
Electrical Engineering Department
Centro Universitário FEI
São Bernardo do Campo, Brazil
Visisng Scholar, NJIT
engfernando@fei.edu.br

Aseel Zeinati
Student Member, IEEE
Electrical and Computer Engineering
Department
New Jersey Institute of Technology,
NJIT
Newark, USA
akz4@njit.edu

Renan Trevisoli
Senior Member, IEEE
Pontifícia Universidade Católica de
São Paulo, PUC-SP
Insper Instituto de Ensino e Pesquisa,
São Paulo, Brazil
renan.trevisoli@ieee.org

D. Misra
Fellow, IEEE
Electrical and Computer Engineering
Department
New Jersey Institute of Technology,
NJIT
Newark, USA
dmisra@njit.edu

Rodrigo T. Doria
Senior Member, IEEE
Electrical Engineering Department
Centro Universitário FEI
São Bernardo do Campo, Brazil
rtdoria@fei.edu.br

Abstract— The main objective of this work is to present an analysis of the switching properties in Resistive Random-Access-Memory devices through the capacitance measurements of the metal-insulator-metal structure. The analysis was carried out in two devices with different insulating layers, one composed of H-plasma-treated HfO$_2$ and the other with a stoichiometric HfO$_2$. The device with a higher quantity of oxygen vacancy related defects in the insulator (HfO$_2$ w/trt) presents a wider spread of the capacitance with the application of a range of varying pulse widths. An increase in the capacitance from 3.904 to 3.917 pF/μm^2 was observed for the same device when it was subjected to a 144 μs pulse width, demonstrating a conductance quantization required for the application in in-memory computing systems.

Keywords— *MIM, ReRAM, Capacitance, Oxygen Vacancies, Defects, H-Plasma treatment.*

I. INTRODUCTION

Since the assumption of a missing device until its conception [1-2], the Resistive Random-Access-Memory (ReRAM) devices have attracted attention, and some applications are upcoming on the academic and industry research where the devices present Multi-Level-Cell (MLC) characteristics [3-5] mimicking the behavior of a neural synapse [6-7]. Through a crossbar array system [8] it allows for the development of a neuromorphic, in-memory computing system [9] where the processing and the storage are located at the same core, reducing latency time. The resistive switching mechanics is based on the thermal activated ion migration [10] through a construction and disruption of a conducting filament (CF) due to the variance of oxygen vacancy distribution in the dielectric. Initially, the device is in its pristine state, where the oxygen defects are on their natural position from the fabrication process. In this condition the device is in a high resistance state (HRS). When the voltage is applied on the devices terminal for the first time, the oxygen vacancies line up due to oxygen ion movement to form the CF, so the device enters in the low resistance state (LRS) and the current reaches the compliance value, required to avoid irreversible permanent dielectric breakdown. The applied

voltage at this point is so-called the forming voltage. Once the CF is formed, the device can be switched back from LRS to HRS by the application of a reverse voltage bias disrupting the filament. After a reset, the device can be switched again to LRS, at this time, a lower voltage than the forming voltage is required to reconstruct the filament, these are the forming, set and reset processes [11] which define the ReRAMs' operating characteristics. For the first-time this work proposes to perform the analysis of switching properties through the capacitance characteristics of two devices with different insulator types, one with higher concentration of oxygen vacancy related defects (plasma treated HfO$_2$) and one with much lower concentration (stochiometric HfO$_2$).

II. DEVICES CHARACTERISTICS AND APPLIED METHODOLOGY

The devices were fabricated at TEL Labs and constitutes two different metal-insulator-metal (MIM) structures as shown in Fig. 1. Both were fabricated on 300 mm silicon wafers where the bottom electrodes (BE) were formed with an initial application of a physical vapor deposition (PVD) of 10 nm of Ti followed by another PVD application of 50 nm of TiN. The dielectric layers were then deposited above the TiN in an atomic layer depositor (ALD) reactor using tetrakis (ethylmethylamido) hafnium as the Hf precursor for HfO$_2$ layer deposition. After HfO$_2$ deposition the top metal electrodes (TE) were deposited as shown in Fig. 1 with 5 nm ALD Ru followed by 5 nm of ALD TiN and 50 nm of PVD TiN with the length and width of the 100 μm. Which respect to the switching layer, the device-A was subjected to hydrogen plasma (H-Plasma) treatment at the middle of the deposition process (after 3-nm of HfO$_2$ deposition) in a clustered microwave plasma chamber [12] to incorporate excess oxygen vacancies, then another 3 nm of HfO$_2$ was deposited, forming a device with a 6 nm-thick dielectric layer (HfO$_2$ w/trt), while the device-B has a 7 nm-thick stoichiometric HfO$_2$ (Fig. 1).

The devices were then characterized at the NJIT VLSI lab

979-8-3503-1191-4/23 $31.00 © 2023 IEEE

using a B1500A semiconductor device analyzer. Different set of pulse widths were applied to the top electrode of the devices. Immediately after each pulse application the capacitances were measured in a voltage range from -0.5 up to 0.5 V at a frequency of 1kHz. For the pulse width variation, pulses with 1, 2, 4, 8, 12 and 24 μs were applied. Subsequently, the pulses were incremented with 20 μs at each step until 144 μs. To assure the devices were working with voltages beyond the set voltage, the amplitude of the pulses was kept at a fixed value of 2.5 V for device A and 3.5 V for device B. It is worth to mention that at the end of each capacitance measurement, a reset process according to [11] was applied at the devices.

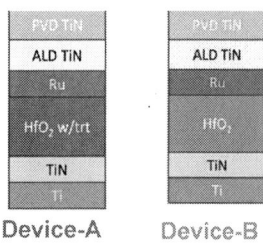

Fig.1. The layers schematics that compose the two studied devices.

III. CAPACITANCE ANALYSIS

Firstly, a sweeping voltage going from 0 to 4 V and going back from 4 to 0 V was applied to the device's top electrode terminal. Fig. 2 presents the curves of the current as a function of the top electrode voltage at a compliance current of 200 nA. It can be observed on the curves of the high resistance states (HRS) that device A presents a smaller forming voltage of 2.5 V i. e. the voltage where the current reaches its compliance value, while device B presents a forming voltage of 3.5 V. In the case of the curves for low resistance states (LRS), the current starts to drop with a TE voltage of 1.5 V for device A and 3 V for device B, indicating that device A presents a larger switching window, which reaches about 1 V whereas device B presents only 0.5 V. It is interesting to note that the curves in device A show a better uniformity dependence of the current on the TE voltage. These effects occur due to the better control of the oxygen vacancies in device A due to the higher number of oxygen vacancies incorporated with the plasma treatment.

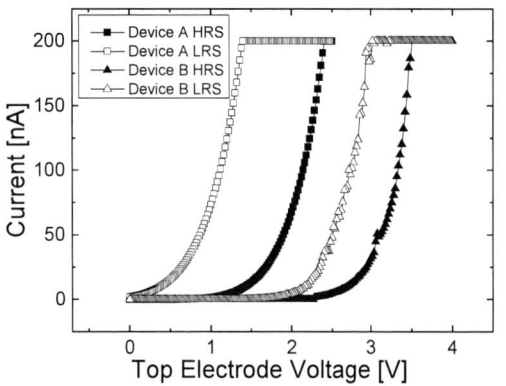

Fig.2. Current as a function of the top electrode voltage for devices A and B.

As mentioned before, without any kind of bias application to its terminals, the insulator is in its natural condition, without any change in its defects properties. In this condition, the device is in pristine state, where the insulating layer is intact of any defect change. So, the capacitances were measured with the device in this state, and then measured after the application of different pulse widths and plotted as a function of the TE voltage as shown in Fig. 3 for both devices. The capacitances with the variation of the pulse widths are shown in Fig. 3 with the curve of the pristine state as a reference. One can observe that all capacitance curves show a parabolic dependence on TE voltage and device A presents higher capacitances than device B due to the final thicknesses of the insulating layers, where devices A and B presents 6 nm and 7 nm respectively and to the larger concentration of oxygen vacancies in device A. With respect to the capacitance's variation in relation to TE voltage increasing, the pristine capacitances vary from 3.923 dropping to 3.905 and increasing to 3.95 pF/μm² in device A while device B presents values of 2.0355, 2.0335 and 2.040 pF/μm² respectively in the same conditions. Also, it is important to note that an increase in pulse width has increased the overall capacitance spread and this effect is much larger in device A, compared to device B. At higher TE voltage the curves in device B tend to converge at a same capacitance value.

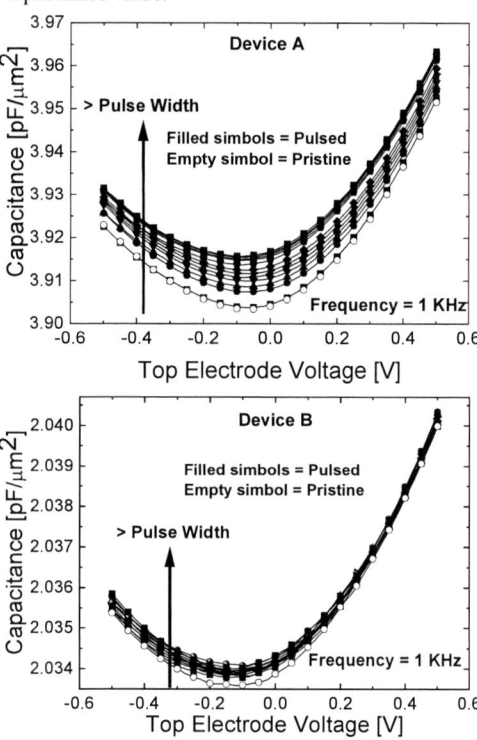

Fig.3. Capacitance as a function of the top electrode voltage at a frequency of 1 KHz for device A (TOP) and device B (BOTTOM) with different pulse widths.

The voltage capacitance coefficients (VCC) are important figures of merit that correlate the dependence of the capacitances on the TE voltage, and they can be obtained through the following expression as performed in [13]:

$$\frac{C(V)-C_0}{C_0}10^6 = [\alpha V^2 + \beta V]_{PPM} \qquad (1)$$

where C_0 is the zero-bias capacitance (pF/μm²), α and β are the quadratic and linear voltage coefficients of capacitance (PPM), and V is the TE voltage (V).

To provide a clear overview of the C-V curves behavior, the VCCs were extracted and plotted in Fig. 4 as a function of the pulse width for devices A and B with the empty circle being the values in pristine state. Firstly, one can observe that device A presents higher values for α and β, which indicates a presence of a large concentration of oxygen vacancy traps near the top electrode [14] in relation to device B, where the slow time constant of traps on the stoichiometric HfO$_2$ reduces the VCCs values [15]. Both devices present a decreasing tendency for α with the pulse width increase as shown by the red lines, which correspond to the linear fitting of α. It can be seen that device A exhibits a higher α drop from 33200 to 31800 PPM/V^2, that occurs due to the improved switching properties [16], providing a better distribution of the capacitances within the pulse width variation with respect to device B, which shows a reduction from 7470 to 7420 PPM/V^2.

The linear voltage coefficient of capacitance β presents different behavior between the devices. While device A shows an increase from 6780 up to 7860 PPM/V, a small decrease in β, from 2180 to 2140 (on the red curve) for 1 and 144 μs, respectively, can be observed with the pulse width variation for device B. This explains the tendency of the capacitance dependence at higher voltage on TE for the different pulse widths as mentioned earlier on this work.

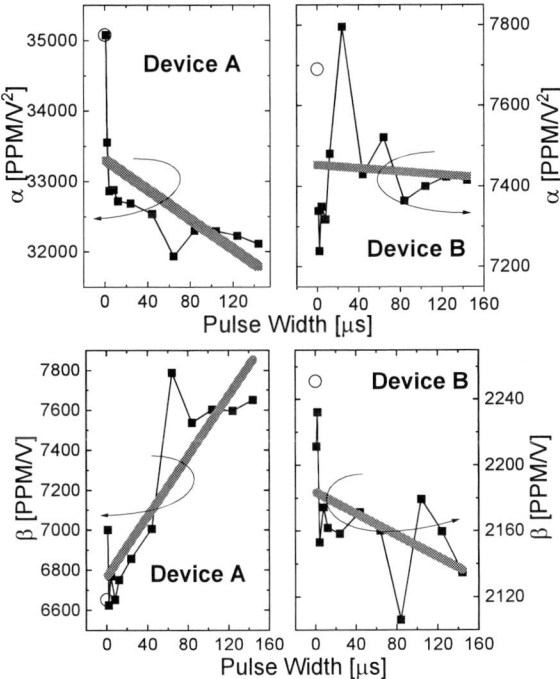

Fig.4. Quadratic capacitance coefficient (TOP) and linear capacitance coefficient (BOTTOM) as a function of the pulse width for devices A and B.

To provide a better overview of the capacitance increasing with the pulse variation, the capacitance values for TE voltage = 0 V were extracted from the curves and plotted as a function of the pulse width for devices A and B as shown in Figs. 5 and 6. The empty circle is the capacitance in pristine state. As one can observe in Fig. 5, device A presents an increase from 3.904 to 3.917 pF/μm^2 with the pulse width increment up to 144 μs. A more uniform distribution of the capacitance states can be seen along the entire pulse width shift in device A in relation to

device B in Fig. 6, which exhibits smaller overall capacitance shift, where the capacitance increases from 2.0339 in the pristine state to 2.0343 pF/μm^2 with the pulse width of 144 μs.

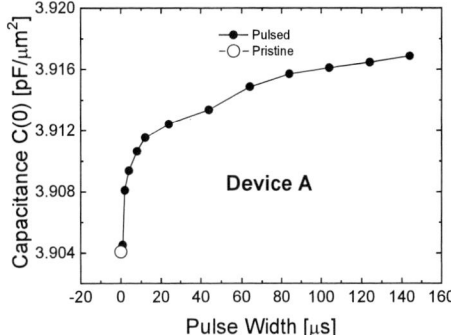

Fig.5. Capacitance C(0) as a function of the pulse width for device A.

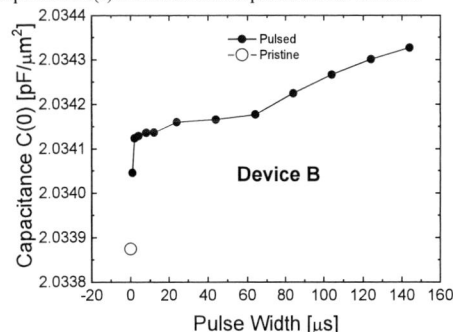

Fig.6. Capacitance C(0) as a function of the pulse width for device B.

The overall capacitance behavior in relation to the pulse widths can be correlated to the quantization of the switching properties of the ReRAM devices [17], and device A, which presents better control of the oxygen vacancies [16-17], exhibits a wider capacitance spreading with the pulse variation. For the device B, a smaller spreading of the capacitance is attributed to a poorer quantization of the switching characteristics of the stoichiometric HfO$_2$ dielectric layer.

The capacitive reactance (Xc) can be described as a resistance of a MIM capacitor, when an alternate current or signal is flowing through its terminals [18]. Thus, in order to better analyze the switching properties of the devices, the extraction of the capacitive reactance as a function of the pulse width has been performed throughout (2) [18]:

$$X_C = \frac{1}{2\pi f C} \qquad (2)$$

where Xc is the capacitive reactance (Ohm), f the frequency (Hz) and C is the C(0) capacitance (F).

In Fig. 7, Xc is presented as a function of the pulse width, and, as both devices present decrease in Xc, exhibiting multiple resistance states. The analysis of the switching properties based on its capacitances indicates the presence of the multi-level cells of the ReRAM devices [19-20] with the quantization of the conductance in relation to the pulse width variation. It is observed that device A presents a reduction from the pristine state with 407.7 to 406.35 kΩ with the pulse width increased up to its maximum value, while device-B

979-8-3503-1191-4/23 $31.00 © 2023 IEEE   107

presents decrease in Xc from 782.54 to 782.37 kΩ for the same analysis.

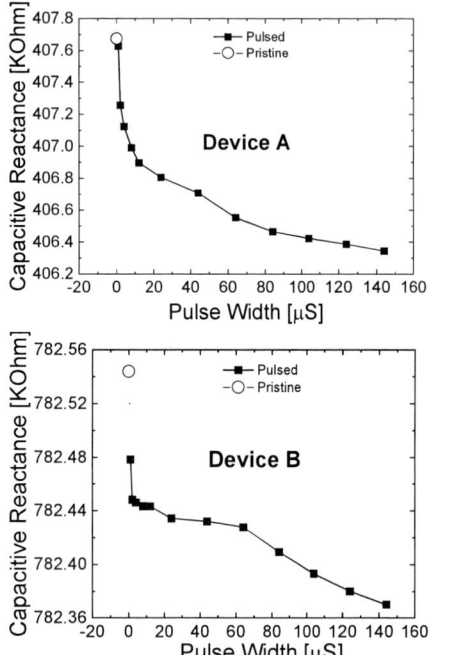

Fig. 7. The capacitive reactance as a function of the pulse width for device A (TOP) and device B (BOTTOM).

IV. CONCLUSIONS

This work has presented an evaluation of ReRAM devices through capacitance measurements of the Metal-Insulator-Metal structure. Devices presenting two kinds of insulating layers were characterized and the quantization of the switching properties could be observed from the behavior of the capacitance with the application of different pulse widths on the devices' top electrode. As wider pulses were applied, the devices showed a higher capacitance value which is attributed to the different switching resistivity promoted by the quantization of the conductive filament responsible for the low resistance state in ReRAM devices. Devices with a higher density of defects based on oxygen vacancies present a wide capacitance spreading in relation to the pulse width increase.

ACKONOWLEDGEMENTS

This work was supported by National Council for Scientific and Technological Development (CNPq) grant #303938/2020-0 and financed in part by the Coordenação de Aperfeiçoamento de Pessoal de Nível Superior - Brasil (CAPES) - Finance Code 001. The authors would like to thank Drs. D.H. Triyoso, K. Tapily, R.D. Clark, S. Consiglio, S. Rogalskyj, K. Imakita, S. Lombardo, C.S. Wajda, and G.J. Leusink of TEL Technology Center, America, Albany, NY, for supplying the devices for this work and helpful discussions.

REFERENCES

[1] L.O. Chua, "Memristor - the missing circuit element", IEEE Trans. Circuit Theory, vol. 18, pp. 507–519, 1971.

[2] D. Strukov, G. Snider, D. Stewart et al., "The missing memristor found", Nature, vol. 453, pp. 80–83, 2008.

[3] Zahoor, F., Azni Zulkifli, T.Z. & Khanday, F.A. "Resistive Random Access Memory (RRAM): an Overview of Materials, Switching Mechanism, Performance, Multilevel Cell (mlc) Storage, Modeling, and Applications" Nanoscale Res Lett 15, 90 2020.

[4] A. Padovani, J. Woo, H. Hwang and L. Larcher, "Understanding and Optimization of Pulsed SET Operation in HfOx-Based RRAM Devices for Neuromorphic Computing Applications," in IEEE Electron Device Letters, vol. 39, no. 5, pp. 672-675, May 2018.

[5] D. Garbin, E et al. "HfO2-based OxRAM devices as synapses for convolutional neural networks," *IEEE Trans. Electron Devices*, vol. 62, no. 8, pp. 2494–2501, Aug. 2015.

[6] Irem Boybat et al. "Multi-ReRAM Synapses for Artificial Neural Network Training" In IEEE Int. Symposium on Circuits and Systems, ISCAS 2019, Sapporo, Japan, May 26-29, 2019. pages 1-5, IEEE, 2019

[7] A. Padovani, J. Woo, H. Hwang and L. Larcher, "Understanding and Optimization of Pulsed SET Operation in HfOx-Based RRAM Devices for Neuromorphic Computing Applications," in IEEE Electron Device Letters, vol. 39, no. 5, pp. 672-675, May 2018.

[8] K.H. Kim et al. "A functional hybrid memristor crossbar-array/CMOS system for data storage and neuromorphic applications" Nano Lett. 12 2012.

[9] Z. Xuan et al "HPSW-CIM: A Novel ReRAM-Based Computing-in-Memory Architecture with Constant-Term Circuit for Full Parallel Hybrid-Precision-Signed-Weight MAC Operation," 2022 IEEE International Symposium on Circuits and Systems (ISCAS), Austin, TX, USA, pp. 3274-3278, 2022.

[10] F. Nardi, S. Larentis, S. Balatti, D.C. Gilmer, D. Ielmini, "Resistive switching by voltage-driven ion migration in bipolar RRAM: I. Experimental study", IEEE Trans. Electron Dev. v. 59, pp. 2461, 2012.

[11] D. Misra et al "Dielectrics and Metal Stack Engineering for Multilevel Resistive Random-Access Memory" 2020 ECS J. Solid State Sci. Technol. V. 9 053004, 2020.

[12] T. Sugawara, et al., "Characterization of ultra thin oxynitride formed by radical nitridation with Slot Plane Antenna plasma", Jpn. J. Appl. Phys, Vol.44(3) pp.1232 2005.

[13] Mu, Jiliang, Xiujian Chou, Zongmin Ma, Jian He, and Jijun Xiong. "High-Performance MIM Capacitors for a Secondary Power Supply Application" Micromachines vol. 9, no. 2, pp 69, 2018.

[14] C. Vallée, P. Gonon, C. Jorel, and F. El Kamel, "Electrode oxygen-affinity influence on voltage nonlinearities in high-k metal-insulator-metal capacitors" Appl. Phys. Lett. 96, 233504 2010.

[15] Xiongfei Yu et al., "A high-density MIM capacitor (13 fF/µm2) using ALD HfO2 dielectrics," in IEEE Electron Device Letters, vol. 24, no. 2, pp. 63-65, Feb. 2003.

[16] Yuvraj Patel et al "RRAM Devices with Plasma Treated HfO2 with Ru as Top Electrode for In-Memory Computing Hardware" ECS Trans. 104 35, 2021.

[17] Zeinati A et al "Process Optimization to Reduce Power in HfO2-Based Rram Devices for in-Memory"242nd ECS Meeting October 2022.

[18] Sedra, A.S. and Smith, K.C. (2004). Microelectronic Circuits. 5th Edition, Oxford University Press, New York, 509.

[19] Bousoulas P et al "Investigating the origins of high multilevel resistive switching in forming free Ti/TiO2−x-based memory devices through experiments and simulations" J Appl Phys 121(9):094501, 2017.

[20] J. -C. Liu, C. -W. Hsu, I. -T. Wang and T. -H. Hou, "Categorization of Multilevel-Cell Storage-Class Memory: An RRAM Example," in IEEE Transactions on Electron Devices, vol. 62, no. 8, pp. 2510-2516, Aug. 2015.

2023 IEEE Latin American Electron Devices Conference (LAEDC)
Puebla, México, July 3-5, 2023

Thermal Evaluation of 28-nm *p*-type FD-SOI MOSFETs

Alan Rossetto
Centro de Desenvolvimento Tecnológico
Universidade Federal de Pelotas
Pelotas, Brazil
alan.rossetto@inf.ufpel.edu.br

Marcelo Pavanello
Departamento de Engenharia Elétrica
Centro Universitário FEI
São Bernardo do Campo, Brazil
pavanello@fei.edu.br

Caroline Soares, Gilson Wirth
Programa de Pós-graduação em Microeletrônica
Universidade Federal do Rio Grande do Sul
Porto Alegre, Brazil
santos.soares, gilson.wirth@ufrgs.br

Ziyi Wang, Dragica Vasileska
School of Electrical, Computer and Energy Engineering
Arizona State University
Tempe, USA
zwang581, vasileska@asu.edu

Abstract—With the advent of quantum computing, MOSFET operation in cryogenic environments is of particular interest. Moreover, the full understanding of the thermal dynamic under these circumstances is key for proper modeling and design of circuits for low-temperature applications. In this work we present the partial results for the ongoing investigation of self-heating effects in 28-nm *p*-type FD-SOI devices aiming cryogenic operation. The simulation framework proved to deliver consistent results for room temperature data, and it is being extended to incorporate ultra-low temperature capabilities.

Index Terms—Self-heating, reliability, CMOS

I. INTRODUCTION

Temperature is a well-known reliability issue. Although it is not directly responsible for the device functionality, it adversely affects its performance, being inherent to the device operation. It is widely acknowledged that the temperature rise in a metal-oxide-semiconductor field effect transistor (MOSFET) degrades the carrier mobility, towards reducing the driving current, transconductance and overall device speed [1]. Such a temperature rise is caused by the device itself by the dissipation of power as heat in a process appropriately called *self-heating*, affecting the device reliability. This phenomenon is known since the '70 [2], but captured significant attention with the emergence of silicon-on-insulator (SOI) devices. In these structures, the conduction channel is separated from the silicon bulk by a buried silicon dioxide layer, which has low thermal conductivity and, thus, hinders the heat flow through the substrate. In addition, the ultrathin silicon films used in SOI technology have enhanced phonon boundary scattering and, thus, poor thermal performance [3]. Consequently, the lattice

This study was financed in part by the Coordenação de Aperfeiçoamento de Pessoal de Nível Superior - Brasil (CAPES) - Finance Code 001, by FAPESP grant 2019/15500-5, CNPq, and FAPERGS grant 19/2551-0002286-7.

temperature of the SOI-based devices increases significantly compared to their bulk counterparts.

Still, fully-depleted SOI (FD-SOI) devices are a well-suited alternative for several applications due to their improved electrostatic and variability characteristics [4], low power consumption [5], adaptive threshold voltage adjustment via back-gate biasing [6], and its resilience against single-event effects [7]. One example of such an application is quantum computing [8], in which the circuits may experience ultra-low temperatures (ULT). In this way, the proper characterization of MOSFET operation in these conditions is of particular interest. On the other hand, self-heating is usually not accounted for in most of the literature on cryogenic effects in CMOS devices, even though it may severely degrade the performance of the qubits or even turn them unusable [9].

Therefore, understanding the thermal dynamic involving FD-SOI devices in these environments is key for the design of high-performance and high-reliability integrated system for cryogenic applications. In this context, this work reports the ongoing investigation of self-heating effects in 28-nm *p*-type FD-SOI technology using a 3-D particle-based electrothermal simulation framework. The developed tool proved to deliver consistent results for room temperature data, and it is being extended to incorporate ULT capabilities.

II. SIMULATION FRAMEWORK

The work presented here was carried out using the 3-D particle-based electrothermal simulator presented in [10]. In short, the tool employs the Ensemble Monte Carlo (EMC) method for solving the Boltzmann Transport Equation (BTE) for holes in silicon self-consistently coupled to a Poisson solver and a thermal module. From the electrical perspective, the simulator accounts for carrier-carrier and carrier-impurity interactions, surface roughness scattering, Coulomb correction force, dynamical screening and simultaneous interaction with

979-8-3503-1191-4/23 $31.00 © 2023 IEEE

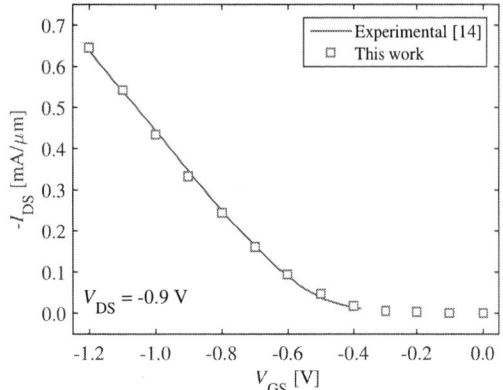

Fig. 1. $I_{DS} \times V_{GS}$ characteristics for the case study device.

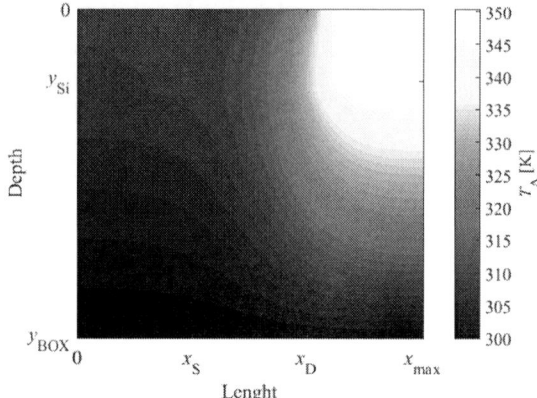

Fig. 2. Lattice temperature evaluated via phonon energy balance model for $V_{DS} = V_{GS} = -1.2$ V. Here, x_S and x_D are the position of the junction between the channel and the source / drain region, respectively, whereas x_{max} is the maximum length of the structure. Along the depth, y_{Si} and y_{BOX} are the extension of the silicon film and buried oxide layer, respectively.

multiple charge centers, being quantum corrections under development [11].

The thermal module, in turn, was developed pursuing the approach of Raleva and co-workers [12], in which the phonon temperatures are obtained by solving the steady-state phonon energy balance (PEB) equations [13], which have the form

$$C_{OP} \frac{\partial T_{OP}}{\partial t} = \frac{3}{2} \rho k_B \left(\frac{T_c - T_{OP}}{\tau_{c-OP}} \right) + \frac{\rho m^* v_D^2}{2\tau_{c-OP}} - C_{OP} \left(\frac{T_{OP} - T_A}{\tau_{OP-A}} \right) \quad (1)$$

and

$$C_A \frac{\partial T_A}{\partial t} = \nabla \cdot (\kappa_A \nabla T_A) + C_{OP} \left(\frac{T_{OP} - T_A}{\tau_{OP-A}} \right) + \frac{3}{2} \rho k_B \left(\frac{T_c - T_A}{\tau_{c-A}} \right), \quad (2)$$

where C stands for the volumetric heat capacity, T is temperature, ρ is the carrier concentration, k_B is the Boltzmann constant, m^* is the carrier effective mass, v_D is the carrier drift velocity, κ is the silicon thermal conductivity, t is the time, and τ is the relaxation time. The subscripts OP, A and c refer to the quantities related to optical phonons, acoustic phonons and carriers, respectively.

To account for the self-heating impact on the device performance, the Monte Carlo kernel and the thermal module are coupled self-consistently. As seen in Eqs. 1 and 2, the thermal model relies on electrical parameters such as ρ, v_D and T_c for determining the acoustic T_A and optical T_{OP} phonon temperatures. These temperatures are fed back into the Monte Carlo phase that, employing the concept of temperature-dependent scattering tables, immediately changes the scattering rate probability for each carrier based on the phonon temperatures in its vicinity. This process is repeated until the desired convergence is achieved and allows the tool to capture the degradation of the device parameters due to the self-heating. Besides the phonon temperatures, the thermal module can also provide insights into the heat generation map and power dissipation.

III. METHODOLOGY

The case study structure in which self-heating effects are investigated in this work is a 28-nm p-type FDSOI whose complete device characterization is presented elsewhere [14]. The transistor has channel length $L_G = 28$ nm, channel width $W = 80$ nm, and oxide thickness $t_{OX} = 1.1$ nm (EOT). The silicon film and the buried oxide layer have thicknesses $t_{Si} = 7$ nm and $t_{BOX} = 25$ nm. Source and drain regions were uniformly doped with $N_A = 5 \times 10^{19}$ cm^{-3}, whereas the silicon film was lightly doped with $N_D = 10^{15}$ cm^{-3} and the substrate with $N_D = 10^{18}$ cm^{-3}. The $I_{DS} \times V_{GS}$ curve of the case study device is shown in Fig. 1. Since the simulation utilizes poly-silicon as gate material and the experimental devices possess metal gate stack, the gate workfunction had to be adjusted by $\Delta\phi = +0.53$ eV.

In order to describe the case study structure in the simulation domain, a mesh size Δ_{ox} of 0.22 nm was used for the

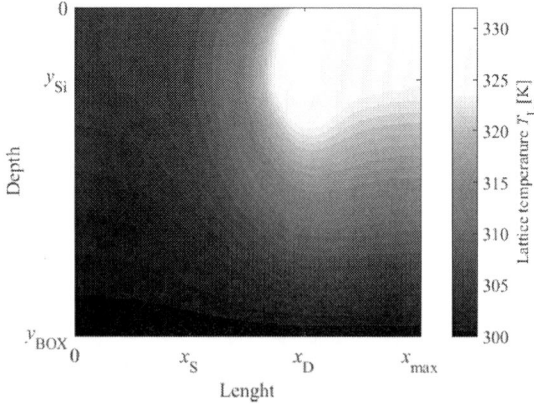

Fig. 3. Lattice temperature evaluated via Joule heating model for $V_{DS} = V_{GS} = -1.2$ V.

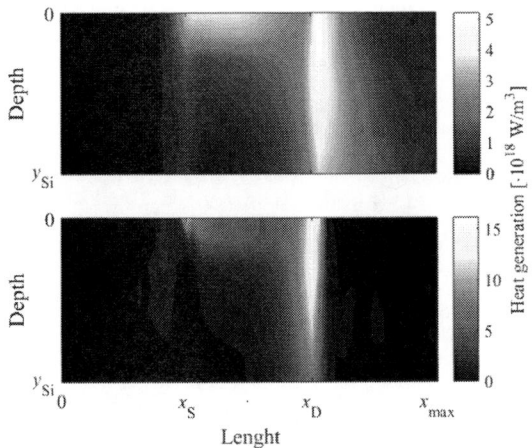

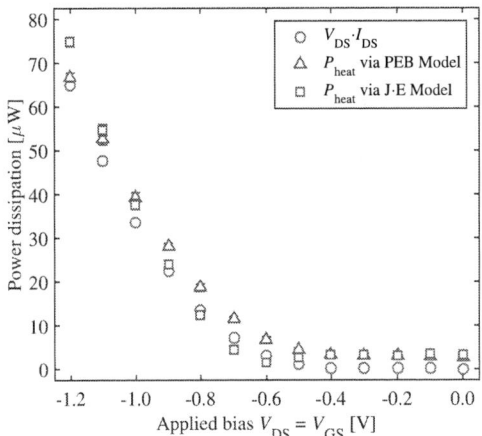

Fig. 4. Heat generation map evaluated via phonon energy balance model (*top*) and via Joule heating model (*bottom*) for $V_{DS} = V_{GS} = -1.2$ V.

Fig. 5. Power dissipation as a function of the applied bias.

discretization of the gate oxide and interface regions, while $\Delta_x = \Delta_z = 2.0$ nm were used respectively along the length and width of the device, and $\Delta_y = 1.0$ nm along the depth. The boundary conditions for the thermal simulation were defined as fixed (Dirichlet) at the bottom of the device and at the gate contact, and adiabatic (Neumann) in the source / drain contact and all the other sides of the device. At the bottom and at the gate contact the temperature was kept at $T_{sink} = 300$ K and $T_{gate} = 300$ K. The latter, however, can be swept to larger values in case the heat removal efficiency through this contact is reduced.

IV. RESULTS AND DISCUSSION

Fig. 2 depicts the lattice (acoustic phonon) temperature profile for the case study device. Note that, for a typical bias condition of $V_{DS} = V_{GS} = -1.2$ V, the hot spot temperature T_{peak} may exceed 350 K. Hence, it peaks at the drain side of the channel, where the carriers lose the energy gained upon crossing the channel by the increased number of scattering events at the drain region. This behavior can be seen in Fig. 4(*top*), where the heat dissipation spreads over the drain region but reducing as the carriers travel further away and thermalize.

In contrast, Joule heating model yields a more localized heat generation profile, as shown in Fig. 4(*bottom*). It peaks exactly at the channel/drain junction, where the electric field and carrier velocity are the highest. Consequently, the hot spot is also very localized, but less intense than the one evaluated via PEB model, as seen in Fig. 3. Note that the hot spot temperature evaluated via Joule heating model roughly exceeds 330 K under the same bias condition.

Regardless the differences between the two heat generation maps, the total power dissipation as heat remains close to the device input power $V_{DS} \cdot I_{DS}$ for the entire range of biases, as expected, and shown in Fig. 5. Besides the hot spot temperature, the average temperature in the channel T_{avg} was

also extracted from the simulation as a function of the applied bias, and it is shown in Fig. 6.

Fig. 7 depicts the temperature increase ΔT in the hot spot temperature and in the average temperature in the channel as a function of the dissipation power, which is also known as *thermal resistance*. Since experimental data for the PMOS case study device were not available in this case, the figure shows the data for its NMOS counterpart. In this study, PMOS and NMOS devices share the same structure, fabrication process and materials, differing solely on the doping type. Therefore, their thermal behavior is expected to be quite similar for a given power dissipation level.

The thermal resistance is somewhat related to the overall device temperature. Nevertheless, basing its extraction on the peak temperature may overestimate its value, since the hot spot occurs solely in a particular point of the device. On the other hand, considering the average temperature in the channel may underestimate it, once the channel region is much colder

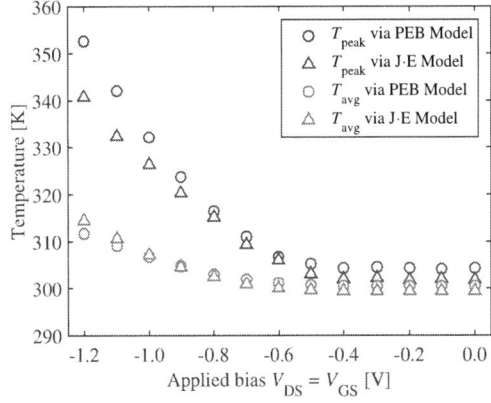

Fig. 6. Hot spot temperature T_{peak} and average temperature in the channel T_{avg} extracted using the phonon energy balance model and the Joule heating model.

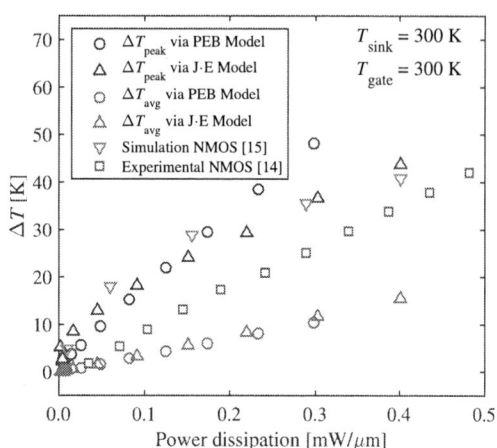

Fig. 7. Hot spot and average channel temperature increase as a function of the power dissipation.

than other active zones (see Fig. 2), such as the drain region, for instance. In this way, Fig. 7 shows that the experimental thermal resistance lies within the desired interval.

V. CONCLUSION

This work presented the partial results for the ongoing investigation of self-heating effects in 28-nm p-type FD-SOI devices for cryogenic applications. We showed that our simulation framework can deliver consistent results for room temperature data. Future works will focus on the extension of the tool to account for low-temperature MOSFET modeling.

REFERENCES

[1] S. Rathore, R. K. Jaisawal, P. N. Kondekar, and N. Bagga, "Design optimization of three-stacked nanosheet FET from self-heating effects perspective," *IEEE Transactions on Device and Materials Reliability*, vol. 22, no. 3, pp. 396–402, 2022.

[2] C. Popescu, "Selfheating and Thermal Runaway Phenomena in Semiconductor Devices," *Solid-State Electron.*, vol. 13, no. 4, pp. 441–450, 1970.

[3] M. Asheghi, K. Kurabayashi, R. Kasnavi, and K. E. Goodson, "Thermal Conduction in Doped Single-crystal Silicon Films," *J. Appl. Phys.*, vol. 91, no. 8, pp. 5079–5088, 2002.

[4] F. Andrieu, O. Weber, J. Mazurier, O. Thomas, J.-P. Noel, C. Fenouillet-Béranger, J.-P. Mazellier, P. Perreau, T. Poiroux, Y. Morand, T. Morel, S. Allegret, V. Loup, S. Barnola, F. Martin, J.-F. Damlencourt, I. Servin, M. Cassé, X. Garros, O. Rozeau, M.-A. Jaud, G. Cibrario, J. Cluzel, A. Toffoli, F. Allain, R. Kies, D. Lafond, V. Delaye, C. Tabone, L. Tosti, L. Brévard, P. Gaud, V. Paruchuri, K. Bourdelle, W. Schwarzenbach, O. Bonnin, B.-Y. Nguyen, B. Doris, F. Boeuf, T. Skotnicki, and O. Faynot, "Low leakage and low variability ultra-thin body and buried oxide (UT2B) SOI technology for 20nm low power CMOS and beyond," in *2010 Symposium on VLSI Technology*, 2010, pp. 57–58.

[5] H. Okuhara, A. Ben Ahmed, J. M. Kühn, and H. Amano, "Asymmetric body bias control with low-power FD-SOI technologies: Modeling and power optimization," *IEEE Transactions on Very Large Scale Integration (VLSI) Systems*, vol. 26, no. 7, pp. 1254–1267, 2018.

[6] O. Weber, "FDSOI vs FinFET: differentiating device features for ultra low power & IoT applications," in *2017 IEEE International Conference on IC Design and Technology (ICICDT)*, 2017, pp. 1–3.

[7] R. Liu, A. Evans, L. Chen, Y. Li, M. Glorieux, R. Wong, S.-J. Wen, J. Cunha, L. Summerer, and V. Ferlet-Cavrois, "Single event transient and TID study in 28 nm UTBB FDSOI technology," *IEEE Transactions on Nuclear Science*, vol. 64, no. 1, pp. 113–118, 2017.

[8] L. L. Guevel, G. Billiot, X. Jehl, S. De Franceschi, M. Zurita, Y. Thonnart, M. Vinet, M. Sanquer, R. Maurand, A. G. M. Jansen, and G. Pillonnet, "19.2 A 110mK 295μW 28nm FDSOI CMOS quantum integrated circuit with a 2.8GHz excitation and nA current sensing of an on-chip double quantum dot," in *2020 IEEE International Solid-State Circuits Conference - (ISSCC)*, 2020, pp. 306–308.

[9] A. A. Artanov, E. A. Gutiérrez-D, A. R. Cabrera-Galicia, A. Kruth, C. Degenhardt, D. Durini, J. Méndez-V, and S. Van Waasen, "Self-heating effect in a 65 nm MOSFET at cryogenic temperatures," *IEEE Transactions on Electron Devices*, vol. 69, no. 3, pp. 900–904, 2022.

[10] A. C. Rossetto, V. V. Camargo, D. Vasileska, and G. I. Wirth, "3-D non-isothermal particle-based device simulator for p-type MOSFETs," *Journal of Computational Electronics*, vol. 20, pp. 1644–1656, 2021.

[11] C. S. Soares, P. K. R. Baikadi, A. C. J. Rossetto, M. A. Pavanello, D. Vasileska, and G. I. Wirth, "Modeling quantum confinement in multigate transistors with effective potential," in *2022 36th Symposium on Microelectronics Technology (SBMICRO)*, 2022, pp. 1–4.

[12] K. Raleva, D. Vasileska, A. Hossain, S.-K. Yoo, and S. M. Goodnick, "Study of self-heating effects in SOI and conventional MOSFETs with electro-thermal particle-based device simulator," *J. Comput. Electron.*, vol. 11, no. 1, pp. 106–117, 2012.

[13] J. Lai and A. Majumdar, "Concurrent thermal and electrical modeling of sub-micrometer silicon devices," *J. Appl. Phys.*, vol. 79, no. 9, pp. 7353–7361, 1996.

[14] M. Cassé, B. C. Paz, F. Bergamaschi, G. Ghibaudo, F. Serra, G. Billiot, A. Jansen, Q. Berlingard, S. Martinie, T. Bedecarrats *et al.*, "FDSOI for cryoCMOS electronics: device characterization towards compact model," in *2022 International Electron Devices Meeting (IEDM)*. IEEE, 2022, pp. 34–6.

[15] Z. Wang, D. Vasileska, C. Soares, G. Wirth, M. Pavanello, and M. Povolotskyi, "Modeling self-heating effects in 28 nm technology node fully-depleted SOI devices," in *2023 22nd International Workshop on Computational Electronics (IWCN)*, 2023, in press.

2023 IEEE Latin American Electron Devices Conference (LAEDC)
Puebla, México, July 3-5, 2023

Fabrication and electrical characterization of Al-based MIM capacitors

Daniel Rocha-Aguilera
Electronics Department, Microelectronics Group
National Institute of Astrophysics, Optics and Electronics
Tonantzintla, Mexico
daniel.rocha@inaoep.mx

Joel Molina-Reyes
Electronics Department, Microelectronics Group
National Institute of Astrophysics, Optics and Electronics
Tonantzintla, Mexico
jmolina@inaoep.mx

Abstract—**In this work, a fabrication process for Al-based metal-insulator-metal capacitors using atomic layer deposition to deposit thin aluminum oxide films with various thicknesses is presented. These devices are used to electrically characterize the thin dielectric films in a structure which is very similar to a Josephson tunneling junction, which is the central element of superconducting qubits. An analysis of the conduction mechanisms through the dielectric and obtention of electrical properties of the materials are also included.**

Keywords— *metal-insulator-metal capacitor, ultra-thin metal oxide, atomic layer deposition, conduction mechanisms in dielectrics, Josephson junction*

I. INTRODUCTION

The fabrication and integration of quantum bits or qubits is a fundamental matter in the development of quantum computers capable of having scientific, and technological applications to perform complex simulations, solve optimization problems and work with vast amounts of data. [1]

Superconducting qubits are superconducting integrated circuits that can be operated as two-level quantum systems and constitute one of the most widely explored platform for the fabrication of qubits. One of the main advantages of these devices is their scalability, as integrated circuit fabrication technology allows for reproducible fabrication and coupling of many-qubit systems. On the other hand, they are vulnerable to diverse decoherence mechanisms that limit the lifetime of the quantum states. These mechanisms are related to material defects and losses as well as external noise. Research and improvement in material choice and processing, fabrication and shielding have resulted in an important improvement on coherence times over the last 25 years. [2-3]

Various architectures for superconducting qubits have been developed, all of them based on Josephson junctions (JJs). JJs are superconducting metal-insulator-metal (MIM) structures in which quantum tunneling of Cooper pairs occurs. [4] They are constituted by two superconducting electrodes separated by an ultra-thin insulator of 1 or 2 nm in thickness. Aluminum and aluminum oxide are the most common materials for JJs. The typical fabrication process for JJs involves deposition of the metallic electrodes by shadow evaporation, a thermal oxidation to grow the insulator from the bottom electrode and e-beam lithography to define the junction geometry and area. [5]

In this work, a fabrication process for JJs is proposed and used to fabricate Al-based integrated MIM capacitors with

insulator thicknesses of 40, 20 and 6 nm. These insulating films are not thin enough to form a JJ, but they are useful to electrically characterize the insulator. The fabrication process uses atomic layer deposition (ALD) to deposit the thin insulator films, allowing a precise thickness control and replacing the typical thermal oxidation of Al. Optimization of the ALD deposition parameters can be useful to minimize defects in the metal-insulator interfaces and in the bulk of the insulator. These defects are one of the main problems for JJ devices, deriving in quantum two-level systems.

II. EXPERIMENTAL

Samples with Al/Al$_2$O$_3$/Al MIM capacitors were processed on silicon substrates. The resulting structure is shown in Fig. 1. Prime grade n-type silicon wafers with 2-5 Ω·cm resistivity and a (100) surface were used as substrates. The silicon wafers were cleaned using a standard RCA clean and HF-last treatment to remove contaminants and obtain an H-terminated surface.

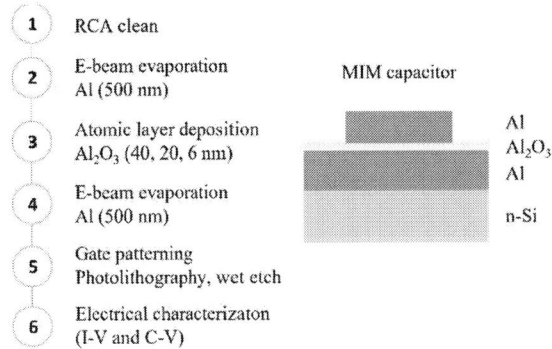

Fig. 1. Fabrication process sequence for integrated MIM capacitors (left). Cross-section schematic of a MIM capacitor (right).

The bottom aluminum electrodes were deposited by e-beam evaporation (Temescal BJD-1800 from Edwards) in ultrahigh vacuum (1E-7 Torr) with a deposition rate of 10 Å/s to obtain a 500 nm thickness. Next, The Al$_2$O$_3$ films were deposited by thermal ALD (Savannah S100 from Cambridge NanoTech) at 200 °C, 200 mTorr using H$_2$O and TMA (trimethylaluminum) as precursors. Immediately after the ALD deposit, the samples were introduced again in the evaporation chamber to minimize the exposure of the Al$_2$O$_3$ to ambient contamination. The top Al electrode was deposited using the same parameters as the bottom electrode. Photolithography and wet etching steps were used to pattern the top aluminum layer in the form of gate electrodes

979-8-3503-1191-4/23 $31.00 © 2023 IEEE
113

with areas of 40x40 and 80x80 um². This fabrication process is summarized in Fig. 1.

Wafer-level characterization was performed using a semiconductor parameter analyzer (Keithley 4200 SCS), to obtain current-voltage (I-V) and capacitance-voltage (C-V) measurements.

III. RESULTS AND DISCUSSION

A. I-V measurements

Capacitors with a gate area of 40x40 um² were characterized. Typical I-V curves for the samples are displayed in Fig. 2. The measured current is the leakage current that flows through the insulator. In Fig. 2b it is evident that this leakage current increases at lower voltages in the devices with thinner dielectrics. Hard dielectric breakdown is also observed as an abrupt increase of the current. In Fig. 2a the curves are shown in the form of current density vs electric field (J-E). In this way, it can be observed that the electric field at which dielectric breakdown occurs (E_{bkd}) is similar for the 3 different thicknesses and is found to be between 5 and 7 MV/cm.

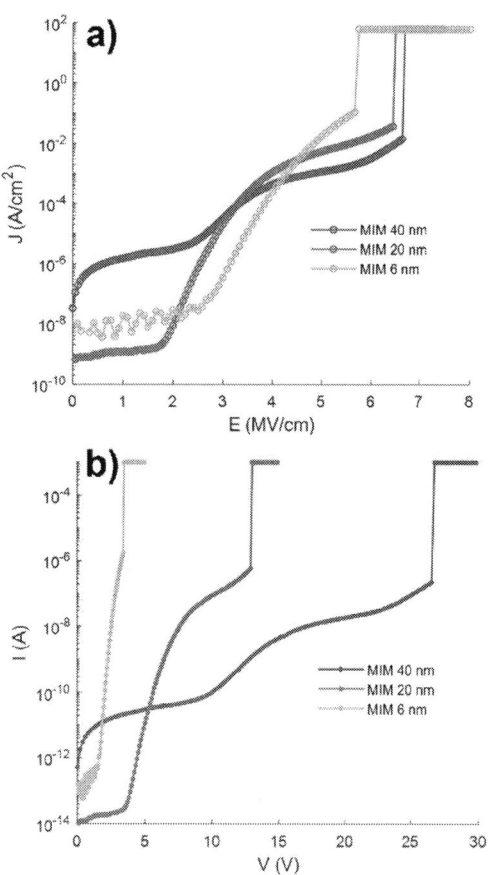

Fig. 2. a) E-J curves, b) I-V curves for Al/ Al_2O_3/Si MIS capacitors.

There is a decreasing trend for E_{bkd} as the Al_2O_3 becomes thinner, even though it is the same material deposited with the same parameters. This can be related to an observed increase in mechanical fragility of the thinner dielectrics. Additionally, these MIM capacitors were found to be more sensitive to

mechanical pressure than similarly fabricated metal-insulator-semiconductor capacitors. This is due to the surface over which the Al_2O_3 is deposited, as the surface of the evaporated Al films has a considerably higher roughness and defect density than the surface of Si wafers, thus affecting the mechanical and electrical quality of the material deposited on top of it. From AFM analysis, an average surface roughness of 3 nm was calculated for the Al surface, in contrast to a 0.7 nm surface roughness for the Si substrates.

B. Analysis of conduction mechanisms

There are different physical mechanisms that allow current flow through a non-ideal dielectric. Some of them are related to energy levels from traps or defects in the bulk of the dielectric and some other take place in the energy barriers at the interfaces between electrodes and dielectric. For these MIM capacitors, various conduction mechanisms were identified and modeled using the semi-empirical equations presented hereafter.

Space Charge Limited Conduction (SCLC) is a conduction mechanism characterized by the appearance of a space charge in the bulk of the dielectric. [6] It is defined by 3 regions:

- An ohmic region in a condition of weak injection. The current follows Ohm's law (1), and the current-voltage relation is I □ V, where q is the electron charge, n_0 is the concentration of carriers in thermal equilibrium, μ is the electron mobility and d is the dielectric's thickness.

$$J = qn_0\mu\frac{V}{d} \qquad (1)$$

- A trap-filled region in which the number of injected carriers becomes greater than the thermally generated carriers, so they can travel across the dielectric and fill traps present in it to create a space charge. The current-voltage relation is I □ V² and follows (2), where ε is the dielectric constant and θ is the ratio of free carriers to total carriers. The transition voltage V_{tr} marks the beginning of this region.

$$J = 9\mu\varepsilon\theta\frac{V^2}{8d^3} \qquad (2)$$

- After all the traps are filled at very strong injection, there is an abrupt increase in current followed by a region described by Child's law (I □ V²) according to (3). The traps-filled limit is defined by the voltage V_{TRL}.

$$J = 9\mu\varepsilon\frac{V^2}{8d^3} \qquad (3)$$

Another mechanism is Fowler-Nordheim (FN) tunneling, which is found at high electric fields that cause the energy bands to bend. In this condition, the dielectric's band gap bends forming a triangular barrier thin enough for charge carriers to tunnel through it. This phenomenon is described by (4), in which ϕ_B is the height of the triangular barrier, h is Planck's constant and m^* is the electron tunneling effective mass. [6]

$$J = \frac{q^3E^2}{8\pi hq\phi_B}e^{-8\pi\sqrt{2m^*}(q\phi_B)^{3/2}/(3qhE)} \qquad (4)$$

979-8-3503-1191-4/23 $31.00 © 2023 IEEE

Poole-Frenkel (PF) emission is a mechanism in which the carriers are thermally emitted from a trap level within the band gap of the dielectric to its conduction band and is described by (5), in which ϕ_τ is the energy of the trapping center measured from the top of the conduction band of the dielectric, N_C is the effective density of states in the conduction band, ε_i is the optical dielectric constant, ε_0 is the vacuum permittivity, k_B is Boltzmann's constant and T is the temperature. [6]

$$ J = q\mu N_C E e^{-q(\phi_t - \sqrt{qE/(\pi\varepsilon_i\varepsilon_0)})/k_BT} \qquad (5) $$

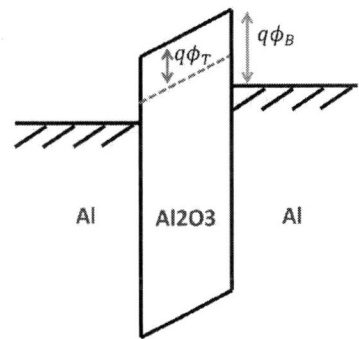

Fig. 3. Simplified energy band diagram for Al/ Al₂O₃/Al MIM capacitor.

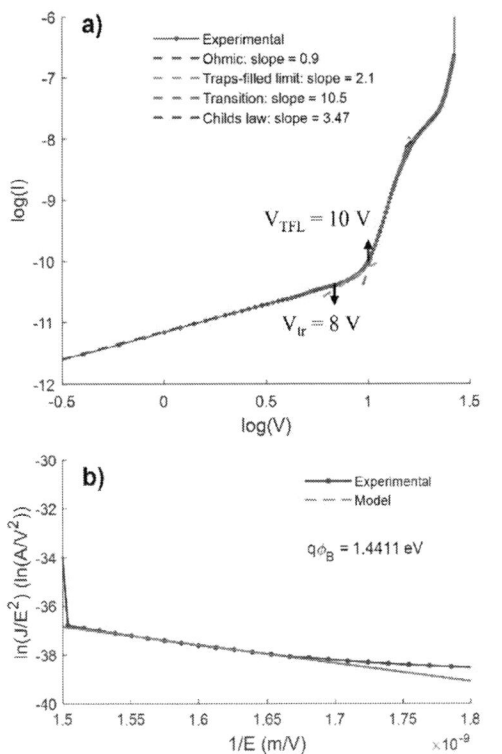

Fig. 4. a) SCLC plot and b) FN plot for MIM capacitor with 40 nm thick Al₂O₃.

Trap-assisted tunneling (TAT) was also observed. This mechanism consists of quantum tunneling of the charge carriers through the dielectric band gap using trap energy levels as intermediate steps to tunnel through a thick energy barrier. The

equation used to model inelastic TAT, in which the electrons lose energy as they leave the trapping center, is shown in (6). ϕ_T is the energy of the trap level, N_t is the trap density in the dielectric and C_t is a function of electron energy. [7]

$$ J = \frac{2C_tN_tq\phi_T}{3E} e^{-\frac{8\pi\sqrt{2qm^*}}{3hE}\phi_T^{3/2}} \qquad (6) $$

The schematic in Fig. 3 is a simplified energy band diagram for the al-based MIM capacitor which shows the location of the energy barrier and a trap energy level such as the ones described in the models of conduction mechanisms.

In the MIM capacitor with 40 nm of Al₂O₃, a SCLC mechanism was observed up to 6 MV/cm, followed by FN tunneling above 6 MV/cm. Fig. 4 includes typical SCLC and FN plots for this device. In Fig. 4a, a transition voltage $V_{tr} = 8$ V and a traps-filled limit voltage of $V_{TRL} = 10$ V were found. The region corresponding to Child's law has a slope of 3.47 instead of the expected value of 2. This is probably a consequence of the contribution of the FN current, which becomes dominant after 6 MV/cm. Fig. 4b corresponds to a typical FN plot from which a barrier height $q\phi_B = 1.44$ eV is extracted.

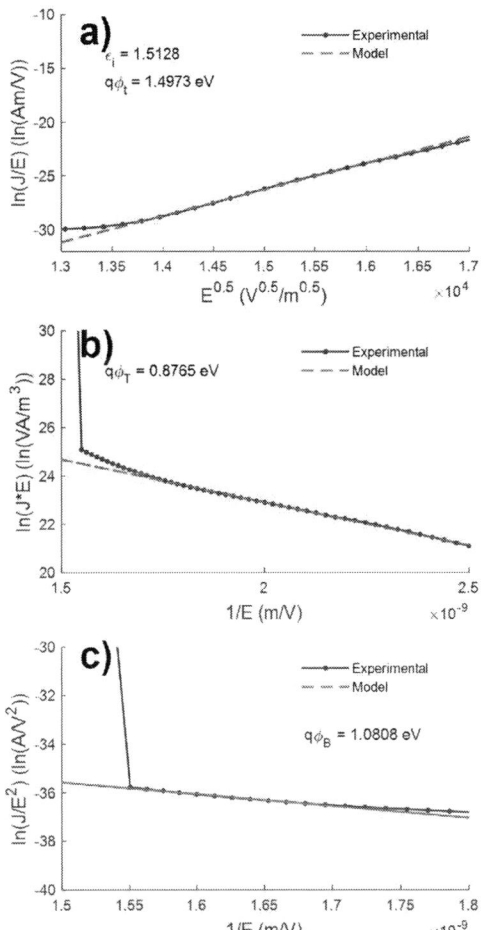

Fig. 5. a) PF plot, b) TAT plot, c) FN plot for MIM capacitor with 20 nm thick Al₂O₃.

In the case of the MIM capacitor with oxide thickness of 20 nm, the observed mechanisms were PF emission from 2 to 3 MV/cm, TAT from 4 to 5 MV/cm and FN tunneling after 5 MV/cm. The extracted trap energy levels were $q\phi_\tau = 1.51$ eV for PF and $q\phi_T = 0.97$ eV for TAT, the obtained FN barrier height is $q\phi_B = 1.08$ eV. Fig. 5 displays the plots from which these values were obtained.

Finally, for the MIM capacitor with 6 nm of Al_2O_3, PF emission from 3 to 4 MV/cm and FN tunneling above 4 MV/cm were identified. PF and FN plots were used to calculate a trap level with an energy of $q\phi_\tau = 1.57$ eV for PF and a FN barrier height of $q\phi_B = 1.39$ eV. This is shown in Fig. 6.

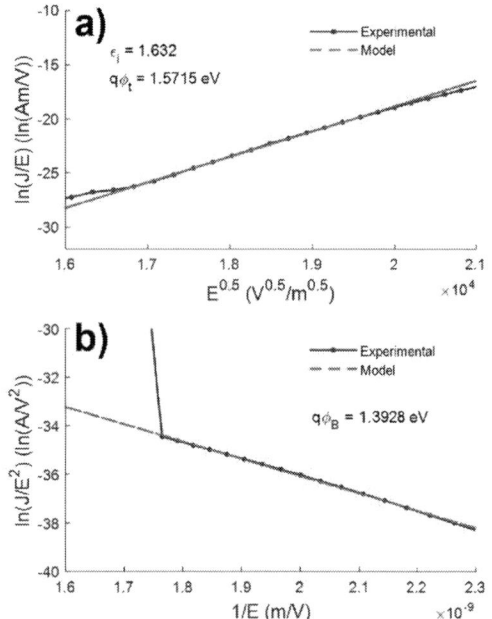

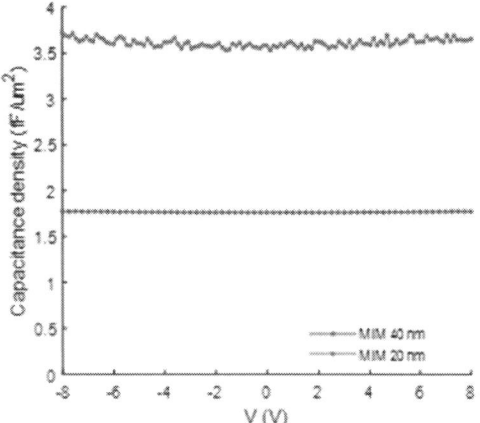

Fig. 6. a) PF plot, b) FN plot for MIM capacitor with 6 nm thick Al_2O_3.

Fig. 7. C-V curves at 1 MHz for MIM capacitors.

For the 6 and 20 nm samples, the trap energy level for PF emission is similar (around 1.5 eV) and agrees with previously reported experiments, corresponding to oxygen vacancies in deep states in the oxide gap. On the other hand, the trap energy

level for the TAT mechanism in the 20 nm sample would correspond to a shallow trap, much closer to the bottom of the conduction band of the oxide, which can be associated to the increase in leakage current when the mechanism changes. [8-9]

The FN barrier heights of the 6 and 40 nm samples show good coincidence (around 1.4 eV), but there is a considerable variation against the 20 nm sample (1.08 eV). The lower value for ϕ_B could be expected to yield a lower breakdown voltage, but this is not the case. Further analysis is necessary to explain this inconsistence or to attribute it to a fabrication issue.

C. C-V Measurements

Fig. 7 includes C-V measurements of the MIM capacitors with 40 and 20 nm of Al_2O_3. The curves have a very flat parabolic shape, yielding a very constant capacitance value even at high voltages. Using the capacitance value at 0 V, the dielectric constant of the Al_2O_3 was calculated, obtaining a value of k = 7.9.

IV. CONCLUSIONS

MIM capacitors based on Al and Al_2O_3 were fabricated using ALD to deposit 6, 20 and 40 nm thick Al_2O_3 layers. I-V and C-V measurements were used to obtain electrical properties of the dielectric and to analyze the conduction mechanisms that result in leakage current in the different devices. It was found that the electrical quality and mechanical integrity of the oxide is limited by the quality of the surface on which it is deposited, which in this case is evaporated Al. It will be necessary to optimize the characteristics of the Al deposit to obtain a higher quality dielectric for quantum devices. The analysis of conduction mechanisms shows that there are traps present in the bulk of the dielectric that can contribute to thermal emission and quantum tunneling of charge carriers. The proposed fabrication process still has work to do regarding optimization, but it appears promising as a feasible and simple way to fabricate integrated Josephson tunnel junctions.

REFERENCES

[1] M. A. Nielsen and I. L. Chuang, "Introduction and overview," in Quantum Computation and Quantum Information, Cambridge: Cambridge University Press, 2022, pp. 1–42.J. Clerk Maxwell, A Treatise on Electricity and Magnetism, 3rd ed., vol. 2. Oxford: Clarendon, 1892, pp.68–73.

[2] J. Clarke and F. K. Wilhelm, "Superconducting Quantum Bits," Nature, vol. 453, no. 7198, pp. 1031–1042, Jun. 2008.

[3] R. Simmonds, K. Lang, D. Hite, S. Nam, D. Pappas, and J. Martinis, "Decoherence in Josephson phase qubits from junction resonators," Physical Review Letters, vol. 93, no. 7, p. 077003, August 2004.

[4] A. Barone and G. Paternò, Physics and applications of the Josephson effect. Ann Arbor, Michigan: UMI Books on Demand, 1997.

[5] F. Lecocq, I. M. Pop, Z. Peng, I. Matei, T. Crozes, T. Fournier, C. Naud, W. Guichard, and O. Buisson, "Junction fabrication by shadow evaporation without a suspended bridge," Nanotechnology, vol. 22, no. 31, p. 315302, July 2011.

[6] F.-C. Chiu, "A review on conduction mechanisms in dielectric films," Advances in Materials Science and Engineering, vol. 2014, pp. 1–18, Feb. 2014.

[7] S. Salas-Rodríguez, J. Molina-Reyes, J. Martínez-Castillo, R. M. Woo-Garcia, A. L. Herrera-May, and F. López-Huerta, "Modeling of conduction mechanisms in ultrathin films of Al2O3 deposited by ALD," Electronics, vol. 12, no. 4, p. 903, Feb. 2023.

[8] J. Molina, H. Uribe-Vargas, R. Torres, P.G. Mani-Gonzalez and A. Herrera-Gomez, "Accurate modeling of gate tunneling currents in Metal-Insulator-Semiconductor capacitors based on ultra-thin atomiclayer deposited Al2O3 and post-metallization annealing", Thin Solid Films, Vol. 638 pp. 48-56 (2017).

[9] H. Uribe-Vargas and J. Molina-Reyes, "Parameter extraction of gate tunneling current in metal–insulator–semiconductor capacitors based on ultra-thin atomic-layer deposited al2o3," Journal of Materials Science: Materials in Electronics, vol. 29, no. 18, pp. 15496–15501, Apr. 2018.

Simulation and Operational Evaluation of Distributed Storage Devices Connected to a Direct Current Distribuition Nanogrid

João Paulo de Andrade Machado
Instituto de Tecnologia
Universidade Federal do Pará
Belém, Brazil
joão.machado@ufpa.itec.br

Felipe Cabral Reis
Instituto de Tecnologia
Universidade Federal do Pará
Belém, Brazil
felipe.cabral.reis@itec.ufpa.br

Arthur Fonseca Côrrea
Instituto de Tecnologia
Universidade Federal do Pará
Belém, Brazil
arthureufpa@itec.ufpa.br

Lucas dos Santos Bulhosa
Instituto de Tecnologia
Universidade Federal do Pará
Belém, Brazil
lucas.bulhosa@itec.ufpa.br

Wilson Negrão Macêdo
Instituto de Tecnologia
Universidade Federal do Pará
Belém, Brazil
ORCID: 0000-0002-6097-8620

Marcos André Barros Galhardo
Faculdade de Engenharias Elétrica e
Biomédica
Universidade Federal do Pará
Belém, Brazil
ORCID: 0000-0001-6248-5187

Abstract—This paper presents the modelling of a DC nanogrid composed of distributed photovoltaic generation and battery banks. The MATLAB/SIMULINK environment is used to simulate the components. The real DC nanogrid, located outside the laboratory of the Grupo de Estudos e Desenvolvimento de Alternativas Energéticas (GEDAE) at the Universidade Federal do Pará (UFPa) in Belém-Pará, Brazil, provided measured data for model validation. The electronic devices had their operational performance and the power quality at connection points of the storage systems evaluated. During the period with the photovoltaic generation, it was verified that during the real operation of the system, there were moments of overvoltage at the terminals of the battery banks and along the DC distribution nanogrid, this was also identified in the simulated case. The simulation results showed good agreement with the measurements, indicating the applicability of the model.

Keywords—nanogrid, direct current, electronic device modeling, simulation

I. INTRODUCTION

In recent years, there has been a significant increase in applications related to the use of renewable energy sources for electricity production through decentralized generation systems integrated with the electrical distribution grid. A large portion of these decentralized systems utilizes DC sources, such as photovoltaic generators and battery banks, and integrates them into DC microgrids.

The concept of microgrids aims to enhance the flexibility of traditional distribution systems. By subdividing a large distribution grid into smaller independent groups, the complexity of the system is reduced. This facilitates the adoption of control and protection strategies, making the system more reliable and robust [1] [2].

Moreover, the operation in DC within the scope of Direct Current Distribution Microgrids (DCDM) and Direct Current Distribution Nanogrids (DCDN) has emerged as a prominent topology for this form of distribution. These systems have gained importance due to their greater efficiency in integrating renewable sources, loads, and storage systems [1] [2] [3] [4]. This integration is increasingly common in residential and commercial buildings, where photovoltaic systems, battery storage systems, electric vehicles, and electronic loads are used. This facilitates the use of DC electricity distribution, eliminating the need for DC/AC conversion stages [1] [5] [6]. Consequently, this configuration proves to be advantageous both technically and economically, as most DCDM and DCDN deployments aim to reduce conversion costs and increase overall efficiency [7].

In the context of small demands in isolated locations, the application of DCDM or DCDN is also promising. It aims to supply a set of buildings and community loads, with the possibility of interconnecting decentralized energy generation and storage. Therefore, studies on this application are crucial for improving the quality, reliability, and efficiency of the electrical energy system.

This paper aims to analyze, both theoretically and practically, the implementation and operation of distributed storage systems in a DCDN, with a focus on modeling and simulation of the system and evaluating its operational performance.

In the theoretical part, the paper considers the use of equivalent mathematical and/or electrical models for the PV generator and battery storage system, as well as the validation of these models through experimental tests. The infrastructure and test performance are conducted in a DCDN installed in the external area of the building of the Grupo de Estudos e Desenvolvimento de Alternativas Energéticas at Universidade Federal do Pará (GEDAE/UFPA), located in the city of Belém, Pará, Brazil.

II. METODOLOGY

A. DC nanogrid

Fig. 1 shows the model of the real nanogrid components simulated in Simulink/Matlab. It is composed of 3 generation and storage systems (GSS), 3 load banks, and conductors connecting these devices represented by wires with resistors. In this case, a GSS consists of a solar photovoltaic (PV) array, Maximum Power Point Tracking (MPPT) Charge Controller taken from [8] with some modifications, and a battery storage system (BSS). The PV array is formed by the series

979-8-3503-1191-4/23 $31.00 © 2023 IEEE

2023 IEEE Latin American Electron Devices Conference (LAEDC)
Puebla, México, July 3-5, 2023

association of two multicrystalline solar modules with 245 W nominal power. The charge controller is a buck converter with the MPPT mechanism implemented and associated with a three-stage charge controller algorithm for lead-acid batteries. Also, the BSS consists of 4 lead-acid solar stationary.

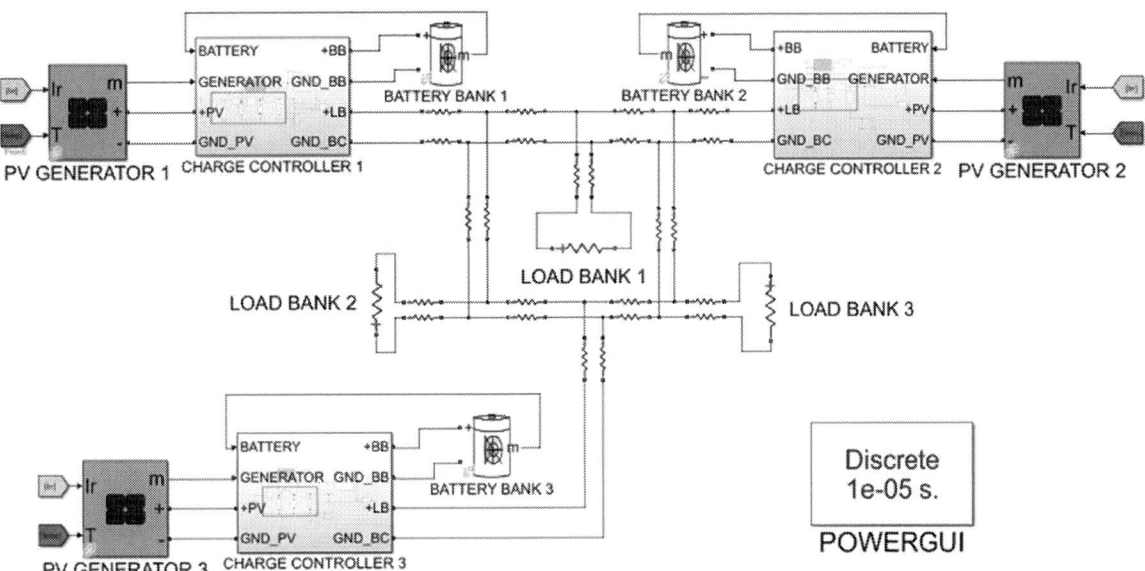

Fig. 1. Nanogrid model simulated in Matlab/Simulink.

(12V/95Ah) batteries in a series-parallel association resulting in a bank of 24V/190Ah. Finally, four 24 V/40 W resistive lamps were used in the load banks: two in the first bank and one in the others.

At the same time, a device is used to store the data from the sensors of irradiance and environmental temperature. It is necessary to collect these environmental data so that the simulation model uses a real generation profile for the place where the nanogrid is installed. Hence, if there is enough irradiance, the generation can feed the load and charge the BSSs. At night, the load is supplied by the energy stored in the BSSs. All the dynamic of the electric parameters of nanogrid devices, such as voltage, current, and power, during this cycle is collected by the MPPT charge controller and stored by a record accessory using the RS485 interface, a device that stores the data of the electrical parameters measured by the charge controller. Thus, from different scenarios of irradiance and temperature collected during the days of operation of the nanogrid, it was possible to simulate its model and evaluate the electric energy quality in the system based on the electric parameters (real and simulated).

B. PV array model

The PV array was modeled based on the five parameters model present in the block PV array in the Simscape library on Simulink to represent the I-V characteristic dependent on the incident irradiance and the PV cell operating temperature. To simulate a day, it was necessary to add some modifications to the model. First, the irradiance data presented noise at night. It was made by some light produced by a lamp near the sensor. To filter this noise, the irradiance data was compared with a constant of around 5 W/m², which is the average noise value. Then, if the irradiance data is greater than this value, the switch is opened, and the PV array is connected to the system. Otherwise, it is short-circuited, and it is disconnected.

Furthermore, during the data analysis, it was noticed that the PV generated power when the irradiance is upper to an average value of 80 W/m². Thus, the irradiance data is compared to a constant block with this value. If it is greater than 80 W/m² the irradiance remains. If not, it is changed to a constant value of 1 W/m².

To validate this model, curves I-V and P-V obtained by this block were compared with the curves obtained from a module test made [1] in the solar simulator Highlight 3c from PASAN MEASUREMENT SYTEMS installed in the GEDAE laboratory. Fig. 2 and Fig. 3 show the measured curves.

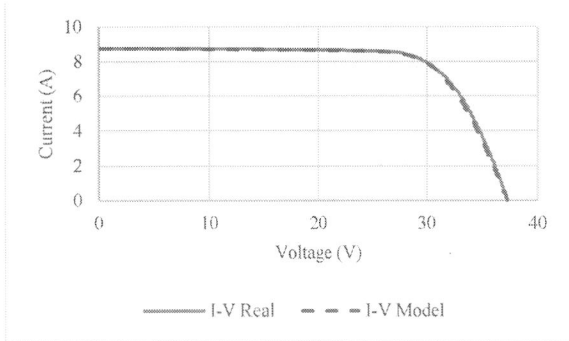

Fig. 2. I-V curves.

979-8-3503-1191-4/23 $31.00 © 2023 IEEE

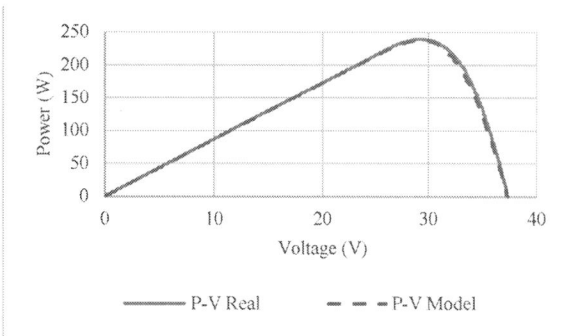

Fig. 3. P-V curves.

Thus, the percent standard deviation of the error (PSDE) was applied to measure the associated error between the curves generated from the PV array block and the real module. The closer to zero, the more accurate the model is. Table I presents the values of PSDE, observing that the model presents a low deviation concerning the measured data.

TABLE I. PSDE OF I-V AND P-V CURVES

Curve	PSDE
I-V	1.84 %
P-V	2.98 %

C. MPPT charge controller

The MPPT charge controller block was based on the work of [8] which consists of a buck converter and an algorithm MPPT associated with a charge control (Fig. 4). The Buck converter circuit was made using a MOSFET, an inductor, a diode Schottky, a capacitor and a blocker diode. The block MPPT Solar Charge Control is responsible to employ the MPPT algorithm and to implement the 3-stages battery charging (bulk, boost, float). By this block, it is possible to define the Boost Voltage, that to these BSS is defined to 28.8 V.

Some modifications were introduced to the Solar Charge Control. At first, the block ABS/Float Control was made to consider the time that the BSSs stay in the boost stage. Also, in series, the MPPT Solar Charge Controller was installed a load protection to disconnect the load when the battery reaches its lower voltage (in this case 22.2 V).

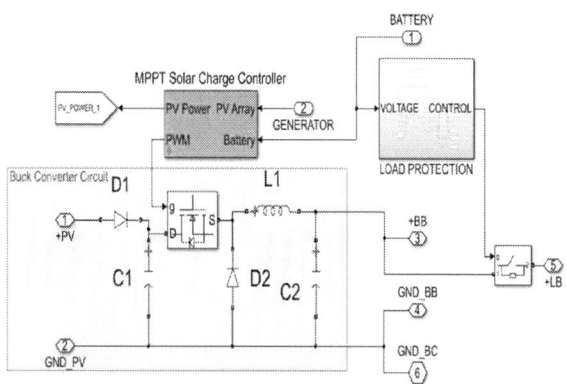

Fig. 4. MPPT Solar Charge Controller, Buck Converter Circuit and load protection system.

D. Battery storage system

The microgrid's BSS block was adapted from the models developed in [8] and [9]. The battery bank is modeled as a voltage source in series with a dynamic internal resistance. In this context, the electric-mathematical model of a rechargeable battery (Fig. 5) from the Simscape library from Simulink was used and adapted to provide a faster simulation of a cycle of discharge/charge. Hence, one hour of a day was considered 1.8 s in simulation. To validate this system, it was simulated a charging where it is verified the three-stages charging method (also known as Constant current/ Constant voltage - CCCV) in BSS's voltage and current. First, the BSS's current is constant, and the voltage increases. It is called the Bulk stage and it continues until the voltage reaches the absorption voltage of 28.8 V, that corresponds the value for the constant voltage charging stage of the battery bank specified in the charge controller settings. When this voltage is reached, it stays constant for 2 hours (or 3.6 s in simulation), and the current decreases. It is known as the Boost stage. Also, in this stage the state of charge almost reaches 100%. After 2 hours, the voltage decreases until a voltage of 27.4 V, and the current is maintained close to 0 A. The charging of the BSSs simulated was compared with the voltage measured. The Fig. 6 shows the voltage curves for the BSS 1 where it is possible to verify the profile for charging method discussed. The error percentage associated with this case is 5%.

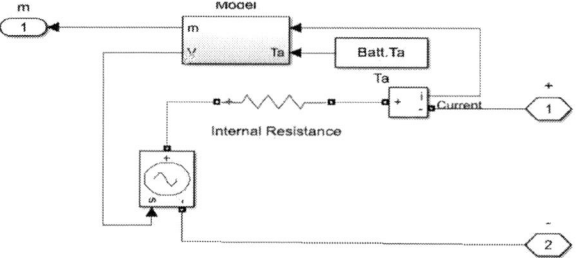

Fig. 5. Electric-mathematical model for the BSSs

As the BSSs are charged in the nanogrid, the current profile varies through the day. The MPPT algorithm defines the current profile during the Bulk stage. At the Boost and Float stages, the MPPT is disabled. It is abled again when the voltage reaches 26.4 V, and the BSS returns to Bulk.

The Table II presents the PSDEs of the BSSs for this validation. Applying the same argument used in the validation of the PV generator, it was understood that the model is a good approximation for the real system.

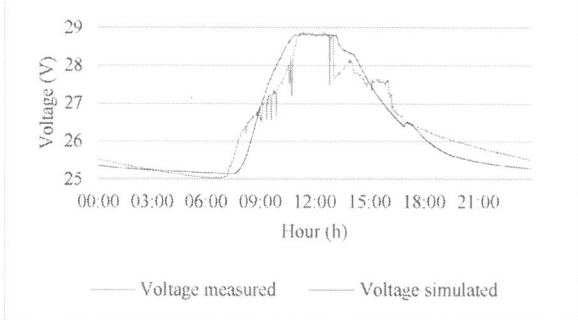

Fig. 6. Voltage simulated and measured curves of BSS 1.

TABLE II. PSDE FOR THE VOLTAGE IN BSS CONNECTION POINTS

PSDE	BSS 1	BSS 2	BSS 3
Voltage	1.35 %	1.19 %	1.07 %

E. Conductors

The distribution grid is formed by two conductors (+Vcc/-Vcc) type XLPE 0.6/1kV aluminium cable. According to the cable manufacturer [11] there is an electric resistance of 0.837 Ω/Km. So, the connections were considered resistors with a resistance calculated with this density.

F. Load banks

The loads used for testing the nanogrid operation were resistive lamps. The nominal grid voltage is 24 V and the lamp's nominal power is 40 W, the resistance calculated was 14.4 Ω. Then, the load bank is the association in parallel of four resistors with this value.

III. RESULTS AND DISCUSSION

The evaluation of the performance of the nanogrid simulated was conducted considering the environmental scenario of a cloudy day. It is because this is the most common in the Brazilian Amazon close to the equator line, where is placed the system studied. The irradiance and the environmental temperature data were collected from the real system. These data were necessary to supply the generation profile along the simulation and to submit the model to the same conditions of the real microgrid.

Fig. 7 shows the profiles of irradiance and environmental temperature for the cloudy day.

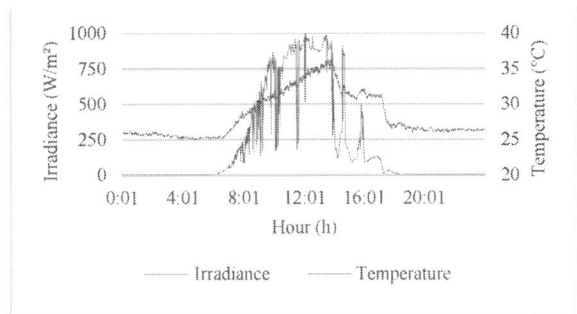

Fig. 7. Irradiance and environmental temperature profile.

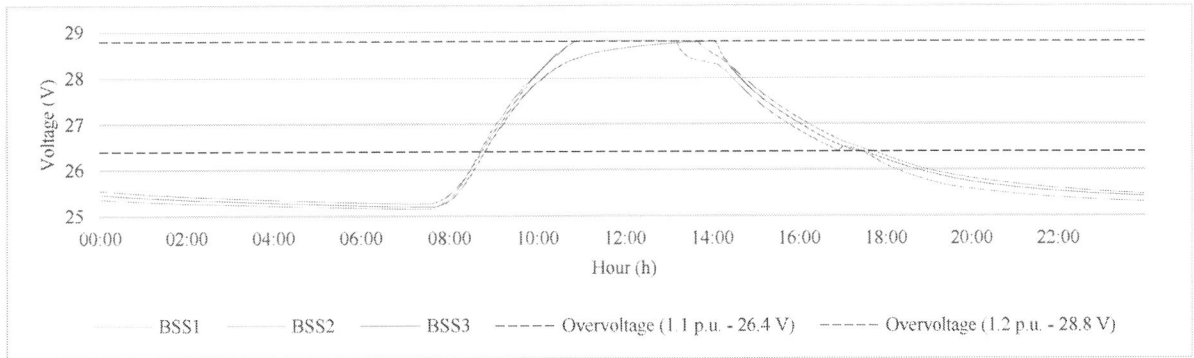

Fig. 8. Simulated voltage curves with the limits of overvoltage.

The Standalone Direct Current nanogrid was simulated without any problem during the time of 43.2 s, which is a day according to the model of the battery, that considered 1.8 s of simulation as one hour of the real system. Fig. 8 shows the simulated voltage curves at connection points of distributed storage systems and the limits of 1.1 p.u. (26.4 V) and 1.2 p.u. (28.8 V) for overvoltage.

The power quality at these connection points was evaluated. It was noticed that occurred moments of overvoltage in p.u. (per unit). With this, [12] defines that the overvoltage is an increase between 1.1 p.u. and 1.2 p.u. in the RMS reference voltage. This standard is AC electrical grid, but relating this definition to DC microgrids, the reference voltage of 24 V for this microgrid is equivalent to 1 p.u. For that reason, to better analyze this long-term event it was necessary to separate these data into two periods with and without PV generation as the voltage in the DC bus depends on the power delivered from the PV arrays. Table III shows the voltage magnitudes in p.u. in the connection points of the distributed storage systems during these two periods.

TABLE III. MAXIMUM VOLTAGE MAGNITUDE IN P.U.

Period	BSS1	BSS 2	BSS 3
With PV generation	1.2	1.2	1.15
Without PV generation	1.04	1.05	1.04

Voltage levels vary throughout the day, reaching lower values during the period without PV generation. It is noticed that the NDCC operated for approximately 9 hours and thirty minutes above 1.1 p.u. Overvoltage occurs during recharging of battery banks and peaks at maximum generation around noon. The occurrence of overvoltage can be characterized by excess generation in relation to the low energy consumption of the loads connected to the DCDN.

Similar results were obtained by [14], where the voltages in the points of connection of the battery banks in the nanogrid stay over 1 p.u. along the day and the lower voltages are present during the absence of PV generation. However, this work verified a maximum voltage of 1.05 p.u. when the PV generation is at its maximum and the battery banks are fully charged.

IV. CONCLUSION

The infrastructure of a standalone DC nanogrid and the modelling of its electronic devices are presented. The models of the PV generator are validated with the results of the I-V and P-V tests and the models of the battery storage system and the MPPT charge controller are both validated with the verification of the voltage levels during the three-stage charging method. All these devices are joined through a DC distribution grid modeled by the resistance equivalent of the conductors. The irradiance and the environmental temperature data of a typical day in the Amazon region give the profile of the generation and are used in the simulation of the nanogrid model. The voltages in different points of connection of the BSSs are obtained. On that account, the operational performance and the power quality are evaluated. Due to the good approximation of the voltage profile of the batteries of the modeled system to the real system, the nanogrid modeled can be flexibly adapted to meet and evaluate the operational performance of another standalone DC micro or nanogrids.

REFERENCES

[1] P. Torres, "Desenvolvimento e modelagem de uma nanorrede de distribuição em corrente contínua em baixa tensão com sistema de geração distribuída", dissertação de mestrado em Engenharia Elétrica, Universidade Federal do Pará, Belém, 2019.

[2] P. Torres, J. Filho, V. Júnior, L. Araújo, S. Williamson, M. Galhardo, J. Pinho, and W. Macêdo, "35th European Photovoltaic Solar Energy Conference and Exhibition," in *Load Flow Simulation of a Low-Voltage PV-Battery Based DC Microgrid to Supply Small Isolated Communities*, 2018, pp. 1636–1640.

[3] J. Filho, "Desenvolvimento de uma Batedeira de Açaí em Corrente Contínua e Monitoramento de sua Aplicação em uma Edificação Ribeirinha na Amazônia Suprida por Sistema Fotovoltaico Isolado", dissertação de mestrado, Universidade Federal do Pará (UFPa), Belém, 2021.

[4] S. Liptak, M. Miranbeigi, S. Kulkarni, R. Jinsiwale, and D. Divan, "Self-organizing NanoGrid (song)," *2019 IEEE Decentralized Energy Access Solutions Workshop (DEAS)*, 2019.

[5] S. Moussa, M. Ghorbal and I. Slama-Belkhodja, "Bus voltage level choice for standalone residential DC nanogrid", Sustainable Cities and Society, vol. 46, 2019.

[6] A. Fonseca, "Qualidade da Energia Elétrica em uma Nanorrede de Distribuição em Corrente Contínua", trabalho de conclusão de curso, Universidade Federal do Pará (UFPa), Belém, 2021.

[7] J. Duan, Z. Li, Y. Zhou and Z. Wei, "Study on the voltage level sequence of future urban DC distribution network in China: A Review", *Electrical Power and Energy System*, vol. 117, 2020.

[8] R. R. Kolluri, I. Mareels, and J. Hoog, "Controlling DC microgrids in communities, buildings and data centers," *IET Smart Grid*, vol. 3, no. 3, pp. 376–384, 2020.

[9] R. Tan, C. Er and S. Solanki, "Modelling of Photovoltaic MPPT Lead Acid Battery Charge Controller for Standalone System Applications", in *Int. Conf. Power, Energy Electrical Engineering*, 2020.

[10] D. Guasch and S. Silvestre, "Dynamic battery model for photovoltaic applications," *Progress in Photovoltaics: Research and Applications*, vol. 11, no. 3, pp. 193–206, 2003.

[11] ALUBAR. Aluminium Electrical Conductors. Alubar Metais e Cabos S.A. [Online] Available: https://alubar.net.br/img/site/arquivo/Cat_Tec_Alubar_Aluminio_2 015.pdf. Acessed: April 14, 2023.

[12] IEEE. IEEE Recommended Practice for Monitoring Electric Power Quality. Manhatam: IEEE Std 1159-2019, 2019.

[13] M. Young, The Technical Writer's Handbook. Mill Valley, CA: University Science, 1989.

[14] P. Torres, J. Filho, V. Junio, L. Araújo, S. Williamson, M. Galhardo, J. Pinho and W. Macêdo, " Load Flow Simulation of a Low-Voltage PV-Bateery Based DC Microgrid to Supply Small Isolated Communities", in 35th European Photovoltaic Solar Energy Conference and Exhibition, 2018.

Optimization of the electrical conductivity and thermal coefficient of temperature (TCR) on hydrogenated amorphous silicon-germanium films doped with nitrogen (a-SiGe:H,N) for applications on high performance infrared detectors

Oscar Velandia
Electronics department
Instituto Nacional de Astrofísica,
Óptica y Electrónica
Puebla, México
oscarveca@gmail.com

Mario Moreno
Electronics department
Instituto Nacional de Astrofísica,
Óptica y Electrónica
Puebla, México
mmoreno@inaoep.mx

Ricardo Zavala
Electronics department
Instituto Nacional de Astrofísica,
Óptica y Electrónica
Puebla, México
ricardojimzav@gmail.com

Alfredo Morales
Electronics department
Instituto Nacional de Astrofísica,
Óptica y Electrónica
Puebla, México
alfredom@inaoep.mx

Alfonso Torres
Electronics department
Instituto Nacional de Astrofísica,
Óptica y Electrónica
Puebla, México
atorres@inaoep.mx

Luis Hernández
Electronics department
Instituto Nacional de Astrofísica,
Óptica y Electrónica
Puebla, México
luish@inaoep.mx

Abstract — **Uncooled microbolometers are infrared detectors used for imaging applications in the long-wave IR spectra, made of semiconductor materials, such as vanadium oxide (VOx), boron doped amorphous silicon (a-Si:H,B) or amorphous silicon germanium (a-SiGe:H), the last two are usually deposited by plasma-enhanced chemical vapor deposition (PECVD). When infrared radiation falls on the thermos-sensor films, they experience changes in their electrical resistance, generating an electrical signal proportional to the amount of incident radiation. In this work, we optimized the absorption of infrared radiation in the long-wave IR range (8 – 14 um) with amorphous silicon germanium films doped with nitrogen (a-SiGe:H,N). Several flow rates of nitrogen (N$_2$) and different SiH$_4$ and GeH$_4$ flow rate ratios have been used for the deposition of two a-SiGe:H,N film series. We have characterized the films electrical and optically, and we have studied the electrical conductivity at room temperature (σ_{RT}), which is in the range of (1.2E-8 to 6.45E-2) (Ω.cm)$^{-1}$ and the Temperature Coefficient of Resistance (TCR), which is in the range (2.67 to 6.23) (%/K). Those σ_{RT} and TCR values of the a-SiGe:H,N films are very attractive for being used in high performance infrared detectors.**

Keywords— *Amorphous, silicon-germanium, conductivity, TCR, LWIR, PECVD, infrared, detector, microbolometer.*

I. INTRODUCTION

Uncooled microbolometers are composed of a thermos-sensing thin film, which detects infrared (IR) radiation in the long-wave IR (8 - 14 um), generally made of semiconductor materials, as vanadium oxide (VO$_X$), boron doped amorphous silicon (a-Si:H,B), or amorphous silicon germanium (a-SiGe:H). Those thermo-sensing films are suspended on a substrate insulated with a metallic reflective film by two legs, at a quarter-wavelength (2.5um) distance to improve absorbance, by forming a Fabry Perot cavity [1]. When infrared radiation falls on the thermos-sensor film, it is heated up and experience changes in its electrical resistance, which translates into an electrical signal proportional to the amount of incident radiation. A significant advantage of microbolometers is that they do not require active cooling like

more traditional infrared sensors, making them more compact, lightweight, and less expensive.

The requirements for high-performance micro-bolometers are a thermo-sensing material with high activation energy Ea, and consequently a high value of the temperature coefficient of resistance, TCR (α(T)), high conductivity room temperature, low noise, fast thermal response time, an excellent IR absorption [2] (silicon nitride, SiN$_x$ is usually used an absorber film at 12µm). As well it must be compatibility with standard integrated circuit (IC) based on silicon (Si) CMOS. The SiN$_x$ is an excellent IR absorber film at 12µm [3],[4].

Fabrication technology of IR cameras with uncooled microbolometers can be used in many applications such as medical, industrial maintenance, agricultural inspection, chemical threats, mineral exploration, and military [5].

The most common thermo-sensor material is vanadium oxide (VO$_x$) and amorphous semiconductors (a-Si:H,B, and a-SiGe:H), those materials present small volume, low-cost production, and non-necessity of cryogenic cooling [6]. Moreover, amorphous silicon-germanium (a-SiGe:H) is compatible with the Silicon CMOS technology and it can be deposited at a low temperature (< 200 °C) as thin film by the plasma enhanced chemical vapor deposition (PECVD) technique [7].

The activation energy value (Ea) of vanadium oxide (VO$_x$), is in the range of 0.2 - 0.28 eV, and the TCR is in the range of 2.7 - 3.4 %/K, given by the expression α(T)=(1/R)(dR/dT)=Ea/KT^2, where T is the temperature, K is the Boltzmann constant, and Ea is the activation energy. In addition, the σ_{RT} of VOx is 8.5 x 10^{-7} (Ω.cm)$^{-1}$ [8]. The thermal and electrical conductivity of the materials used in a microbolometer plays an important role in its performance and sensitivity. However, one of the main disadvantages of VO$_X$ is that it is not compatible with standard Si CMOS process.

Hydrogenated amorphous silicon (a-Si:H) is a standard semiconductor material in microelectronics. This material has a high Ea (0.8 - 1 eV) and a high TCR value (α(T) $\approx$10-

979-8-3503-1191-4/23 $31.00 © 2023 IEEE

12 %/K), with σ_{RT} of 1×10^{-9} $(\Omega.cm)^{-1}$. However, the low conductivity of a-Si:H results in a mismatch with the input impedance of the CMOS readout circuit (ROIC) [9]. To increase the low conductivity of a-Si:H films, boron doping has emerged as a solution but also showed decreased Ea (~0.22 eV) and TCR (2.8%/K) [10]. Additionally a-Si:H,B, it is not a very stable material and will undergo radiation degradation.

Hydrogenated amorphous silicon-germanium alloys, a-SiGe:H, have been studied to modify the optical and electrical properties, without doping by adjusting the Ge and Si contents in the alloy. Interesting values of Ea (0.36 eV), TCR (4.6%/K) and σ_{RT} (2×10-5 $(\Omega.cm)^{-1}$) have been reported [11].

Recently, polymorphous semiconductors have been developed for been used as thermo-sensig films. This material consists of an amorphous matrix with embedded small nanocrystals (size range 2 - 5 nm). These nanocrystals improve the structural quality of the films, reducing the stress and defect density, making them more stable to radiation. Ea of pm-Si:H is 0.69 eV, TCR is 8%/K, σ_{RT} is $7 \times 10^{-9}(\Omega.cm)^{-1}$; while Ea of pm-Ge:H is 0.39 eV, TCR is 5%/K, σ_{RT} is 2×10^{-4} $(\Omega.cm)^{-1}$ [12].

In this work we report in two parts, our study on the development of a-SiGe:H films doped with N_2 (a-SiGe:H,N), since the incorporation of N_2 into an amorphous semiconductor film can improve the IR absorption. Therefore in the first part we varied the N_2 flow rate in a wide range in order to improve the σ_{RT}, maintaining a high TCR value. In the second part, we studied the effect of the variation of the SiH_4 and GeH_4 flow rate ratios in order to improve even more the σ_{RT} and TCR, since our aim is to produce thermosensing films with high TCR, high σ_{RT} and high IR absorption.

II. EXPERIMENTAL DETAILS

A. Film series number 1.

The samples were fabricated in a low-frequency capacitively coupled PECVD reactor operating at 110 kHz, and the films were deposited at a low temperature (200 °C). SiH_4, GeF_4,and N_2 are used as gas sources, and H_2 is used as diluent. The gas purity of GeF_4 is 99.995%, that of SiH_4 is 99.995%, and that of H_2 is 99.999%. A variety of substrates were used, including Corning 1737 glass for the determination of optical properties, and Corning glass substrates with aluminum (Al) contacts for the determination of electrical properties, dark conductivity, and Ea. Crystalline silicon (c-Si) wafers were used for characterization by Fourier transform infrared spectroscopy (FTIR).

The following gas flow rates were used for thin film deposition, Q_{SiH_4} = 25 sccm, Q_{GeH_4} = 25 sccm, and Q_{H_2} = 1000 sccm, chamber pressure (P) was 1200 mTorr, HF - power is 300 W, and the deposition time is 60 minutes [14]. N_2 was also added to the gas mixture using eight different gas flow rates Q_{N_2} = 0, 20, 30, 40, 60, 80, 100 and 1200 sccm, with film thicknesses ranging from ~3000–3950 Å (see Table I).

The temperature dependence of the dark conductivity (σ_T) in the above samples was measured in the range of 300–400 K. Measurements are performed in a vacuum chamber with a pressure of 60 mTorr , the use of vacuum when measuring the conductivity of a material provides a controlled environment, free from external thermal interferences, maintaining a stable temperature, allowing for precise and reliable measurements. A model MMR Inc. K20 temperature controller and a model 6517A electrometer, Keithley Inst, were used for the temperature dependence of conductivity. These measurements allowed us to obtain the temperature dependence of (σ_T) and then determine Ea, TCR and room temperature (σ_{RT}) conductivity. Finally, the infrared absorption was analyzed using FTIR (Bruker Vector 22 spectrometer) and checked for hydrogen and nitrogen bonds in the film. For this analysis, thin films were deposited on c-Si substrates and measured with a resolution of 5 cm −1 in the 400–2500 cm-1 range.

Table I. Parameters for a-SiGe:H,N films deposition, where pressure was 1200 mTorr, RF power was 300 W and the deposition temperature was 200 °C.

| Film | Time (min) | Gas flow (sccm) | | | | thickness (nm) |
		SiH_4	GeH_4	H_2	N_2	
Series 1 - 1	60	25	25	1000	0	317.38
Series 1 - 2	60	25	25	1000	20	327.54
Series 1 - 3	60	25	25	1000	30	339.61
Series 1 - 4	60	25	25	1000	40	361.93
Series 1 - 5	60	25	25	1000	60	379.99
Series 1 - 6	60	25	25	1000	80	390.14
Series 1 - 7	60	25	25	1000	100	376.24
Series 1 - 8	60	25	25	1000	120	392.78

B. Film series number No. 2: dependence of SiH_4/GeH_4 flow rates.

The samples were same fabricated in a low-frequency capacitively coupled PECVD reactor operating at 110 kHz, and the films were deposited at a low temperature (200 °C). A variety of substrates were used, including Corning 1737 glass for the determination of optical properties, and Corning glass substrates with aluminum (Al) contacts for the determination of electrical properties, dark conductivity, and Ea. Crystalline silicon (c-Si) wafers were used for characterization by Fourier transform infrared spectroscopy (FTIR). The N_2 gas flow has been left constant in Q_{N_2} = 100 sccm, rates were used for thin film deposition, Q_{SiH_4} = 45, 40, 35, 30, 25, 20, 15, 10 and 5 sccm, Q_{GeH_4} = 5, 10, 15, 20, 25, 30, 35, 40 and 45 sccm, and Q_{H_2} = 1000 sccm, chamber pressure (P) was 1200 mTorr, HF - power is 300 W, and the deposition time is 70 minutes [14], with film thicknesses ranging from ~2500–4000 Å (see Table II). The characterization was the same as described for series number 1.

979-8-3503-1191-4/23 $31.00 © 2023 IEEE

Table II. Parameters for a-SiGe:H,N films deposition, where pressure was 1200 mTorr, RF power was 300 W and the deposition temperature was 200 °C.

| Film | Time (min) | Gas flow (sccm) | | | | thickness (nm) |
		SiH₄	GeH₄	H₂	N₂	
Series 2 - 1	60	45	5	1000	100	251.97
Series 2 - 2	60	40	10	1000	100	327.54
Series 2 - 3	60	35	15	1000	100	247.67
Series 2 - 4	60	30	20	1000	100	361.93
Series 2 - 5	60	25	25	1000	100	319.96
Series 2 - 6	60	20	30	1000	100	390.14
Series 2 - 7	60	15	35	1000	100	370.58
Series 2 - 8	60	10	40	1000	100	392.78
Series 2 - 9	60	5	45	1000	100	401.33

III. RESULTS

A. Film series no. 1: dependence of N_2 flow rate.

The conductivity temperature dependence can be well represented by $Ln(\sigma) = Ln(\sigma_0) - E_a/kT$, where σ_0 is a pre-factor, Ea is the activation energy, k is the Boltzmann constant and T is the temperature, the $TCR = E_a/kT^2$, and the room temperature conductivity (σ_{RT}) is calculated at T = 300 K. The TCR is an important feature when manufacturing a microbolometer as it determines the sensitivity of the device. It indicates how the resistance of the microbolometer changes in relation to temperature. A high TCR means that the resistance of the microbolometer changes more significantly with temperature variations, enabling better detection of infrared radiation. Table 2 shows those values of a-SiGe:H,N, films doped with different N₂ flow rates, of 0, 20, 30, 40, 60, 80, 100 and 120 sccm. As well these values can be seen in Figure 2, where is shown the Eₐ dependence with the N₂ flow rate used for dope the a-SiGe:H films, while figure 3 shows the TCR and σ_{RT} dependence of the films also with N₂ flow rate.

In figure 2 is observed that TCR and Eₐ decrease as the N₂ flow rate increases for the a-SiGe:H films deposition. With an N₂ flow of 60 sccm is observed the lower Eₐ value of 0.22 eV. However, the most interesting result is for the film deposited with an N₂ flow rate of 100 sccm, which has an Eₐ value of 0.23 eV, corresponding to a TCR of 3.0 %/K.

Figure 3 also is observed that the σ_{RT} significantly increases as the N₂ flow increases. the most interesting result is for the film deposited with an N₂ flow rate of 100 sccm, which has σ_{RT} the highest value of 6.73E-03 ($\Omega.cm^{-1}$).

The sample was characterized using an FTIR spectrometer in the range of 700 cm⁻¹ to 1300 cm⁻¹. The sample is exposed to infrared light, and the amount of light absorbed at different wavelengths is analyzed based on the characteristics of the chemical bonds presented. The absorbed light at each wavelength is detected and converted into an absorption spectrum. The obtained absorption spectrum displays characteristic peaks corresponding to the molecular vibrations of the chemical bonds presented in the sample.

For microbolometer termo-sensing a-SiGe:H,N films absorption in the L-W IR region, it is of interest the absorption coefficient in the wavenumber range of 700 - 1250 cm⁻¹.

Figure 4, it shows how much radiation is absorbed by the sample rather than being transmitted or reflected, based on the wavenumber of the incident radiation. It illustrates how the absorption capacity of the material changes when nitrogen dopant is introduced.

Doping is the process of adding controlled impurities to a material to alter its properties. In this case, nitrogen is being added to the amorphous film of silicon germanium, significantly impacting the structure and optical properties of the samples. It enhances the material ability to absorb radiation of interest for the microbolometer to some extent. In Figure 4, changes in absorbance are observed as nitrogen doping occurs, where the peaks compared to the undoped sample remain the same, indicating an alteration in the material optical properties. The chemical bonds and composition remain unchanged, indicating that the presence of doping improves the material absorbance.

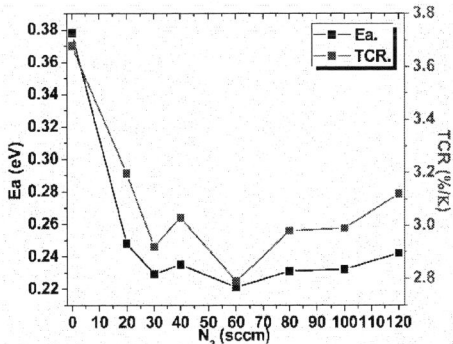

Figure 2. Activation energy, Eₐ, and Temperature Coefficient of Resistance, (TCR).

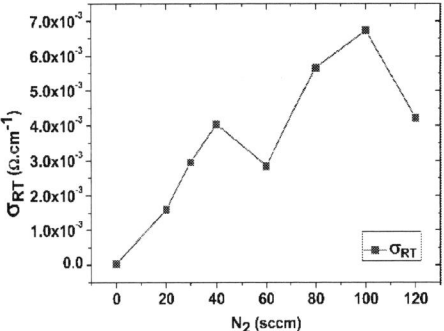

Figure 3. Room temperature conductivity (σ_{RT}).

The peaks could be related to the vibration of Si-N or Ge-N bonds. However, it is important to note that the precise interpretation of absorption at a specific frequency requires a more detailed analysis and depends on the exact composition and structure of the nitrogen-doped material. Considering other factors such as dopant concentration and interactions between different components of the material can also be helpful.

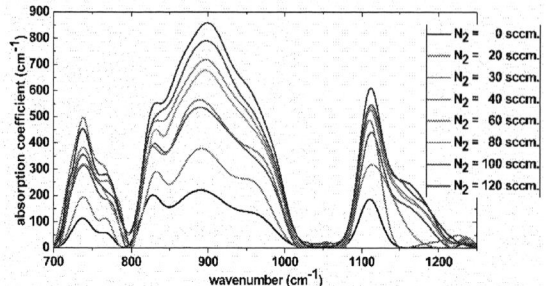

Figure 4. Absorption coefficient spectra of the a-SiGe:H,N films deposited with different N_2 flow rates.

Figure 4 shows the absorption coefficient spectra, corresponding to the spectra in the wavelength range of 8-14 μm. In those spectra is observed that the absorption coefficient increases as the N_2 flow rate increases as well, during the a-SiGe:H,N films deposition. The films deposited with N_2 flow rates of 100 sccm are the larger infrared absorption.

Therefore, coupling with the TCR and σ_{RT} results, the a-SiGe:H,N film deposited with an N_2 flow rate of 100 sccm, seems to be very suitable to be used in the second part of this study, with an improved IR absorption. (TCR of 3.0 eV and σ_{RT} of 6.73 E-3 $(\Omega\ cm)^{-1}$).

Table III. E_a, TCR, and the room temperature conductivity (σ_{RT}) values of the film series number 1.

Film	SiH$_4$ (sccm)	GeH$_4$ (sccm)	E_a (eV)	TCR (%/K)	σ_{RT} $(\Omega\ cm)^{-1}$
Series 2 - 1	45	5	0.48	6.23	1.20E-08
Series 2 - 2	40	10	0.36	4.65	7.56E-06
Series 2 - 3	35	15	0.26	3.35	9.96E-05
Series 2 - 4	30	20	0.24	3.15	1.47E-03
Series 2 - 5	25	25	0.22	2.84	5.40E-03
Series 2 - 6	20	30	0.21	2.70	1.16E-02
Series 2 - 7	15	35	0.2	2.61	3.26E-02
Series 2 - 8	10	40	0.17	2.23	3.24E-02
Series 2 - 9	5	45	0.21	2.67	6.45E-02

B. Film series no. 2: dependence of SiH$_4$/GeH$_4$ flow rates.

The conductivity temperature dependence can be well represented at the same form by $Ln\ (\sigma) = Ln\ (\sigma_0) - E_a/kT$, Table 4 shows Ea, TCR and (σ_{RT}) values of a-SiGe:H,N, films doped with different Q_{SiH_4} = 45, 40, 35, 30, 25, 20, 15, 10 and 5 sccm, Q_{GeH_4} = 5, 10, 15, 20, 25, 30, 35, 40 and 45 sccm and Q_{N_2} = 100 sccm. As well these values can be seen in Figure 5, where is shown the E_a and TCR dependence with the SiH$_4$ and GeH$_4$ flow rate used for develop the a-SiGe:H films, while figure 6 shows the σ_{RT} dependence of the films also with Q_{SiH_4} and Q_{GeH_4} flow rate.

In Figure 5 is observed the TCR and Ea, sample 4 is an intrinsic film Q_{SiH_4} = 25sccm, Q_{GeH_4} = 25sccm and Q_{N_2} = 100 sccm, the sample in the first part and the second part have the same flow rate but are different samples, the TCR = 3.0 %/K and Ea = 0.23 eV corresponds to the sample of the first part and the TCR = 3.15 %/K and Ea = 0.24 eV correspond to the sample of the second part. TCR and Ea

decreases when Q_{SiH_4} decrease and Q_{GeH_4} increases in the a-SiGe:H films deposition. With Q_{SiH_4} = 10sccm, Q_{GeH_4} = 40sccm and Q_{N_2} = 100 sccm is observed the lower E_a and TCR values of 0.17 eV and 2.67%/K. However, the most interesting result is the range of Ea from 0.48 eV to 0.17 eV and the variation of the TCR from 2.23%/K to 6.23%/K. Figure 6 also is observed that the σ_{RT} significantly increases when Q_{SiH_4} decrease and Q_{GeH_4} increases in the a-SiGe:H films deposition, showing higher values than those seen with the film intrinsic. The higher σ_{RT} is 6.45E-03 $(\Omega.cm^{-1})$.

Figure 7 shows the spectra of absorbance as a function of the wavenumber of a nitrogen-doped amorphous silicon germanium films, with a constant nitrogen flow rate of 100 sccm, while varying the silicon and germanium flow rates. Initially, a silicon flow rate of 45 sccm and a germanium flow rate of 5 sccm are used. Then, changes in the silicon and germanium flow rates are performed, as is shown in Table IV.

It can be observed that the absorbance improves when the silicon flow rate is 45 sccm and the germanium flow rate is 5 sccm, and it decreases as the silicon flow rate decreases and the germanium flow rate increases.

Table IV. E_a, TCR, and the room temperature conductivity (σ_{RT}) values of samples deposited with different Q_{SiH_4} and Q_{GeH_4} flow rates.

Film	N_2 flow Rate (sccm)	E_a (eV)	TCR (%/K)	σ_{RT} $(\Omega\ cm)^{-1}$
Series 1 - 1	0	0.38	3.68	3.68E-05
Series 1 - 2	20	0.25	3.20	1.60E-03
Series 1 - 3	30	0.23	2.92	2.95E-03
Series 1 - 4	40	0.24	3.03	4.04E-03
Series 1 - 5	60	0.22	2.79	2.85E-03
Series 1 - 6	80	0.23	2.98	5.67E-03
Series 1 - 7	100	0.23	3.00	6.73E-03
Series 1 - 8	120	0.24	3.12	4.22E-03

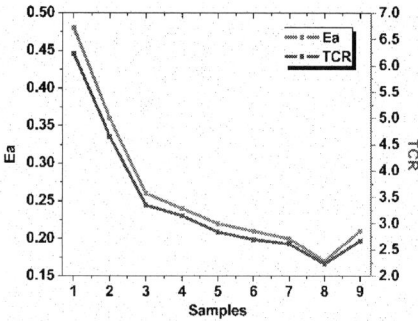

Figure 5. Activation energy, E_a and Temperature Coefficient of Resistance, (TCR) deposited with different Q_{SiH_4} and Q_{GeH_4} flow rates.

This suggests that there is a relationship between the silicon and germanium composition in the film and its ability to absorb radiation. The specific combination of 45 sccm of silicon and 5 sccm of germanium appears to enhance the absorbance, while changing the ratio towards 5 sccm of silicon and 45 sccm of germanium decreases the absorbance. Additionally, the gas flow rates used during the film deposition processes also play an important role in shaping the final optical properties of the material.

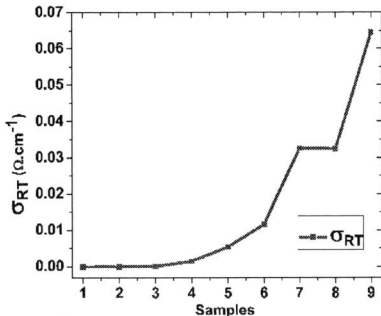

Figure 6. Room temperature conductivity (σ_{RT}) deposited with different $Q_{SiH_4} - Q_{GeH_4}$ flow rates.

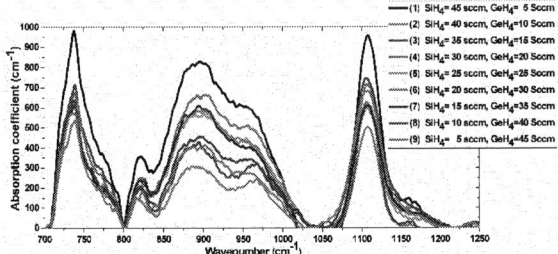

Figure 7. Absorption coefficient spectra of the a-SiGe:H,N films deposited with different Q_{SiH_4} and Q_{GeH_4} flow rates.

IV. CONCLUSIONS

Two series of hydrogenated amorphous silicon-germanium doped with nitrogen (a-SiGe:H,N) thin films have been deposited by LF-PECVD at 200 °C. In the first seriesthe N_2 flow rate has been varied from 0 to 120 sccm, keeping the SiH_4, GeH_4 and H_2 flow rates constant at 25 sccm, 25 sccm, and 1000 sccm, respectively. In the sencond series, the N_2 flow rate was maintained constant to 100 sccm, which produced the film with the best IR absorption and the SiH_4 and the GeH_4 flow rates were varied from 5 to 45 sccm and 45 to 5 sccm respectively.

The objective for the incorporation of N_2 in the gas mixture was to dope the films, in order to improve the LW-IR absorption in the range of 8 - 14 μm, while the objective fto vary the SiH_4 and GeH_4 flow rates was to improve the σ_{RT}. Also we have observed that the Ea and TCR were afected, leaving a range of possibilities to use this type of film as a faithful candidate for a thermos-sensing film for the fabrication of high performance uncooled microbolometers.

Moreover, one of the films presented an improved σ_{RT}, which is 2 orders of magnitude larger than that of the intrinsic film. The above result is very important because meets the requirements for the input impedance of the CMOS readout integrated circuit (ROIC).

Having a variation in activation energy from 0.21 eV to 0.48 eV is important in the fabrication of microbolometers because it allows for tailoring the material properties to different applications. A larger activation energy means that the

material can more easily respond to temperature changes, resulting in higher thermal sensitivity and improved capability to detect small temperature variations.

A variation in TCR (temperature coefficient of resistance) from 2.67 %/K to 6.23 %/K implies that the resistance of the microbolometer have a larger change with temperature. A higher TCR enables more precise detection of temperature changes, which is essential for obtaining high-quality thermal images.

The conductivity at room temperature, is ranging from 1.2×10^{-8} $(\Omega cm)^{-1}$ to 6.4×10^{-2} $(\Omega cm)^{-1}$, which is important for ensuring optimal performance of the microbolometers. Excessively low conductivity values can lead to a mismatch with the input impedance of the ROIC.

In conclusion, the variation in activation energy, TCR, and conductivity at room temperature are crucial factors in the fabrication of microbolometers. These parameters influence thermal sensitivity, precision, response time, and heat dissipation capability of the device, directly impacting its performance and thermal imaging quality. Therefore, having an optimal range in these parameters is essential for achieving efficient and reliable microbolometers.

V. REFERENCES

[1] A. Abdullah, A. Koppula, O. Alkorjia, y M. Almasri, «Uncooled two-microbolometer stack for long wavelength infrared detection», *Sci. Rep.*, vol. 13, n.º 1, Art. n.º 1, mar. 2023, doi: 10.1038/s41598-023-30328-1.
[2] M. Moreno, R. Ambrosio, A. Torres, A. Kosarev, M. García, y J. Mireles, «Measurements of thermal characteristics in silicon germanium un-cooled micro-bolometers», *Phys. Status Solidi C*, vol. 7, n.º 3-4, pp. 1172-1175, 2010, doi: 10.1002/pssc.200982739.
[3] L. V. Mercaldo, P. D. Veneri, E. Esposito, E. Massera, I. Usatii, y C. Privato, «PECVD in-situ growth of silicon quantum dots in silicon nitride from silane and nitrogen», *Mater. Sci. Eng. B*, vol. 159-160, pp. 77-79, mar. 2009, doi: 10.1016/j.mseb.2008.09.029.
[4] T. Drüsedau, «The role of nitrogen in sputtered a-Si:H, a-Ge:H and a-Si56Ge44:N:H», *J. Non-Cryst. Solids*, vol. 137-138, pp. 821-824, ene. 1991, doi: 10.1016/S0022-3093(05)80246-7.
[5] C. Marshall, T. Parker, y T. White, «Infrared sensor technology», en *Proceedings of 17th International Conference of the Engineering in Medicine and Biology Society*, sep. 1995, pp. 1715-1716 vol.2. doi: 10.1109/IEMBS.1995.579906.
[6] L. Yu, Y. Guo, H. Zhu, M. Luo, P. Han, y X. Ji, «Low-Cost Microbolometer Type Infrared Detectors», *Micromachines*, vol. 11, n.º 9, Art. n.º 9, sep. 2020, doi: 10.3390/mi11090800.
[7] C. Ascencio-Hurtado, A. Torres, M. Moreno, y R. Ambrosio, «High conductivity intrinsic a-SiGe films deposited at low-temperature», en *2021 IEEE Latin America Electron Devices Conference (LAEDC)*, abr. 2021, pp. 1-4. doi: 10.1109/LAEDC51812.2021.9437924.
[8] V. Y. Zerov, V. G. Malyarov, I. A. Khrebtov, Y. V. Kulikov, I. I. Shaganov, y A. D. Smirnov, «Uncooled membrane-type linear microbolometer array based on a VOx film», *J. Opt. Technol.*, vol. 68, n.º 6, p. 428, jun. 2001, doi: 10.1364/JOT.68.000428.
[9] R. Jimenez *et al.*, «Fabrication of Microbolometer Arrays Based on Polymorphous Silicon–Germanium», *Sensors*, vol. 20, n.º 9, Art. n.º 9, ene. 2020, doi: 10.3390/s20092716.
[10] A. J. Syllaios, T. R. Schimert, R. W. Gooch, W. L. McCardel, B. A. Ritchey, y J. H. Tregilgas, «Amorphous Silicon Microbolometer Technology», *MRS Online Proc. Libr. OPL*, vol. 609, ed 2000, doi: 10.1557/PROC-609-A14.4.
[11] A. Torres, M. Moreno, A. Kosarev, y A. Heredia, «Thermo-sensing silicon–germanium–boron films prepared by plasma for un-cooled micro-bolometers», *J. Non-Cryst. Solids*, vol. 354, n.º 19-25, pp. 2556-2560, may 2008, doi: 10.1016/j.jnoncrysol.2007.09.112.
[12] M. Moreno *et al.*, «Boron doping compensation of hydrogenated amorphous and polymorphous germanium thin films for infrared detection applications», *Thin Solid Films*, vol. 548, pp. 533-538, dic. 2013, doi: 10.1016/j.tsf.2013.08.102.

2023 IEEE Latin American Electron Devices Conference (LAEDC)
Puebla, México, July 3-5, 2023

A proposed STEM program to make institutions more inclusive for people with visual and physical disabilities in Panama

1st Victoria Serrano
Research Group on Emerging Computing Technologies
Universidad Tecnologica de Panama
David, Panama
victoria.serrano@utp.ac.pa

2nd Vladimir Villarreal
Research Group on Emerging Computing Technologies
Universidad Tecnologica de Panama
David, Panama
vladimir.villarreal@utp.ac.pa

3rd Lilia Muñoz
Research Group on Emerging Computing Technologies
Universidad Tecnologica de Panama
David, Panamá
lilia.munoz@utp.ac.pa

4th Konstantinos Tsakalis
Ira A. Fulton Schools of Engineering
Arizona State University
Arizona, USA
tsakalis@asu.edu

Abstract—**STEM programs have attracted popularity among educational institutions in the recent years. Not only because of their interest in seeking new strategies to teach Science, Technology, Engineering and Mathematics, but also because there is constant demand of professionals in those areas. In this paper, a STEM program is proposed to involve the participation of university and high school students and professors, a non profit organization with people with disabilities, and community service to solve the most basic problems that disable people face in educational institutions in Panama.**

Index Terms—**STEM programs, community service, disabilities**

I. INTRODUCTION

According to the 2010 census in Panama, people with reduced mobility accounted for 30.1% of all disabilities in the country [1]. That is the person who, because of an accident, illness, stroke or amputation, has difficulty performing activities of daily living with skill or moving independently. Also, those who have malformations that limit them physically or those who were born with a lack of a limb. Additionally, in Panama 87,000 people have a condition of visual impairment, ranging from individuals with congenital disabilities, others with total disabilities or with low vision, which represents 11.3% of the total population with disabilities.

In addition, most of the educational institutions in Panama lack of inclusive spaces where people with physical and visual impairments feel welcomed. That is also the reality at the Technological University of Panama in Chiriqui where there is not automated doors at all.Therefore, this project seeks to focus on the Panamanian population that has visual disabilities and/or physical disabilities to provide them with tools that

allow them to attend educational institutions that have almost non-existent spaces for inclusion.

This project is going to be developed by having a co-ordinated program where high school students will work together with university students a professors to find the most suitable solutions to the most basic problems that people with disabilities face. To be able to do that, these group of students and professors will be interviewing and having the cooperation of a non-profit organization. They are comprised of a group people with different disabilities, so that students will have a first-hand information of how convenient are the solutions proposed. High school students will be attending the university to learn about Science, Technology, Engineering and Mathematics (STEM), so that these concepts may be used to propose the possible solution to problems faced by people with disabilities.

II. RESEARCH BACKGROUND

As seen throughout Latin America, it has been very difficult to motivate students and guide them towards engineering careers. In an effort to meet the specific STEM needs of high school students in Panama, Parts I and II of the STEM Beyond Borders program were proposed in 2015 and 2016, respectively [2]. This program impacted fifteen 12th grade students and twenty 11th grade students (over 50% female) over a six-week period during the second semesters of 2016 and 2017. Participating schools were from David district , in the province of Chiriqui, Panama: Felix Olivares Contreras High School, San Agustín High School, San Francisco de Asis High School, Beatriz Miranda de Cabal High School and Bilingual Adventist High School. These projects motivated high school students to enter engineering careers and

979-8-3503-1191-4/23 $31.00 © 2023 IEEE

some of them have already graduated from the Technological University of Panama.

Other initiatives have also emerged to spark interest in STEM areas. That is the case of the use of small robots to teach mathematical concepts to preschool and first grades students [3] or the application of augmented reality in teaching-learning natural sciences [4].

Additionally, the use of Arduino-based cars and mechanical flapping birds have been used to educate youth from 7th to 9th grades in Arizona [5].

Although the previous work has been focused in promoting STEM areas among primary and secondary school, there has been also an interest in using similar tools to teach concepts at the university. In that sense, a proposal was made using Lego Mindstorms to teach robotic concepts [6] and multivariable controller design [7] through the use of a lego robotic arm.

III. PROBLEM STATEMENT

In an effort to strengthen the inclusive education of people with visual disabilities and physical disabilities, the project "Closing the gap in engineering education for people with disabilities" has been proposed. This project seeks to provide an environment in which people with visual disabilities feel welcomed in the educational institution in Panama by having spaces where the location signs of the different classrooms and service places have Braille language, as well as points of auditory information. None of them are actually existent at the Technological University of Panama in Chiriqui.

Additionally, as part of the inclusion plan, the program seeks to provide spaces where an educational institution has automated doors to access classrooms and corridors for a wheelchair to go through. In addition, to that a custom wheelchair is going to be designed and built according to the disable person limbs' measures.

Although wheelchairs are already available on the market, the most economical ones are generic which are not suitable for all the people with physical disabilities since they do not face all the same problem and/or have the same limbs' measures. Moreover, the customed ones are able to provide a better postural support, but they are usually most expensive that not many people may afford.

A. Objectives

The objectives of this project are:

- To contribute with education in Science, Technology, Engineering and Mathematics among university and high school students through community service.
- To generate ideas that stimulates the autonomy and individual development of people with moderate to severe visual impairment and/or physical disabilities.
- To contribute to the development of quality of life in a comprehensive and inclusive manner for people with visual disabilities and/or physical disabilities.
- To create a more inclusive environment in educational institutions by generating Braille signs using 3D printing.
- To develop auditory information points with low-cost equipment.

IV. METHODOLOGY

This project will focus on the Panamanian population that has visual disabilities and/or physical disabilities, especially in the Chiriqui region. All this work will be carried out through an after-school program with university students together with students from four different schools in the province of Chiriquí, Panama: Beatriz Miranda de Cabal High School, San Francisco de Asis High School, Cambridge Bilingual School and San Agustin High School. High school students will be attending university to learn how to design, manufacture, and assemble both electronic and mechanical parts; as well as to learn about programming to facilitate inclusion for people with disabilities at educational institutions. A total of six students from each high school will attend the weekly sessions starting from June through November, 2023 for a total of twenty four students (50% female). Additionally, a total of fifteen university students will be involved in the development of the program. This will mainly be students from the Electrical Engineering, Mechanical Engineering and Computer Science Deparments.

Continuing insight will be sought through regular online meetings with professors at Arizona State University to improve the proposed solutions for the problems that the program is trying to solve. The results of this project will be presented at international conferences and indexed journals, as well as through local media and community outreach sessions.

A. Resources needed

To be able to develop the STEM program, the following resources are necessary:

1) Personnel: twenty four high school students and fifteen university students.
2) Materials: for the construction of a personalized wheelchair and to automate the door access. Also, 3D filament+resin and support are necessary to make haptic maps and Braille signs. Additionally, Arduino boards and circuit devices are needed for electronics. Also, a CNC milling machine will be purchased to create PCB's. T-shirts for students and professors involved in the project are going to be designed and distributed to use them in all the sessions of the STEM program. A banner will be displayed at each of the meetings and promotion of the program. Additionally, office supplies will be purchased and distributed among the participants.
3) Arduino software will be installed to program the boards to provide auditory information.
4) FREECAD or TinkerCAD will be used to teach and perform 3D design.
5) Promotional items will be acquired to invite local authorities to let them know about the project.
6) Results presentation will be performed at the end of the program to present the insights about the project.

B. Project Activities

The project will have a duration of twelve months distributed as follows:

979-8-3503-1191-4/23 $31.00 © 2023 IEEE

1) **April-May, 2023: Application to the Bioethical committee.** Since this project involves the participation of human beings, application to the bioethical coommitee is necessary to understand the impact of each of the activities and how people with disabilities are not put at risk. It is expected to get the application approved at the end of May.

2) **June, 2023: Understanding Braille language and first report.** Sessions to learn and understand Braille language are going to take place in this month, so that the haptic map may be built using Braille to provide instructions for people with visual disabilities. First report will be delivered.

3) **July, 2023:** 3D design. Free software will be used such as: FreeCAD or TinkerCAD to create signs with Braille language.

4) **August, 2023: 3D printing of Braille signs and haptic maps.** A scaled haptic map will be created using Braille signs to define where the location of each building is?

5) **September, 2023: programming the Arduino board and building PCB's to provide auditory information for people with visual disabilities and second report.** Arduino-based points of information will be developed using sound for people with visual disabilities. A second report delivered.

6) **October, 2023: automation of door access for wheelchairs.** One or two doors will be selected to automate them so that a wheelchair may go through. For that, either sensors and/or buttons will be used together with a motor and a special mechanism to open and close hinged doors.

7) **November-December, 2023: designing and building a personalized wheelchair and third report.** A personalized wheelchair will be designed and built according to the person's limbs. A third report delivered.

8) **January, 2024: Results presentation** Final presentation will take place to share the results with the local community.

9) **February, 2024: Final results publication** Final results will be shared through a journal and/or international conference.

10) **March, 2024: Final Report.** A final report will be delivered.

The main activities of the STEM program may be summarized as shown in Fig.

V. SUSTAINABILITY PLAN

To ensure a sustainable plan, part of the materials used in this program will be donated to the Universidad Tecnologica Panama in David, Chiriqui and participating schools. This will allow local groups to repeat and develop new projects while they enhance the program to satisfy the needs of people with disabilities.

The final results of this project will benefit directly members of the NGO who are the core of the development of this project. Additionally, we will keep communication with

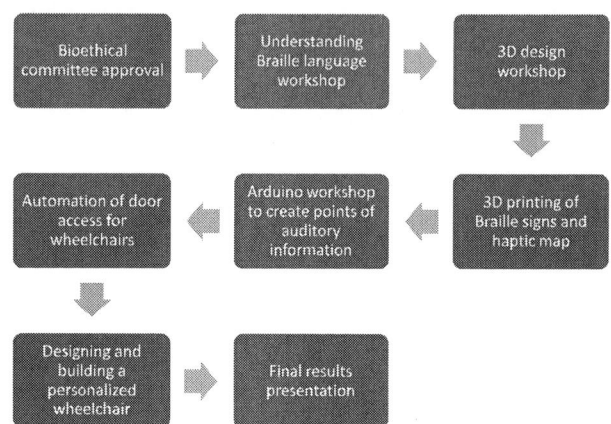

Fig. 1. Activities flowchart of the STEM program proposed.

the NGO and participating schools via email, phone, in-person meetings and video-calls to provide insights about the methodology and new requirements for new needs. This will enhance the experience in all institutions and will allow the future collaboration between all communities taking part in this program. Future efforts will seek funding from the National Secretariat of Science, Technology and Innovation (SENACYT) to help involve more student and help sustain the program in Panama.

VI. PROJECT IMPACT

Since this program involve the participation of human beings, there will be an application to the Bioethical committee at the Technological University of Panama. A total of twenty four high school students (9th-12th grades) and fifteen university students will benefit by developing new tools and methodologies to solve problems for people with disabilities, enabling Arizona State University and the Technological University of Panama to establish continuing international collaboration as shown in Fig. 2.

A broader impact will be achieved by helping an estimated of 300 people with visual and/or physical disabilities in the Chiriqui province wich is the second one in the country affected by those conditions.

Therefore, this project will provide a most efficient method to improve the quality of education in developing countries. This will bring the possibility to develop new applications that seeks to improve the learning process of people with disabilities as well as making them more independent.

VII. THREE MAIN COMPONENTS OF THE STEM PROGRAM

To achieve the goal, there are three main components of this STEM program: university and high school students, the non-profit organization and community service as shown in Fig. 3.

A. Non-profit organization

The Association of People with Disabilities and Problems of Locomotion of Chiriqui (APEDIPROLOCHI) is constituted

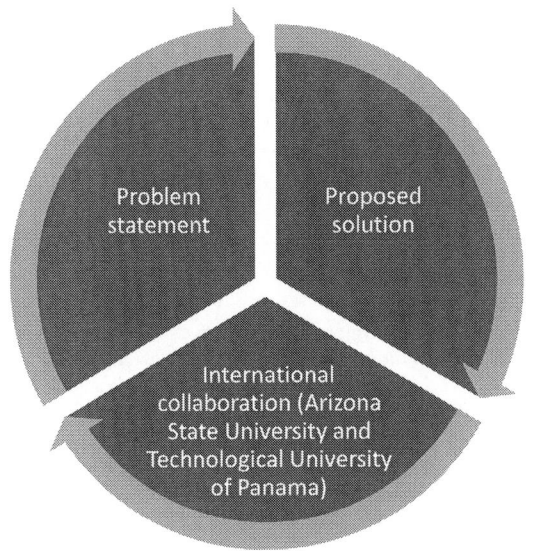

Fig. 2. Continuing international collaboration the STEM program proposed.

Fig. 3. Three main components of the STEM program proposed.

as a civic, non-political, non-profit and non-religious organization with legal status constituted on April 14th, 1996. The Association groups people with different disabilities and family members; achieving with hard work, fighting to become visible and have a space in society, which has grown from the union and personal movement. The goal of the association is to make social inclusion effective, continue in the fight against discrimination, active inclusion and improve employability, improve the employment of people with disability in a situation or risk of social exclusion, through social activation or entrepreneurship. The purpose is to promote the creation

of self-employment, promote their skills, the entrepreneurial spirit through a viable, close and sustainable project.

B. University and High school collaborators

An after-school program will be conducted with university students together with students from four different schools in the province of Chiriquí, Panama: Beatriz Miranda de Cabal High School, San Francisco de Asis High School, Cambridge Bilingual School and San Agustin High School. Twenty four high school students will be working with a team of university students who will be developing workshops about 3D design, programming, mechanical design, budget and circuit analysis. Each month a different topic will be covered and a challenge will be given to students to brainstorm about the best solution to issues for people with disabilities. The sessions will start in April after regular classes have started in all high schools and universities and is expected to finish in November with all the solutions implemented.

C. Community service plan

Since 2018 all university students at the Technological University of Panama will have to complete a total of 120 hours of social work. This time may be divided as follows: 40 hours may be completed by having any work that help their community and 80 hours should be closely related to the program that they are studying at the university. Therefore, this project will provide another opportunity for them to contribute with the community as well as fullfill their graduation requirement. Although students from any department may participate in the STEM program, this may directly benefit students from the Mechanical, Electrical and Computer Science department

ACKNOWLEDGMENT

The authors would like to thank EPICS in IEEE for their financial support in the development of this STEM program. Vladimir Villarreal and Lilia Muñoz are members of SNI in SENACYT.

REFERENCES

[1] D. Muñoz Rivera, "Cifra actualizada de personas con discapacidad se desconoce en Panamá," 2020.
[2] V. Serrano and M. Thompson, "Aprendizaje Tecnológico," *El Tecnológico*, vol. 17, no. 1, pp. 9–10, 2017.
[3] L. Muñoz, V. Villarreal, I. Morales, J. Gonzalez, and M. Nielsen, "Developing an interactive environment through the teaching of mathematics with small robots," *Sensors (Switzerland)*, vol. 20, no. 7, pp. 1–23, 2020.
[4] L. Muñoz, R. Montenegro, and B. Aparicio, "Uso de la Realidad Aumentada en la enseñanza-aprendizaje de ciencias naturales Use of Augmented Reality in teaching-learning natural sciences," in *4to Congreso Internacional AmITIC 2017*, (Popayán, Colombia), pp. 4–9, 2017.
[5] M. Thompson, V. Serrano, D. Ixtabalan, V. Garcia, A. Godinez, A. Rodriguez, and K. Tskalis, "Building a Mechanical Flapping Bird and Arduino Robotic Cars for Educating Youths in 7th, 8th and 9th Graders at Arizona State University," in *Proceedings of 14th Annual Hawaii International Conference on Education*, (Honolulu, HI), pp. 2203–2220, 2016.
[6] M. Thompson, V. M. Serrano R., J. E. Willem Noriega, and V. Martinez, "Learning Robotic Concepts with a 3R Lego NXT Robotic Arm," *Journal of Automation and Control Research*, vol. 1, pp. 38–44, 2014.
[7] V. Serrano, M. Thompson, and K. Tsakalis, "Learning Multivariable Controller Design a Hands-on Approach with a Lego Robotic Arm," *Advances in Automation and Robotics Research in Latin America, Lecture Notes in Networks and Systems*, vol. 13, pp. 271–278, 2017.

2023 IEEE Latin American Electron Devices Conference (LAEDC)
Puebla, México, July 3-5, 2023

Development of Drop-On-Demand Inkjet Process for the Fabrication of Thin-Film Printed Devices

Salvador Ivan Garduño*
Institute of Basic Sciences and Engineering
CONACyT – Autonomous University of Hidalgo State
Pachuca, Hidalgo, Mexico
orcid.org/0000-0003-0101-6285

Angel Sacramento-Orduño
Section of Solid-State Electronics
Electric Engineering Department
Center of Research and Advanced Studies – IPN
CDMX, Mexico
orcid.org/0000-0001-5207-3154

Magaly Ramírez-Como
Interdisciplinary Professional Unit in Engineering and Advanced Technologies
National Polytechnic Institute
CDMX, Mexico
orcid.org/0000-0002-9313-8337

María Isabel Reyes-Valderrama
Institute of Basic Sciences and Engineering
Autonomous University of Hidalgo State
Pachuca, Hidalgo, Mexico
orcid.org/0000-0002-1912-7998

Ventura Rodríguez-Lugo
Institute of Basic Sciences and Engineering
Autonomous University of Hidalgo State
Pachuca, Hidalgo, Mexico
orcid.org/0000-0001-8767-032X

Magali Estrada
Section of Solid-State Electronics
Electric Engineering Department
Center of Research and Advanced Technologies
CDMX, Mexico
orcid.org/0000-0002-6244-6492

Abstract— The main benefit of inkjet printing (IJP) lies in its ability to eliminate the photolithography and etching processes that are fundamental to silicon manufacturing technology. This is due to the direct deposition of solution-processed materials in pre-defined patterns. Although IJP has been employed in the production of various thin-film devices (TFD), there are still significant challenges in achieving fully printed TFDs. These include adjusting the rheological properties of formulated inks with drop-on-demand (DoD) printers and developing a comprehensive understanding of the parameters and conditions required for the IJP process to yield a desired morphology, film thickness, and pattern area uniformity. This investigation proposes to develop a process for the IJP of solution-processed materials, considering not only the ink rheological properties but also the parameters and setup conditions of the DoD printer employed. Commercial inks with specific rheological properties for IJP are used to print materials patterns with different resolutions, under continuous ejection conditions. The morphology and film thickness variations of the printed material are characterized by employing atomic force microscopy, ellipsometry, and optical microscopy. The obtained results depend on the rheological properties and solid content in the ink, as well as the configured parameters of the DoD printer. These findings can provide valuable evidence for attaining optimal conditions to enhance the surface wettability and develop the stacking of printed patterns of different materials. The focus of this work is on the fabrication of TFDs through a complete IJP process and improving their performance.

Keywords—Inkjet printing, Drop-On-Demand, Printability, Morphology, Film thickness, Thin-film electronic devices.

I. INTRODUCTION

Numerous studies have devoted significant attention to inkjet printing (IJP) owing to its advantages for producing low-cost thin-film devices (TFD) [1]. These benefits include the ca-

pacity for the direct deposition of materials in patterns, which circumvents the need for photolithography and etching, processes employed in the manufacture of integrated circuits using silicon technology. The advantages of the IJP technique render it compatible with the development of flexible and transparent electronics; moreover, it is an environmentally sustainable technique that minimizes materials waste. Nonetheless, several pertinent challenges persist in the fabrication of fully printed TFDs [2]. The most noteworthy is the detailed description of the IJP process and the variations yielded in the morphology, film thickness, pattern area uniformity and adherence. Generally, such variations are not reported and would allow the enhancement of device performance. Zinc oxide (ZnO) has established its position as a crucial and versatile metal oxide semiconductor, particularly for the development of applications based on solution processed materials and printing fabrication techniques. This can be attributed to the different advantages that ZnO offers, such as the capacity to obtain a crystalline structure even at low deposition temperatures [3]. In contrast to ZnO, the utilization of Indium-Gallium-Zinc Oxide (IGZO) as an active layer in thin-film transistors (TFTs) has demonstrated superior field-effect mobility and polarization/illumination stress stability. However, the synthesis of IGZO in solution can be very intricate and requires a post-annealing process >300°C [4]. Aluminum-doped ZnO (Al:ZnO) has been considered as an alternative for tin-doped indium oxide as a transparent conductive oxide. Al:ZnO has garnered notable interest for its non-toxic properties and it can exhibit a resistivity of 25.4 mΩ·cm [5]. The impact of the annealing temperature and atmosphere on the electrical resistivity and morphology of Al:ZnO printed patterns has been reported; however, any potential modifications in pattern characteristics resulting from the conditions and parameters of the IJP process have not been taken into account [5]. There have been reports on the utilization of ink containing dispersed silver (Ag) nano-

---

This work was supported by CONAHCYT program *Investigadoras e Investigadores por México*, and the postdoctoral grants BP-PA-20220711103224 986-2580943 and BP-PA-20220624083033039-2364083.

979-8-3503-1191-4/23 $31.00 © 2023 IEEE

particles (nPs) as contact electrodes for TFD and interconnection lines between circuits [6]. This has led to the IJP being used for the direct definition of source and drain contacts in the fabrication of IGZO-TFT, presenting a cost-effective option for electrode formation. However, the IJP optimization is required to accurately define both the channel geometry and contact resistances [6]. To assess the suitability of an ink for use in DoD printers and its behavior upon contact with a solid surface, several dimensionless numbers must be calculated [7, 8]. These include Reynolds ($Re = v\rho a/\eta$), Weber ($We = v^2\rho a/\gamma$), Ohnersorge ($Oh = (We)^{1/2}/Re$) and $Z = 1/Oh$. These numbers are determined on the ink's density (ρ), viscosity (η), and surface tension (γ), as well as the velocity (v) at which it is ejected from a nozzle with a defined dimension (a). At present, there are few solution-processed materials with appropriate rheological properties for IJP. Furthermore, reported works that detail the fabrication of TFDs, partially using IJP, fail to explore the printing conditions of materials and their impact on performance. This study aims to address this gap by offering a comprehensive account of the conditions and parameters of the IJP process for three commercial inks. The study underscores the significance of proper configuration to ensure a continuous, uniform, and repeatable ejection of ink drops, which in turn enables control over the morphology and film thickness of printed materials.

II. EXPERIMENTAL PROCEDURE

The inks used were purchased from Nanograde Ltd. and Genes Ink. They consist of dispersed solid content in the form of nPs in a solvent mixture of alcohols. The average size of these nPs ranges from 8 nm to 80 nm. The inks have suitable rheological properties for Dimatix DoD inkjet printing (DMP-2800). Before filling the cartridges of 1 pL (ZnO and Al:ZnO) and 10 pL (Ag), the inks were ultrasonicated for 10 min to break up agglomerates [9]. The inks were filtered through polytetrafluoroethylene membranes to prevent large aggregates [10]. Square patterns of 5×5 mm² were printed on silicon substrates for mechanical support, which were cleaned with an RCA standard process for optimal surface wetting. It is feasible to increase film thickness through pattern stacking, however, for this research, each pattern only was printed once to evaluate the impact of drop spacing (DS) on film thickness. The IJP process was performed continuously and uniformly without cleaning cycles to avoid compromising pattern homogeneity due to clogging mechanisms [10]. By varying the DS from 20 to 100 μm, diverse material patterns resolutions were obtained. The patterns were an-

nealed at 120±30 °C during 27.5±17.5 min to sinter the nPs and evaporate the solvent. For the selected DS, the thickness was measured using a variable angle ellipsometer (L116S300) at two wavelengths (λ). Atomic force microscopy (AFM) was used to examine printed materials in semi-contact mode, at a scanning frequency between 0.5 and 1 Hz on a molecular viewer (Pico+ SPM II). Micrographs were acquired in ambient air, using standard silicon probes with an elasticity constant of 1-5 nN/m and a resonant frequency of 75 kHz.

III. ANALYSIS AND DISCUSSION OF THE OBTAINED RESULTS

Table I summarizes the rheological properties and dimensionless numbers of the used inks. The dotted lines in Fig. 1(a) are physical criteria suggested by Derby in [11]. The inks in the figure meet the criteria for printability. Their rheological properties allow for the inertial force to overcome viscosity and surface tension at the cartridge nozzles. This force prevents satellite droplets and spatter during impact with the substrate surface. The dimensionless numbers of ZnO based inks are close to the limit where $Z = 1$, indicating the importance of viscosity for printability of solution processed materials. Figure 1(b) illustrates the criteria proposed by Schiaffino in [12], which describes the physical processes influencing an injected drop's behavior upon impacting the deposition surface. The used inks are in the upper/left area, suggesting that ink spreads on the surface because of its impact speed and the only resistance to dispersion is its own inertia. The reduction of kinetic energy should increase viscosity's influence, but in analyzed cases, this has minimal effect on drop dispersion on substrate surface. Ink droplets are formed

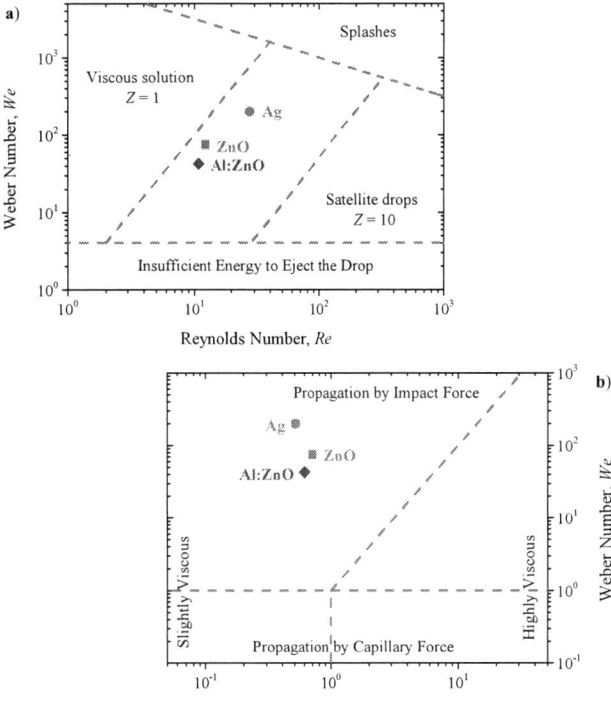

Fig. 1. a) Printability map with the positions of commercial inks, representing the predicted mechanism for drops formation and ejection with a piezoelectric DoD printer. b) Diagram that depicts the interactions of a drop impacting on a solid surface, where the ink position exhibits a drop propagation dominated by the impact force.

TABLE I.	RHEOLOGICAL PROPERTIES OF USED COMMERCIAL INKS AND CALCULATED ADIMENSIONAL NUMBERS

Material	ZnO (v~16 m/s)	Al : ZnO (v~16 m/s)	Ag (v~16 m/s)
ρ [kg/m³]	940	900	1070
η [mPa·s]	11	12	13
γ [mJ/m²]	29[a]	48.6[a]	29
Re	12.31	10.8	27.66
We	74.68	42.67	198.36
Oh	0.7	0.6	0.51
Z	1.42	1.65	1.96

[a.] Values taken from literature.

and ejected from the nozzle using an electrical signal or waveform (WF) that stimulates the piezoelectric membrane in the cartridge header. Although the rheological properties are crucial for ink printability, the WF also affects the droplet formation and ejection mechanism, as well as drop speed. The WFs had a peak amplitude between 14 and 37 V, a duration of 11 μs, and an injection frequency of 5 kHz. This resulted in a velocity of 16 m/s for all inks. The insets of Fig.2 display the WF for each material. The DIMATIX printer's drop watcher captured a series of individual images of the drops produced by the WFs. The delay time capturing ranged from 6 to 62 μs. A drop ejection is achieved in ZnO based inks by configuring a nearly identical WF, as seen in Fig. 2(a) and 2(b). The lower part of the drop is formed along with ejecting a fluid column or filament. This column thins as it falls and forms an upper satellite drop. Although the same v is achieved with ZnO based inks, Al:ZnO drops presented a contracting fluid filament due to slightly higher ink viscosity. Figure 2(c) displays the drop ejection sequence of Ag ink with a varied WF designed based on its printability map position. In this last case, drop ejection occurs via a main drop and a fluid filament which separate from the nozzle at 30 μs. The drop formation and ejection mechanisms are similar for all inks. However, there is a noticeable increase in volume for Ag drop when using a cartridge with 10 pL ejection capacity. The behavior is like Notz's analysis on fluid filament stability [13]. Notz modeled the expulsion of a fluid filament from an injector and found that for $Z > 10$, the filament thickness decreased, and satellite drops were generated. Despite the potential to avoid satellite drops, nozzle clogging remains a challenge. Only the WFs designed according to this study allow continuous, uniform, and repeatable droplet ejection, while also preventing clogging. The horizontal axis resolution is provided by DS and the cartridge angle handles the

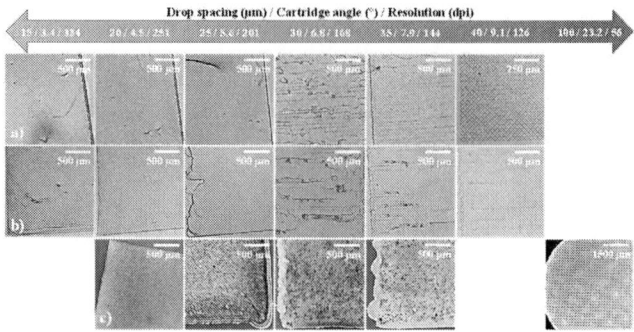

Fig. 3. Optical micrographs depicting the evolution of square pattern of the printed material, by varying the DS from 15 μm to 100 μm, which required the change in the mounting angle of the cartridge and an increase in the linear resolution with: a) the ZnO ink, b) the Al:ZnO ink, and c) the Ag ink.

vertical axis [14]. Morphology of printed materials is affected by DS variation and resolution, while layer thickness is compromised. The optical micrographs (OMs) in Fig. 3 depict the printed patterns obtained by using above-mentioned printing parameters and annealing conditions It is noted that the pattern morphology strongly is influenced by the configured DS, which can result in either individual droplets (DS≥40 μm) or a homogeneous surface by reducing the DS≤25 μm. In patterns with higher DS, controlled printing of aligned drop arrangements is noticeable. For both ZnO and Al:ZnO inks, reducing the DS to 35 μm produce some overlap between drops but not enough to achieve vertical coalescence. This issue also persisted at DS=30 μm, even a less visible homogeneity of the pattern was achieved. The complete incorporation of drops in both axes was seen from a DS of 25 μm, which improved the morphology of the printed material and resulted in uniform films. Samos reported in [15] that as the contact angle increases, deposited drops do not fully

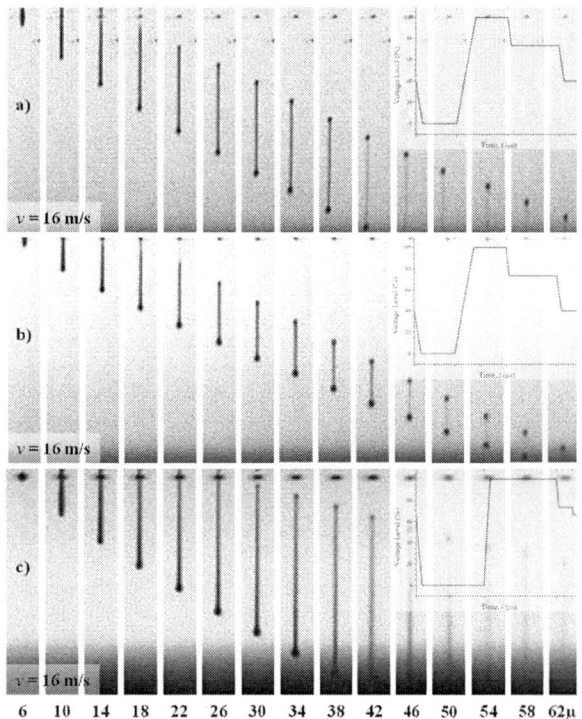

Fig. 2. Images of the droplets expulsion, using the WF depicted on inset figures: a) for ZnO ink, b) for Al:ZnO ink and c) for Ag ink.

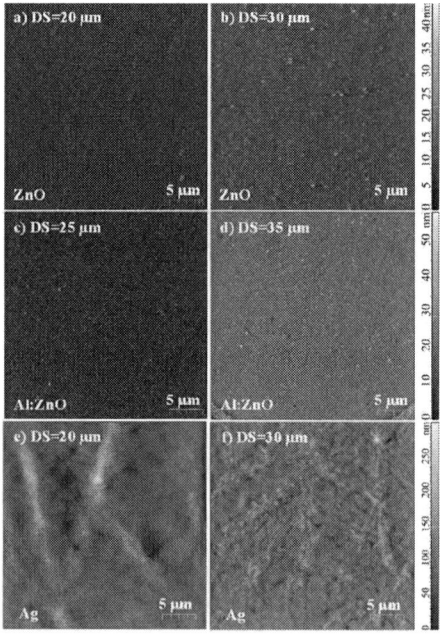

Fig. 4. 2D AFM images using the semi-contact mode over a 30×30 μm² scan area at the center of the pattern, which were printed with a DS of a) 20 μm and b) 30 μm, for the ZnO ink; with a DS of c) 25 μm and d) 35 μm, for the Al:ZnO ink; finally, with a DS of e) 20 μm and f) 30 μm, for the Ag ink.

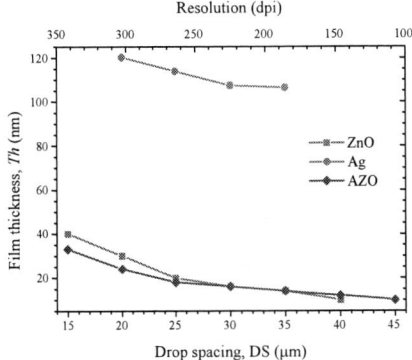

Fig. 5. Average thickness measured in materials patterns printed with different DS, from 15 μm to 45 μm, according to the used commercial inks.

TABLE II. REFRACTIVE INDEX (n), EXTINCTION COEFFICIENT (k) AND REFERENCE VALUES USED TO DETERMINE THE THICKNESS OF THE PRINTED MATERIALS PATTERN AT DIFFERENT WAVELENGTHS (λ)

Material		ZnO	Al : ZnO	Ag
λ=632.8 nm	n	1.63±0.05	1.73±0.07	0.78±0.09
	k	0.16±0.1	0.09±0.04	2.53±0.17
λ=824.3 nm	n	1.62±0.1	1.62±0.07	1.44±0.58
	k	0.55±0.3	0.03±0.02	3.27±0.16
Reference [16, 17, 18]	n	1.63	1.8	0.07
	k	0.02	0.003	3.2

integrate with others on the substrate, so increasing the resolution is necessary to achieve a more homogeneous surface. AFM topographic images, showed in Fig. 4(a) to 4(f), were obtained within a 30 μm per side scanning area. The images were captured in the central region of patterns with a DS ranging from 20 μm to 35 μm. The roughness in ZnO patterns increased with DS, resulting in droplets overlapping and a RMS roughness of 1.4-2.6 nm when DS was between 15-35 μm. For Al:ZnO, the DS of 25 μm to 40 μm resulted in an RMS roughness range of 2.7 nm to 8.6 nm. The variation in printed patterns at DS=40 μm between Al:ZnO and ZnO inks is due to the higher surface tension of the first one. This causes a difference in drop diameter and affects the propagation of the drop upon impact on the substrate surface. This is also illustrated in Fig. 1(b) and Table I. It is observed that there is an inverse relationship between morphology and DS for Ag ink in Fig. 4(e) and 4(f). Material roughness increases with DS reduction, resulting in an RMS value range of 24.6 nm to 78.8 nm. The minimum RMS value of 24.6 nm was obtained with a DS=35 μm, whereas the maximum of 78.8 nm was produced for a DS value of 20 nm. In comparison to ZnO based inks, this difference is due to the formation of large Ag-nPs agglomerates caused by sintering and DS increase, as observed in Fig. 4(e) AFM micrography. Figure 5 displays the mean thickness in relation to changes in resolution of printed patterns. In general, thickness decreases as resolution decreases, regardless of the printed material. The thickness variation shows two slopes, with the larger one indicating complete incorporation of drops in the IJP process. This is seen in patterns with visible surface homogeneity. The smooth slope is linked to a limit on minimum thickness due to less overlap between drops and nanoparticle size. Mean values of refractive index and extinction coefficient used to determine film thickness are shown in Table II. As can be seen, the considered values agree well with those previously reported for bulk materials [16,17,18]. Discrepancies may be attributed to the fact that the material in nPs form was printed only once and to the presence of ink solvent residues, where carbon at Ag-nPs surface specially can affect their properties [6].

IV. CONCLUSIONS

The rheological properties of the used commercial inks agree with the suggested criteria for printability and ejected droplet/substrate surface interaction, indicating that they are suitable for inkjet printing. When using the DIMATIX equipment in drop-on-demand mode, it is necessary to establish conditions that ensure continuous, uniform, and repeatable drop ejection and prevent nozzle clogging. This involves preparing both inks and cartridges beforehand, as well as configuring the printer settings, specifically the waveform design, despite the potential formation of satellite drops. The morphology results indicated that homogeneous and dense films can be achieved by decreasing the drop spacing (DS) >30 μm. A DS of at least 40 μm resulted in aligned single drops in both axes. The AFM analysis exhibited that the surface roughness depends on the overlap between drops, which varies with the DS ranging from 15 μm and 40 μm. The thickness of printed materials is affected by resolution, ejected volume, and ink surface tension. To achieve a homogeneous surface, adjacent drops must be integrated in both printing directions based on DS configuration and nanoparticles size. It is feasible that the findings on controlling morphology and film thickness can be useful in fabricating fully printed thin-film semiconductor devices.

ACKNOWLEDGMENT

The authors appreciate the support of Eng. Luis Abad for the preparation and cleaning of the used substrates. Likewise, thanks to M. Sc. Georgina Ramírez for AFM measurements.

REFERENCES

[1] J. W. Park, et al., "A Review of Low-Temperature Solution-Processed Metal Oxide Thin-Film Transistors for Flexible Electronics," Adv. Func. Mater. 1904632, pp. 1–40, 2019.

[2] X.-Z. Chen, Q. Luo, and C-Q. Ma, "Inkjet-Printed Organic Solar Cells and Perovskite Solar Cells: Progress, Challenges, and Prospect," Chinese J. Polym. Sci., pp. 1–29, March 2023.

[3] G. Arrabito, Y. Aleeva, R. Pezzilli, V. Ferrara, P. G. Medaglia, B. Pignataro, G. Prestopino, "Printing ZnO Inks: From Principles to Devices," Crystals 10, 449, pp. 1–34, May 2020.

[4] G. H. Kim, H. S. Shin, B. D. Ahn, K. H. Kim, W. J. Park, and H. J. Kim, "Formation Mechanism of Solution-Processed Nanocrystalline InGaZnO Thin Film as Active Channel Layer in Thin-Film Transistor," J. of The Electrochem. Soc. 156, 1, pp. H7–H9, October 2009.

[5] K. Vernieuwe, , "Thermal processing of aqueous AZO inks towards functional TCO thin films," J. of Alloys and Comp. 690, 5, pp. 360–368, January 2017.

[6] H. Ning, J. Chen, Z. Fang, R. Tao, W, Cai, R. Yao, and J. Peng, "Direct inkjet printing of silver source/drain electrodes on an amorphous InGaZnO layer for thin-film transistors," Materials 10, 1, pp. 1–7, January 2017.

[7] J. E. Fromm, "Numerical calculation of the fluid dynamics of drop-on-demand jets," IBM J. of Research and Develop. 28, 3, pp. 322–333, 1984.

[8] G. D. Martin, et al., "Inkjet printing: The physics of manipulating liquid jets and drops," J. of Physics: Conference Series 105, pp. 1–14, 2008.

979-8-3503-1191-4/23 $31.00 © 2023 IEEE

[9] Sigma-Aldrich, (2021). Zinc oxide nanoparticle ink. Safety Data Sheet, available at: https://www.sigmaaldrich.com

[10] Y. Li, *et al.*, "Deposited Nanoparticles Can Promote Air Clogging of Piezoelectric Inkjet Printhead Nozzles," Langmuir 35, 16, pp. 5517–5524, 2019.

[11] B. Derby and N. Reis, "Inkjet Printing of Highly Loaded Particulate Suspensions," MRS Bulletin: Inkjet Printing of Functional Materials 28, 11, pp. 815–818, 2003.

[12] S. Schiaffino and A. A. Sonin, "Molten droplet deposition and solidification at low Weber numbers," Physics of Fluids 9, 11, pp. 3172–3187, 1997.

[13] P. K. Notz and O. A. Basaran, "Dynamics and breakup of a contracting liquid filament," J. of Fluid Mechanics 512, pp. 223–256, July 2004.

[14] Fujifilm Dimatix, Inc., "Dimatix Materials Printer DMP-2850 User Manual," 2016.

[15] A. Samos-Puerto, G. Rodríguez-Gattorno, and M. A. Ruiz-Gómez, "Fine tuning of inkjet printability parameters for NiO nanofilms fabrication," Colloids and Surf. A: Physicochemical and Engineering Aspects 583, pp. 1–9, September 2019.

[16] R. E. Treharne, A. Seymour-Pierce, K. Durose, K. Hutchings, S. Roncallo, and D. Lane, "Optical design and fabrication of fully sputtered CdTe/CdS colar cells," J. of Phys.: Conference Series 286, pp. 1–8, December 2010.

[17] C. Stelling, C. R. Singh, M. Karg, T. A. F. König, M. Thelakkat, and M. Retsch, "Plasmonic nanomeshes: their ambivalent role as transparent electrodes in organic solar cells," Sci. Rep. 7, 1, pp. 1–13, February 2017.

[18] K. Stahrenberg, T. Herrmann, K. Wilmers, N. Esser, W. Richter and M. J. G. Lee, "Optical properties of copper and silver in the energy range 2.5–9.0 eV", Phys. Rev. B 64, 11, pp. 115111, September 2001.

Empirical DC compact model for source-gated transistors using TCAD simulation data

Patryk Golec
Advanced Technology Institute,
School of Computer Science and Electronic Engineering,
University of Surrey
Guildford, UK
pg00334@surrey.ac.uk

Dr Radu A. Sporea
Advanced Technology Institute,
School of Computer Science and Electronic Engineering,
University of Surrey
Guildford, UK
r.a.sporea@surrey.ac.uk

Dr Eva Bestelink
Advanced Technology Institute,
School of Computer Science and Electronic Engineering,
University of Surrey
Guildford, UK
e.bestelink@surrey.ac.uk

Prof. Benjamin Iñiguez
Universitat Rovira i Virgili
Tarragona, Spain
Department of Electronic Engineering
benjamin.iniguez@urv.cat

Abstract— **A source-gated transistor (SGT) is a type of thin film transistor (TFT) with unique behaviors. Functionally, the SGT is designed to decrease the saturation voltage at a lower and often more stable saturation current in comparison to a standard TFT, which makes it preferable in some circuit applications. This is achieved by introducing additional effects especially around the source region which differentiate the SGT enough that it cannot be described by a standard TFT compact model. The work presented aims to create the first empirical SGT compact model which may be implemented in DC circuits. This compact model has been developed using the data obtained from an SGT TCAD model.**

Keywords—SGT, Compact modelling, Schottky barrier

I. INTRODUCTION

A source gated transistor (SGT) has a unique structure differentiating it from conventional thin film transistors (TFT) [1]. The adjustments mostly concern replacing the ohmic contacts with a non-linear Schottky contact. The goal is to make the contact injection mechanism the dominant effect controlling the characteristics of the device as opposed to the channel effects. This is achieved by extending the gate such that it overlaps the source contact, as well as choosing materials which result in a significant Schottky barrier at the source contact – semiconductor interface[2]–[4]. This results in the bulk of the source contributing to the current injection[5], as well as a much greater and non-Ohmic contact resistance[6]. Notably, these changes only affect the current injected at the source. The physical channel behaviour of an SGT and a TFT are the same, although here it plays a much lesser role. As the contact resistance is greater than the channel resistance, which is typically the limiting factor in a TFT[7], [8].

The main benefit of using an SGT is that it has a lower saturation voltage, a higher gain and a lower power consumption than a TFT[1], [8], [9]. However, this comes at the cost of a lower saturation drain current, in most cases by multiple orders of magnitude. As such an SGT should only be applicable to specific circuits, such as a dc inverter[10], current mirror[11], and others where the low current is not a disadvantage.

The behaviour of the TFT has already been implemented in compact modelling[12]. Therefore, since the SGT shares channel behaviour with a TFT, the effects in the source region need to be investigated [13]. In an SGT, the current is injected from the bulk of the source contact to the accumulation layer through thermionic-field emission, in a similar way to a Schottky diode.

Due to this behaviour, a depletion region forms between the contact and the accumulation layer. This is dependent on the potential in the accumulation layer, which itself decreases further into the source. Thus, the depletion region is highest at the source edge as illustrated in Fig. 1. Increasing the gate voltage (V_G) will increase electron accumulation in the accumulation region, requiring a greater drain voltage (V_D) for the depletion region to pinch-off. At this point, the first saturation voltage (V_{Dsat1}) [14] is reached, as increasing V_D further will only increase the depletion layer further into the source, with the potential in the accumulation layer remaining constant. A similar effect occurs when the channel pinches off at V_{Dsat2}, when V_D is equal to V_G. However, V_{Dsat1} is reached at a comparatively lower voltage. As in saturation, the depletion region forms two dielectric capacitors in series at the source edge, which act as a potential divider for the gate voltage [8], [15].

For the contact effects to become dominant, the contact resistance must be much greater than the resistance of the channel[8]. This causes the saturation current of SGTs to be much lower than that of TFTs of similar structures[1]. It should be noted that this only occurs as the current injected at the source edge is limited by the depletion envelope limiting its bias potential to V_{Dsat1}[16] Additionally, the structure has two points of pinch off. This allows for a design with especially flat output characteristics when the device is in full saturation, which leads to SGTs having a potentially higher gain compared to TFTs[9].

979-8-3503-1191-4/23 $31.00 © 2023 IEEE

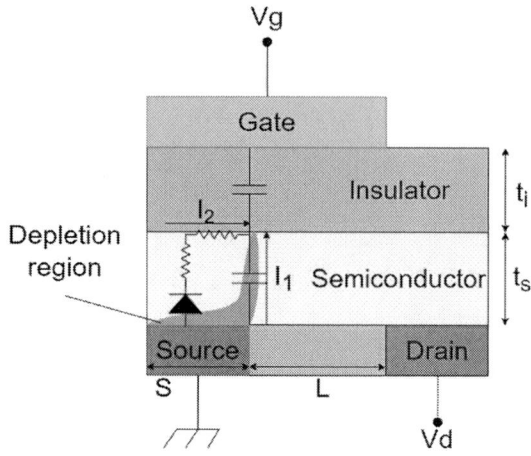

Fig. 1. Structure of an SGT in saturation. There are two current injection mechanisms present, I1 represents the current injected at through thermionic-field emission in high field mode. I2 also represents thermionic-filed emission, but in low field mode. Although I1 is greater per unit area, I2 is dominant due to the much greater area over which it is injected, as well the bias of I1 being limited due to the pinch-off effect.

II. METHODS

The channel effects present in an SGT can be derived from existing compact models of TFTs, therefore the analysis of the contact effects is prioritized. In order to obtain the data required to develop an empirical compact model, a Silvaco Atlas TCAD model is used, an example of which can be seen in Fig. 2. The TCAD model is used to extract data on parameters at areas of interest. The output characteristics of the TCAD model have also been taken, shown in Fig. 3.

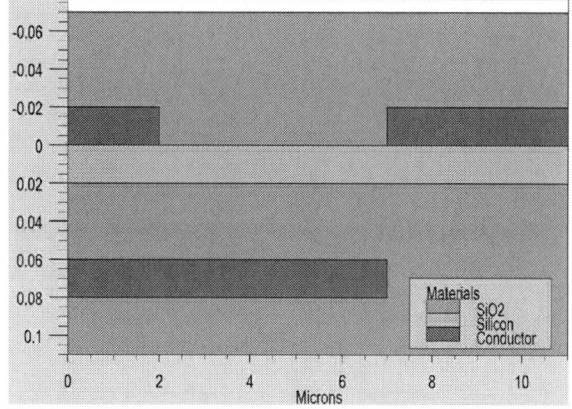

Fig. 2. One of the structures analyzed in TCAD. In this case, the V_D and V_G were both set to 20V, S is 2 microns, WF is 4.67 eV corresponding to a Schottky barrier of approximately 0.75 eV. The cutlines were taken from (0,0.019) to (11,0.019) for J_x and U and (0,001) to (11,0.001) for J_y.

The analysis has been conducted on an amorphous silicon (a-Si) SGT, the structure of which is provided in Fig. 2[13]. The parameters under investigation are the current density (J_x) and potential (U) in the accumulation layer above the source, as well as the current density (J_y) from the source to the accumulation layer. Those can then be used to calculate the resistance throughout the

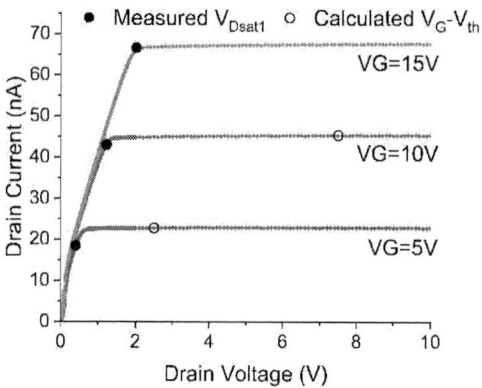

Fig. 3. Output characteristics obtained from an SGT TCAD model where S is 2μm and WF is 4.67eV. The plot demonstrates how the drain current saturates at a lower V_{Dsat1} which is a lower voltage than a TFT would saturate at.

accumulation layer (R_x) and between points on the source to the corresponding point on the accumulation layer (R_y).

The purpose of extracting the data for the TCAD structure was to create a set of empirical equations to represent J_x, J_y, U, R_x and R_y. To do this, two cutlines were taken, one from just below the source from (0,0.001)μm to (11,0.001)μm; and the other from just above the semiconductor-insulator interface from (0,0.019)μm to (11,0.019)μm, Fig. 2. The former was used to obtain J_y and the latter to obtain both J_x and U. These cutlines were repeated for a set of values within the ranges: work function (WF) 4.6-4.8eV; source length (S) 0-100μm; V_G 0-40V; and V_D 0-40V. This also demonstrates some of the differences in approach between an SGT and TFT compact model. Parameters such as WF are often considered negligible as even in organic TFTs[6] the Schottky barrier is of a much lower importance than for an SGT. Source length also becomes an important parameter as a major amount of the current is injected from the bulk of the source. The current injection also decreases further into the source, eventually reaching a point where increasing S becomes redundant (S_{sat})[3]. The final parameter to investigate is temperature, as it is involved in effects such as the current injection mechanism.

III. RESULTS

Using the extracted datasets from the TCAD model, the results at the point on the accumulation layer directly above the source have been analyzed. This is because properties such as the current are expected to be approximately equal at that point and the drain edge. The data has been used to produce fitting lines, as seen in Fig.4 to express the observed dependencies as an equation. This has been done using the fitting app available on MATLAB. Equation (1) showcases the dependence of J_x on V_G, which represents the line seen in Fig. 4.

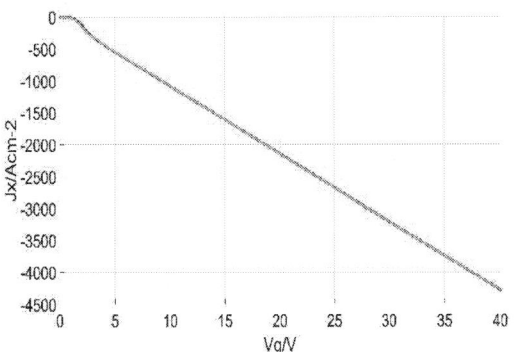

Fig. 4. A plot of J_x against V_G for V_D=20V, S=2μm, WF=4.67eV at the accumulation layer above the source edge. The coloured points represent the data extracted from the TCAD simulation and the yellow line is the fitting equation used to represent the dependance of J_x on V_G.

$$J_x(V_G) = -4.265e^{(1.62V_G)}$$
$$\times \left(0.5 + 0.5\tanh\left(\frac{-V_G + V_{th}}{0.5}\right)\right)$$
$$+ (-106.8V_G - 4.217)$$
$$\times (0.5 + 0.5\tanh(V_G - V_{th}))(Acm^{-2})(1)$$

The equation above accounts for two regimes, the first being exponential and the other linear. This occurs as the threshold voltage is approximately 2.5V (V_{th}), below this value increasing the gate voltage works towards forming the accumulation layer. Once V_G is greater than V_{th}, increasing the gate voltage works towards the feasibility of thermionic-filed emission between the source and the accumulation layer in a linear relationship. The hyperbolic tangent functions are used to allow for a smooth transition between the two regimes. It should be noted that the units for voltages in this and future expressions are in volts but have been omitted for simplicity. Since the equation is empirical, the coefficients are kept as numerical values. In a physical model, those would be replaced by constant device parameters such as width (W). However, in this branch of application, parameters do not need to be implemented, as instead a numerical implementation is sufficient.

The same method as above can be implemented for the other parameters as seen in (2) − (4).

$$J_x(V_D) = (-731.9V_D - 234.7)$$
$$\times \left(0.5 + 0.5\tanh\left(\frac{-V_D + V_{Dsat1}}{0.85}\right)\right)$$
$$+ (-0.111V_D - 2135)$$
$$\times (0.5 + 0.5\tanh(V_D - V_{Dsat1}))(Acm^{-2})(2)$$

$$J_x(S) = (-6000(Acm^{-3}) + 6000e^{-0.1S})(Acm^{-2}) \quad (3)$$

$$J_x(WF) = -1.695e^{77} \times e^{-37.72WF}(Acm^{-2}) \quad (4)$$

Equation (1) − (4) can be combined for a complete representation of J_x involving all the relevant parameters.

This allows for the investigated dependencies to be reflected within a single equation, which can be applied to a compact model. Although the final equation also needs to be rescaled to fit the data and for the units to match.

$$J_x = J_x(V_G) \times J_x(V_D) \times J_x(S) \times J_x(WF) \ (A^4cm^{-8}) \quad (5)$$

This method can be repeated for J_y and U, however, in order to obtain the values of resistance R_x and R_y, the values must be calculated mathematically. This can be done using the current density and potential. If we define that the voltage at the source metal is 0V, then the potential drop from the source to the accumulation layer should be equal to the U at that point on the accumulation layer. Thus, R_y can be expressed by the current density going towards a point on the accumulation layer and the U at that point. Similarly, R_x can be expressed by the change in U between two different points on the accumulation layer and an approximate current density between them.

$$R_y(\Omega cm^2) = \frac{U\ (V)}{J_y(Acm^{-2})} \quad (6)$$

$$R_{\frac{x1+x2}{2}}(\Omega cm^2) = \frac{U_{x1}(V) - U_{x2}(V)}{J_{\frac{x1+x2}{2}}(Acm^{-2})} \quad (7)$$

However, it should be noted that R_x should be calculated for as short a distance as possible. This is because J_x and U do not change linearly which is assumed to be the case for a given range. This is particularly important to consider, as the total resistance of the accumulation layer is the parameter being measured. This is done mainly in order to understand how the resistance of the accumulation layer changes with S. As otherwise the resistance of the accumulation layer would be greatly skewed towards the depletion layer at the source edge. The resulting expression is shown in (8).

$$R_x(\Omega cm^2) = \sum_{i=0}^{S} R_{xi} \ (\Omega cm^2) \quad (8)$$

Fig. 5 provides an example of the dependence of R_y on a device parameter V_G. The fitting is shown in (9), it follows the pattern set in (6) and it fits the data almost perfectly. This is shown to be the case from (10) − (11) which are the fitting equations for J_y and U respectively.

$$R_y(V_G) = \frac{(-0.001V_G^2 + 0.26V_G + 0.48)\ (V)}{(0.0018V_G^2 - 0.39V_G + 0.06)\ (Acm^{-2})} \quad (9)$$

$$J_y(V_G) = (0.0018V_G^2 - 0.39V_G + 0.06)(Acm^{-2}) \quad (10)$$

$$U(V_G) = (-0.001V_G^2 + 0.26V_G + 0.48)\ (V) \quad (11)$$

979-8-3503-1191-4/23 $31.00 © 2023 IEEE

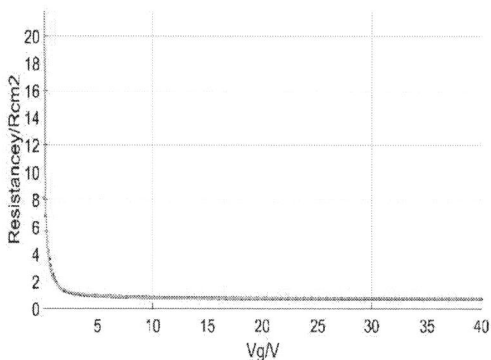

Fig. 5. Plot of R_y against V_G for V_D=20V, S=2μm, WF=4.67eV between the source edge and the accumulation layer. The data extracted for the TCAD model is represented by the coloured points and the fitting equation is shown as a yellow line.

IV. CONCLUSION

In conclusion, five parameters have been investigated in a TCAD model, in terms of V_G, V_D, S, WF; those being J_x, J_y, U, R_x, R_y. This resulted in 20 equations, which may be combined into five, one for each measurement. These five expressions are only accurate while conditions, such as temperature, are kept constant. Also, all bias dependent parameters need to be kept within their respective range.

The next step would be to assign variables to all the numerical values in the equations. This could be done by implementing the model in Verilog-A and comparing the performance of the model and predefined devices. This would allow for a comparison against a real SGT device. To do this, the model would function as a resistive network in series with a TFT compact model [8].

The results would allow for the development of a 2D compact model of the distributed resistor and diode network which forms in the source region[8], [17]. However, this would require an investigation across the full length of the source. This would be necessary in order to fully define the physical behavior of an SGT.

This would require the analysis of a point on the source (S_p), which would also be necessary in determining S_{sat} [5]. Another variable of interest is the temperature, which is highly impactful on many of the effects present in an SGT[10]; making it necessary for a physical model.

V. REFERENCES

[1] J. M. Shannon, R. A. Sporea, S. Georgakopoulos, M. Shkunov, and S. R. P. Silva, "Low-Field Behavior of Source-Gated Transistors," *undefined*, vol. 60, no. 8, pp. 2444–2449, 2013, doi: 10.1109/TED.2013.2264547.

[2] R. A. Sporea, M. Overy, J. M. Shannon, and S. R. P. Silva, "Temperature dependence of the current in Schottky-barrier source-gated transistors," *J Appl Phys*, vol. 117, no. 18, p. 184502, May 2015, doi: 10.1063/1.4921114.

[3] R. A. Sporea and S. R. P. Silva, "Design considerations for the source region of Schottky-barrier source-gated transistors," *Proceedings of the International Semiconductor Conference, CAS*, vol. 2017-October, pp. 155–158, Nov. 2017, doi: 10.1109/SMICND.2017.8101185.

[4] F. Balon, J. M. Shannon, F. Balon, and J. M. Shannon, "Analysis of Schottky barrier source-gated transistors in a-Si:H," *SSEle*, vol. 50, no. 3, pp. 378–383, Mar. 2006, doi: 10.1016/J.SSE.2005.12.020.

[5] R. A. Sporea, X. Guo, J. M. Shannon, and S. R. Silva, "Source-Gated Transistors for Versatile Large Area Electronic

Circuit Design and Fabrication," *ECS Trans*, vol. 37, no. 1, pp. 57–63, Jun. 2011, doi: 10.1149/1.3600724.

[6] J. Pruefer *et al.*, "Compact Modeling of Nonlinear Contact Effects in Short-Channel Coplanar and Staggered Organic Thin-Film Transistors," *IEEE Trans Electron Devices*, vol. 68, no. 8, pp. 3843–3850, Aug. 2021, doi: 10.1109/TED.2021.3088770.

[7] P. Migliorato, M. D. H. Chowdhury, J. G. Um, M. Seok, M. Martivenga, and J. Jang, "Characterization and modeling of a-IGZO TFTs," *IEEE/OSA Journal of Display Technology*, vol. 11, no. 6, pp. 497–505, Jun. 2015, doi: 10.1109/JDT.2014.2328335.

[8] A. Valletta, L. Mariucci, M. Rapisarda, and G. Fortunato, "Principle of operation and modeling of source-gated transistors," *J Appl Phys*, vol. 114, no. 6, Aug. 2013, doi: 10.1063/1.4817502/372621.

[9] J. Zhang *et al.*, "Extremely high-gain source-gated transistors," *Proc Natl Acad Sci U S A*, vol. 116, no. 11, pp. 4843–4848, Mar. 2019, doi: 10.1073/PNAS.1820756116/SUPPL_FILE/PNAS.1820756116.SAPP.PDF.

[10] R. A. Sporea, M. Trainor, N. Young, J. M. Shannon, and S. R. P. Silva, "Temperature effects in complementary inverters made with polysilicon source-gated transistors," *IEEE Trans Electron Devices*, vol. 62, no. 5, pp. 1498–1503, May 2015, doi: 10.1109/TED.2015.2412452.

[11] E. Bestelink *et al.*, "Compact Source-Gated Transistor Analog Circuits for Ubiquitous Sensors," *IEEE Sens J*, vol. 20, no. 24, pp. 14903–14913, Dec. 2020, doi: 10.1109/JSEN.2020.3012413.

[12] X. Cheng, S. Lee, G. Yao, and A. Nathan, "TFT Compact Modeling," *Journal of Display Technology*, vol. 12, no. 9, pp. 898–906, Sep. 2016, doi: 10.1109/JDT.2016.2556980.

[13] N. F. Mott, "Conduction in non-crystalline materials," *Philosophical Magazine*, vol. 19, no. 160, pp. 835–852, 1969, doi: 10.1080/14786436908216338.

[14] R. A. Sporea, M. J. Trainor, N. D. Young, J. M. Shannon, and S. R. P. Silva, "Intrinsic gain in self-aligned polysilicon source-gated transistors," *IEEE Trans Electron Devices*, vol. 57, no. 10, pp. 2434–2439, Oct. 2010, doi: 10.1109/TED.2010.2056151.

[15] J. M. Shannon, D. Dovinos, F. Balon, C. Glasse, and S. D. Brotherton, "Source-gated transistors in poly-silicon," *IEEE Electron Device Letters*, vol. 26, no. 10, pp. 734–736, Oct. 2005, doi: 10.1109/LED.2005.855404.

[16] E. Bestelink, U. Zschieschang, I. Bandara R M, H. Klauk, and R. A. Sporea, "The Secret Ingredient for Exceptional Contact-Controlled Transistors," *Adv Electron Mater*, vol. 8, no. 4, Apr. 2022, doi: 10.1002/AELM.202101101.

[17] S. Jahn and M. E. Brinson, "Interactive compact device modelling using Qucs equation-defined devices," *International Journal of Numerical Modelling: Electronic Networks, Devices and Fields*, vol. 21, no. 5, pp. 335–349, Sep. 2008, doi: 10.1002/JNM.676.

2023 IEEE Latin American Electron Devices Conference (LAEDC), Puebla, México, July 3-5, 2023

DC Biased Field Plate RESURF for Further R_{DSON} Reduction of LDMOS Transistors

Wendi Wang, Z. John Shen, *Fellow, IEEE* and Ian P. Brown, *Senior Member, IEEE*

Abstract— A new RESURF variant concept termed DC Biased Field Plate RESURF (BFP-RESURF) is proposed and studied through TCAD simulation in this work. The new LDMOS device structure features multiple field plates over the drift region that are biased at constant voltages. Significant reduction of R_{DSON} can be achieved by a more ideal electric field profile and an accumulation channel induced by the BFPs. Simulation study indicates that the new BFP-LDMOS offers a specific R_{DSON} of 58 mΩ·mm² comparing to 91 mΩ·mm² of the conventional LDMOS, a 1.6X reduction at a BV of 100V. The concept could be further extended to a wider range of BV ratings, the Rdson benefit becomes more pronounced as BV goes higher. The switching performance shows no obvious difference between the BFP and standard LDMOS. The BFP-RESURF concept is completely compatible with conventional BCDMOS processes and scalable over a wide range of voltage ratings.

Index Terms— Biased Field Plates, Accumulation Mode, Specific on-Resistance

I. INTRODUCTION

RESURF (Reduced Surface Field) technology enables design and fabrication of LDMOS transistors with a good tradeoff between breakdown voltage and on-resistance [1][2][3]. The RESURF concept suppresses the peak surface electric fields by diverting it into multiple directions, and serves as the basis of the BCDMOS IC industry. Over the past few decades, numerous efforts have been made to further improve the effectiveness of the original RESURF concept, including but not limited to double or triple or multi-dimensional RESURF [4][5], homogenization field technology (HOFT) RESURF [7]and floating RESURF [6]. All these RESURF variants deviate considerably from the baseline BCDMOS process with significantly increased process complexity. Furthermore, LDMOS transistors are inherently limited by their poor silicon utilization efficiency and specific R_{DSON} several times higher than the vertical discrete MOSFETs (which also benefit from some forms of the RESURF principle in the vertical direction). Fig. 1 shows such an R_{DSON} gap (modified from [8] [9] [10]). This is mainly because LDMOS devices consume a large silicon surface area to support the blocking voltage. It is highly desirable to investigate novel RESURF concepts to narrow the R_{DSON} gap between LDMOS and VDMOS transistors while maintaining the baseline BCDMOS processing flow. One recent concept of using a separately controlled field plate (FP)

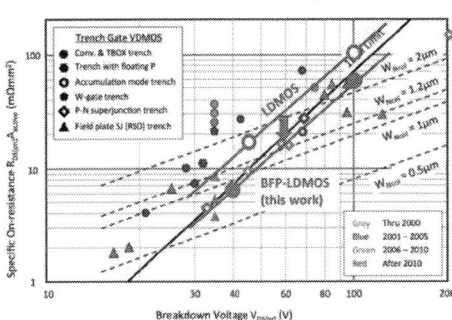

Fig. 1. Specific R_{DSON} vs. breakdown voltage for VDMOS, LDMOS and the proposed BFP-LDMOS (modified form [6]).

to induce an accumulation channel in the drift region to reduce R_{DSON} was proposed in [11]. However, the FP must switch simultaneously with the MOS gate to maintain the BV capability, resulting in control complexity and dynamic robustness concerns.

In this work, LDMOS structures with one or multiple FPs with DC biasing voltage(s) are investigated with TCAD simulation at various voltage ratings. Unlike the prior work, the FPs in this work are biased at fixed DC voltages all the time, benefitting not only on-state accumulation enhancement but also off-state RESURF optimization. They do not need any switching control either. The BFP-RESURF concept leverages the flexible electrode arrangement in power ICs which is unavailable in vertical discrete MOSFETs, and requires no additional masks or process changes.

II. BASIC CONCEPT

A. DC Operation

Fig. 2 conceptually compares two LDMOS transistors targeting 100V BV based on the conventional RESURF and new BFP-RESURF concepts, respectively. Table.I summarizes the device structural parameters used in TCAD simulation. The two devices have similar structural parameters except the BFP-

TABLE I: LDMOS DEVICE PARAMETERS

Parameters	RESURF	BFP-RESURF
Cell pitch (μm)	5.5	5.5
N-drift depth (μm)	1.3	1.3
N-drift length (μm)	4.8	4.8
N-drift doping (cm⁻³)	3.8e16	3.8e16
P-well doping (cm⁻³)	1.5e16	1.5e16
P-sub doping (cm⁻³)	8e14	8e14
STI thickness (nm)	300	150
TOX (nm)	30	30
FP width (um)	N/A	0.9
BV (V)	103	100
R_{DSON} (mΩ·mm²)	110	72.9

979-8-3503-1191-4/23 $31.00 © 2023 IEEE

141

LDMOS has two additional field plates over the STI, which are constantly biased at 15V and 43V, respectively. The STI of the BFP-LDMOS is thinner than the conventional LDMOS, allowing the two field plates to be more effective. The BFPs help improve device performance in both the ON and OFF states through two different mechanisms: 1) to force a more uniform electric field distribution in the OFF state and thus more optimal R_{DSON} vs. BV tradeoff, and 2) to induce an accumulation channel under the STI to enhance conductivity in the ON state. The BFP-RESURF concept uses the baseline BCDMOS process flow and takes advantage of flexible metal interconnect arrangement in power ICs. The BFP biasing voltages can be provided with a simple capacitor voltage divider circuit from the main DC voltage because there is no need of any active control during device operation. The biasing voltage for each field plate is optimally chosen for the LDMOS voltage rating. The BFP-RESURF is completely compatible with conventional BCDMOS processes with the only drawback of requiring additional voltage busses to the BFPs from the external IC power supply pins. Fig. 3 compares the E-field profiles along the drift region under the STI/Si interface of the conventional LDMOS and the new BFP-LDMOS structures in OFF state. The proposed BFP structure exhibits more ideal RESURF contour along the interface as the E-field peaks distribute evenly under the FP corners instead of a single peak under the poly gate in the conventional LDMOS. Fig. 4 shows the device ID-VG characteristics where the dual-BFPs structure exhibits a transconductance 1.5X that of the conventional LDMOS. This is due to two factors: thicker drift region and strong accumulation channel, both enabled by the positive FP biasing. Fig. 5 compares the potential distribution and E-field intensity of the two structures in avalanche, three hotspots are found in the BFP-LDMOS instead of one in the conventional one. Fig. 6 shows the ON-state electron current density of the two devices at $V_{GS}=5V$ and $V_{DS}=1.5V$. It is clearly shown that the current density is intensely enhanced under the STI in the BFP-LDMOS because of the strong accumulation channel

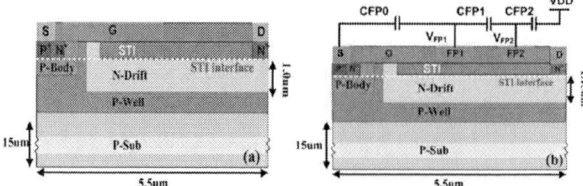

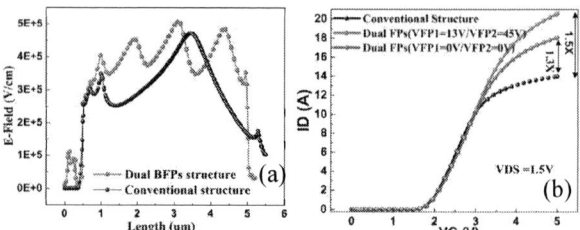

Fig. 2. LDMOS structure based on (a) conventional RESURF and (b) the proposed BFP-RESURF concept.

Fig. 3. (a) E-field distribution along STI/Si interface of conventional and BFP structure and (b) ID-VG curves of conventional and BFP LDMOS structure in ON state.

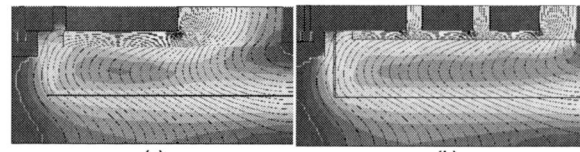

Fig. 4. Potential and E-field distributions in (a) conventional RESURF and (b) BFP-RESURF in the OFF state.

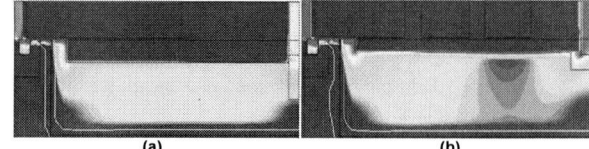

Fig. 5. eCurrent density distribution in (a) RESURF and (b) BFP-RESURF in the ON state, showing the accumulation layer in (b).

induced by the two positively biased BFPs, whereas the current density is rather uniformly distributed in the conventional one. The specific R_{DSON} of the new BFL-LDMOS is 72.9 m$\Omega \cdot$mm^2 comparing to 110 m$\Omega \cdot$mm^2 of the conventional LDMOS (33% reduction) while both devices exhibit BV of around 100V. Note that the specific R_{DSON} of the simulated LDMOS is consistent with the experimental data reported in the literature [7][8].

B. Dynamic Switching

One may wonder if the additional field plates will negatively impact the dynamic behavior of the LDMOS transistor. The switching performance of 85V BFP-LDMOS is studied with mixed-mode TCAD simulation using the standard double pulse inductive switching test circuit. Fig. 6 (a) and (b) show very similar Vg, Id and Vd waveforms of both devices except the BFP-LDMOS exhibits a faster turn-on phase during the second pulse due to its higher transconductance.

The field plates are charged and discharged in the form of displacement currents when the BFP-LDMOS switches between ON and OFF states. Theoretically there should be no power losses for the capacitive charging/discharging process. Ideally, the field plates are biased at the optimal voltages supplied by a capacitor voltage divider circuit which is connected to the IC power supply pin. The field plates simply transfer charge back and forth among the capacitors without

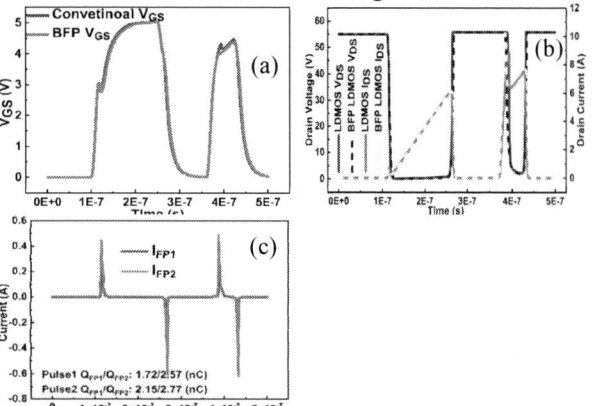

Fig. 6.(a) V$_{gs}$ waveforms of the conventional and new LDMOS transistors during double pulse switching. (b) I$_{DS}$ and V$_{DS}$ waveforms of the conventional and new LDMOS transistors during double pulse switching. (c) Charging and discharging current waveforms of FP1 and FP2 during double pulse switching.

TABLE II: 40/150/300V DEVICE PARAMETERS

Voltage Rating	Structure	TOX (nm)	Drift doping	BV (V)	R_{DSON} (mΩ·mm²)	FPs bias (V)
40V	standard	300	5.6e16	67	**38**	N/A
40V	1 BFP	150	5.2e16	63	29	25
85V	Standard	300	3.8e16	103	**110**	N/A
85V	2 BFP	150	3.8e16	100	73	15/43
150V	standard	650	2.4e16	178	**416**	N/A
150V	2 BFPs	300	2.2e16	175	255	40/85
300V	standard	1000	6.4e15	360	**3040**	N/A
300V	3 BFPs	600	7.6e15	360	1320	55/120/200

drawing power from the IC power supply. Fig. 6 (c) shows the charging and discharging displacement current waveforms of FP1 and FP2 during the double pulse switching. Note that the transient charging/discharging charges of the two field plates during the turn-on and turn-off of the BFL-LDMOS are very small (i.e., 1.7-2.8 nC). Therefore, small decoupling capacitors of a few nF can be used to hold the DC bias voltages constant in the capacitor voltage divider circuit, making the adoption of the BFP-RESURF concept easier from circuit application's perspective. It is also possible to integrate the decoupling capacitors on the IC chip or in the package to further simplify the power IC pin requirement.

III. CONCEPT EXTENSION AND DESIGN OPTIMIZATION

The concept of DC biased FP assisted LDMOS can be easily extended to a wider range of BV by adjusting the field plates design accordingly. Details of the design rules will be discussed in the following section. Table II summarizes the key parameters of 4 sets of BFP/standard LDMOS rated at 40V, 85V, 150V and 300V. The Rdson improvement factors are 1.3X, 1.5X, 1.6X and 2.3X as voltage rating goes up. This is because bigger device dimension can provide more space to accommodate additional field plates (FPs), and result in a stronger RESURF enhancement. Thus, the drift layer design (doping concentration or thickness) can be more aggressive with the benefit of better E-field contour. Moreover, with increased voltage rating the FPs bias will increase accordingly and result in stronger accumulation at silicon surface. In summary, the Rdson reduction bonus is more pronounced in high voltage BFP LDMOS than low voltage ones.

The remaining part of this section will evaluate several key parameters that are unique to BFP-LDMOS structure by simulation. Although it may be impractical to determine the optimal value for each design parameter based solely on mathematical solutions to the electrostatic boundary condition of a structure, an empirical range of values can be identified by analyzing simulation data. Those principles can help a designer to fine tune any specific device with ease and achieve preferable performance.

A. Number of Field Plates

Selection of appropriate number of FPs should be a delicate balance between device performance, process margin and circuitry complexity while taking into account the specific BV range. Using 85V LDMOS as an example where 3 different structures implementing 1, 2 and 3 FPs are compared in Table III. The R_{DSA} improvement compared to standard structure is 1.32X, 1.51X and 1.58X for 1, 2, and 3 FPs, respectively. The

On-Resistance as well as the RESURF gain from the third FP will be very minimum due to the limitation of small structure dimension. The benefit of adding a third FP becomes prominent only when the voltage rate rises to 300V. As the length of the MOSFET drift region increases, the number of required FPs to achieve optimal RESURF effect also increases. And the dimension of a 300V LDMOS can easily accommodate 3 FPs into its STI. Characteristic comparison of 2&3 FPs structures is presented in Table IV. The structure with 3 FPs can achieve a 23% reduction in RdsA compared to the structure with 2 FPs, and an overall improvement of 2.3X over conventional LDMOS.

B. Field Plates Placement and Its Critical Dimension

Similar to the shield gate in vertical trench MOSFET where the E-field peaks at the bottom part closest to the drain node, the electric field also spikes at the right edge of each FP in the BFP structure. Given that the overall E-field profile is primarily influenced by its peaks, it is natural to position the right edge of the FPs uniformly across the length of the drift region to achieve the maximum RESURF effect. Table V summarizes the coordinates of the FPs right edges used in every case. In most cases, the coordinate of each FP (right edge) is located either right at or close to the equidistant points of the drift layer, depending on the number of FPs been used. Misplacing the FP edges away from their optimal points, however, would lead to degraded breakdown voltage. Fig.7 shows the BV window of 85V LDMOS with varying FP1 and FP2 edge location. Point (0, 0) is the B/L condition where the BV is the highest; positive offset stands for narrower spacing between the drain side and negative value means the wider spacing. In both scenarios the BV is more strongly affected by FP2 variation than those in FP1.

TABLE IV: 2/3 FPs 300V DEVICE

Rate	Design	TOX (um)	FP CD (um)	FPs bias (V)	BV (V)	R_{DSA} (Ω·mm²)
300V	STD	1	-	-	360	3.0
300V	2 BFPs	0.65	3.5	50/150	350	1.7
300V	3 BFPs	0.6	2.7	55/120/200	360	1.3

TABLE V: FPs RIGHT EDGE LOCATION

Rate	Drift length(um)	Gate (um)	FP1(um)	FP2(um)	FP3(um)
40V	2.3	0.95	1.8	-	-
85V	4.5	1.25	2.45	3.65	-
150V	8.7	2.35	4.65	6.95	-
300V	24.3	5.95	10.2	15.45	20.4

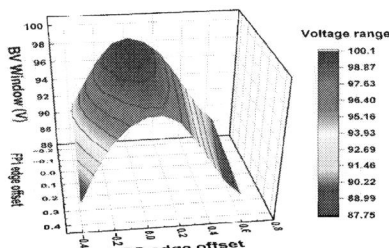

Fig. 7. BV window of 85V BFP LDMOS with FPs edge location variation.

This because FP2 plays a direct role in determining the boundary of the E-field contour on the drain side. Overall, the FP nearest to the drain terminal in any structure is most critical for advantageous RESURF effect.

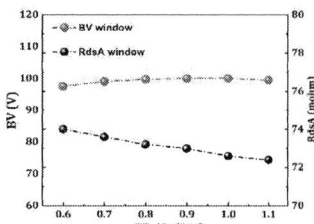

Fig. 8. BV window of 85V BFP LDMOS with FPs CD variation.

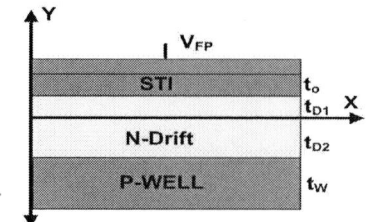

Fig. 9. BV window of 85V BFP LDMOS with FPs CD variation.

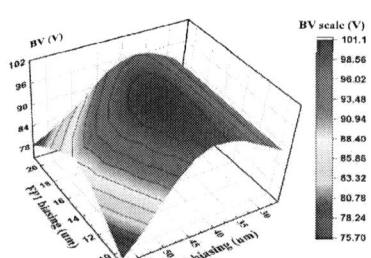

Fig. 10. BV window of 85V BFP LDMOS with FPs biasing variation.

The length/CD of the FP on the other hand, has less impact on BV than its edge coordinates. The BV fluctuation is very minimum as shown in Fig. 8. Both FP length and CD will have limited influence on RdsA.

C. Field Plates Voltage Biasing

After the FP number and its structure dimension are fixed, each FP has to be given a certain positive voltage bias to activate the RESURF effect. The exact biasing voltage has to be obtained by sufficient trials of simulation, though an initial value to start the experiment can firstly be estimated by solving 2D Poisson equation based on the study of [12]:

$$\frac{dE_x(x,y)}{dx} + \frac{dE_y(x,y)}{dx} = \frac{qN_D(x)}{\varepsilon_s} \quad (1)$$

The BFP-LDMOS structure can be simplified as a model shown in Fig. 9 where the drift region is divided into two quadrants by X-axis. Assuming ideal and symmetrical RESURF in both quadrants, the following equations have to meet:

$$E_x = \frac{-BV}{L} \quad (2)$$

$$N_D \cdot t_{D2} = N_W \cdot t_W \quad (3)$$

$$V_{FP}(x) = \frac{BV}{L}x - \frac{qN_D}{\varepsilon_s}t_{D1}\left(\frac{t_{D1}}{2} + \frac{\varepsilon_s}{\varepsilon_o}t_o\right) \quad (4)$$

$$\frac{qN_D}{2\varepsilon_s}t_{D1}^2 + \frac{qN_D}{\varepsilon_o}t_{D1}t_o = \frac{qN_D}{2\varepsilon_s}t_{D2}^2 + \frac{qN_W}{2\varepsilon_s}t_W^2 \quad (5)$$

Equation (3) is the simplified charge balance condition in the lower half structure and equation (4) is the FP voltage function to satisfy ideal RESURF in the upper half structure. Equation (5) dictates symmetrical vertical voltage drop from y=0 to y=t_{D1}+t_o and y=- t_{D2}-t_w in both sections. BV can be estimated as $BV_{ideal} = \frac{b_n \cdot L}{\ln(b_n L)}$. Solving above equations using conditions of 85V BFP-LDMOS structure in Fig.2 will yield the V$_{FP}$ biasing function as VFP=22.4X-21.72. Inserting the X coordinates of the center of the 2 FPs gives V$_{FP2}$=50V and V$_{FP1}$=23.1V, which are higher than the actual optimized values verified by simulation. The simulated BV window of 85V device with respect to FP biasing is plotted in Fig. 10. Similar to FPs edge location in previous section, the FP2 biasing has larger impact on BV than that of FP1. Additionally, V$_{FP1}$ has wider window with under biased V$_{FP2}$ than over biased V$_{FP2}$. It is important to not excessively bias FP2 so that the impact ionization hotspot is not prematurely clamped at body junction or under FP1 edge.

IV. CONCLUSION

In summary, this work proposes a new RESURF variant concept termed Biased Field Plate RESURF (BFP-RESURF) to significantly reduce R$_{DSON}$ of LDMOS using baseline BCDMOS processes. This concept is scalable over a wide range of voltage ratings simply by changing the FP design, such as the FP number, FP placement and FP biasing, without any other

TABLE III: 1/2/3 FPs 85V DEVICE

Rate	Structure	TOX (nm)	FP CD (um)	FPs bias (V)	BV (V)	R$_{DSA}$ (mΩ·mm²)
85V	Standard	300	-	-	100	110
85V	1 BFP	200	1.3	25	100	83.1
85V	2 BFPs	150	0.9	15/43	100	72.9
85V	3 BFPs	150	0.6	2/17/40	100	69.6

process modifications. This methodology is relatively easy to implement in power ICs since all FPs are laid out on the chip surface. Also, the BFP RESURF requires no extra control scheme since all the FPs are biased at constant voltage all the time. TCAD simulation shows R$_{DSON}$ reduction of 1.3X-2.3X in 60V-300V LDMOS transistors. The R$_{DSON}$ benefit becomes more pronounced at higher BV ratings. The drawbacks of the BFP-RESURF include IC layout complexity of the BFP busses and additional IC pins allocated for them.

REFERENCES

[1] J. A. Appels and H. M. J. Vaes, "High voltage thin layer devices (RESURF devices)," in IEDM Tech. Dig., Washington, DC, USA, Dec. 1979, pp. 238–241, doi: 10.1109/IEDM.1979.189589.

[2] A. W. Ludikhuize, "A review of RESURF technology," ISPSD, 2000.

[3] B. Zhang, W. Zhang, L. Zhu, J. Zu, M. Qiao and Z. Li, "Review of technologies for high-voltage integrated circuits," in Tsinghua Science and Technology, vol. 27, no. 3, pp. 495-511, June 2022,

[4] Jie Wu, Jian Fang, Bo Zhang and Zhaoji Li, "A novel double RESURF LDMOS with multiple rings in non-uniform drift region," ICSICT, 2004

[5] Zhuo Wang, Muting Lu, Xin Zhou, Jun Wang and Bo Zhang, "A novel Triple-RESURF SON LDMOS and its analytical model," 2014 IEEE International Conference on Electron Devices and Solid-State Circuits, 2014.

[6] V. Khemka; V. Parthasarathy; Ronghua Zhu; A. Bose, "A floating RESURF (FRESURF) LD-MOSFET device concept," IEEE Electron Device Letters, Vo. 24, No. 10, 2003.

[7] Bo Zhang; Wentong Zhang; Jian Zu; Ming Qiao; Sen Zhang; Zhili Zhang; Boyong He; Zhaoji Li, "Novel Homogenization Field Technology in Lateral Power Devices," IEEE Electron Device Letters, Vo. 41, No. 11, 2020.

[8] Darwish, R. A. Blanchard, R. Siemieniec, P. Rutter and Y. Kawaguchi, "The Trench Power MOSFET: Part I—History, Technology, and Prospects," in IEEE Transactions on Electron Devices.

[9] M. N. Chil et al., "Advanced 300mm 130nm BCD technology from 5V to 85V with Deep-Trench Isolation," ISPSD. 2016.

[10] H. Cha, K. Lee, J. Lee and T. Lee, "0.18μm 100V-rated BCD with large area power LDMOS with ultra-low effective specific resistance," ISPSD, 2016

[11] A. Ferrara et al., "The boost transistor: A field plate controlled LDMOST," ISPSD15.

[12] A. Ferrara et al., "Design optimization of field-plate assisted RESURF devices," ISPSD13

2023 IEEE Latin American Electron Devices Conference (LAEDC)
Puebla, México, July 3-5, 2023

Multi-level Operation in Ultra-scaled MRAM

Viktor Sverdlov
Christian Doppler Laboratory for
Nonvolatile Magnetoresistive Memory
and Logic
Institute for Microelectronics, TU Wien
Vienna, Austria
sverdlov@iue.tuwien.ac.at

Mario Bendra
Christian Doppler Laboratory for
Nonvolatile Magnetoresistive Memory
and Logic
Institute for Microelectronics, TU Wien
Vienna, Austria
bendra@iue.tuwien.ac.at

Wolfgang Goes
Silvaco Europe Ltd.
Cambridge, United Kingdom
wolfgang.goes@silvaco.com

Simone Fiorentini
Christian Doppler Laboratory for
Nonvolatile Magnetoresistive Memory
and Logic
Institute for Microelectronics, TU Wien
Vienna, Austria
fiorentini@iue.tuwien.ac.at

Abel Garcia-Barrientos
Faculty of Science, UASLP
San Luis Potosí 78217, Mexico
abel.garcia@uaslp.mx

Siegfried Selberherr
Institute for Microelectronics, TU Wien
Vienna, Austria
selberherr@iue.tuwien.ac.at

Abstract—**Magnetoresistive random access memory (MRAM) prototypes demonstrate fast operation and are suitable for the last level caches. MRAM possesses high endurance, long retention, and requires less masks for fabrication than its competitor flash memory. MRAM is nonvolatile and scalable. Strong perpendicular magnetic anisotropy in most advanced single-digit nanoscale footprint devices is enhanced by elongating the magnetic layers. To facilitate the switching and to increase the interface-induced magnetic anisotropy even further, the free magnetic layers are made of several elongated pieces separated by tunnel barriers with multiple interfaces. To properly model such devices, accurate evaluation of the spin-transfer torques is required. The interfacial and bulk-torques are not independent, and the use of a spin-charge transport approach coupled to the magnetization dynamics allowing to treat the torques on equal footing in magnetic tunnel junctions with elongated layers becomes mandatory. By employing this advanced modeling approach a multi-level memory operation in an ultra-scaled MRAM cell with a composite free layer is demonstrated.**

Keywords—*Spin and charge drift-diffusion, spin-transfer torque, magnetic tunnel junctions, STT-MRAM, composite free layer, shape-induced magnetic anisotropy, double-spin MTJ*

I. INTRODUCTION

Spin-transfer torque magnetoresistive random access memory (STT-MRAM) is nonvolatile, scalable, and fast. It possesses high endurance and long retention [1-3]. Compared to flash memory, MRAM also requires less additional masks, which significantly reduces fabrication costs [4]. STT-MRAM can be used for stand-alone and embedded applications, and it can compete with SRAM [5] provided it is written with a sub-nanosecond speed and scaled to a few nanometers footprint. Although an MRAM cell consists of many layers, the main element is a magnetic tunnel junction (MTJ) typically composed of a CoFeB reference layer (RL) and free magnetic layer (FL) separated by an MgO tunnel barrier (TB). An MTJ with perpendicular RL and FL orientations (pMTJ) of the magnetization allows to reduce the cell footprint and increase the memory density. Thanks to the interface-induced effects, thin CoFeB layers grown on MgO become perpendicularly magnetized [6]. To increase the perpendicular magnetic anisotropy, a thin FL is interfaced with a second MgO layer [7]. The FL becomes composite as it now consists of two thin CoFeB films with a normal metal

spacer in-between. By introducing additional MgO layers along the FL [8] and elongating the FL out-of-plane in the magnetization direction one benefits from additional CoFeB/MgO interfaces and from the one-dimensional out-of-plane shape anisotropy [9]. It allows increasing the perpendicular anisotropy even further by simultaneously reducing the diameter of the FL, which results in a single digit nanometer cell footprint [9].

To comprehensively model advanced MRAM cells, an accurate evaluation of spin currents, spin accumulations, and spin-transfer torques in MTJs with composite FLs is required. As the STT-MRAM cell includes several elongated CoFeB magnetic layers separated by MgO TBs, it is necessary to evaluate the charge and spin currents and the corresponding torques acting on the magnetization within the same simulation approach. The approach must be capable of treating on equal footing the traditional interfacial Slonczewski torques [10,11] acting close to the interfaces in thin magnetic films and the torques acting on the nonhomogeneous magnetization along the elongated FLs.

Here we extend the recently proposed spin-charge drift-diffusion transport approach coupled to the magnetization dynamics [12,13] to investigate a multi-level operation in an ultra-scaled memory cell with a free layer consisting of two elongated parts separated by a tunnel barrier (Fig.1). We demonstrate the complete multi-level switching cycle by only employing the voltage bias of a fixed polarity.

II. SIMULATION METHODOLOGY

We implemented a fully three-dimensional finite element method (FEM) based modeling and simulation tool [14] which includes all physical phenomena important for proper ultra-scaled MRAM operation (Fig.2). μ_B is the Bohr magneton, e is the elementary charge, λ_φ is the transverse spin dephasing length, λ_J is the exchange length, and λ_{sf} *is* the spin-flip length. We numerically solve the Landau-Lifshitz-Gilbert (LLG) equation describing the normalized magnetization **m** dynamics. The effective field $\mathbf{H}_{eff}$ includes the magnetic anisotropy field, the exchange field, as well as the demagnetization and stray fields. The latter contributions are evaluated only on the disconnected magnetic domain by using a hybrid boundary element/finite element method [15]. The

979-8-3503-1191-4/23 $31.00 © 2023 IEEE

2023 IEEE Latin American Electron Devices Conference (LAEDC)
Puebla, México, July 3-5, 2023

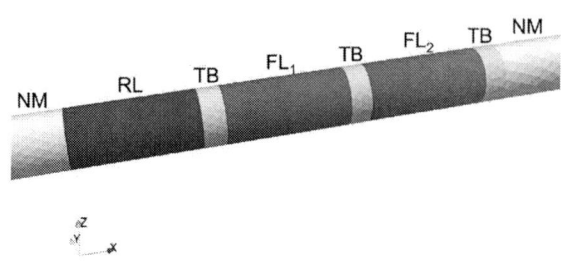

Fig.1 Ultra-scaled memory cell with a composite elongated FL consisting of the two parts FL$_1$ and FL$_2$ separated by a TB. Two TBs are introduced between the FL and the RL and the FL and the top contact.

coupled spin $\mathbf{J_S}$ and charge $\mathbf{J_C}$ drift-diffusion (DD) transport accurately describes the spin accumulation $\mathbf{S}$ and the corresponding torques $\mathbf{T_S}$ acting on the magnetization. The continuous boundary conditions for the spin accumulation and the spin current densities [12] are employed at the interfaces between ferromagnetic parts and normal metal contacts. To describe the charge current in an MTJ, we model the TB as a poor conductor, with a local conductivity $\sigma(\theta)$ depending on the angle θ between the local magnetization directions across the TB [13]. The tunneling magnetoresistance ratio (TMR) is used as a parameter to accurately describe the modulation of the conductivity and the charge current density.

To describe the spin transport through a TB correctly, the spin current density at the TB must satisfy [13]:

$$\mathbf{J_{S,TB}} = -\frac{\mu_B}{e} \frac{\mathbf{J_C} \cdot \mathbf{n}}{1 + P_{RL}P_{FL}\mathbf{m_{RL}} \cdot \mathbf{m_{FL}}} \left(P_{RL}\mathbf{m_{RL}} + P_{FL}\mathbf{m_{FL}} \right.$$
$$\left. + \frac{1}{2}\left(P_{RL}P_{RL}^{\eta} - P_{FL}P_{FL}^{\eta}\right)\mathbf{m_{RL}} \times \mathbf{m_{FL}} \right) \quad (1)$$

The in-plane Slonczewski interface polarization parameters P_{RL}, P_{FL} supplemented with the out-of-plane factors $P_{RL}^{\eta}, P_{FL}^{\eta}$ define the spin current density floating through the TB, and $\mathbf{m_{RL}}$ and $\mathbf{m_{FL}}$ are the normalized magnetization vectors of the RL and FL, respectively. We apply the method to an MTJ structure with a FL composed of the two elongated

LLG :
$$\frac{\partial \mathbf{m}}{\partial t} = -\gamma \mathbf{m} \times \mathbf{H}_{eff} + \alpha \mathbf{m} \times \frac{\partial \mathbf{m}}{\partial t} + \frac{1}{M_S}\mathbf{T_S}$$

Torque :
$$\mathbf{T_S} = -\frac{D_e}{\lambda_J^2}\mathbf{m} \times \mathbf{S} - \frac{D_e}{\lambda_\varphi^2}\mathbf{m} \times (\mathbf{m} \times \mathbf{S})$$

$\mathbf{S}$:
$$D_e\left(\frac{\mathbf{S}}{\lambda_{sf}^2} + \frac{\mathbf{S} \times \mathbf{m}}{\lambda_J^2} + \frac{\mathbf{m} \times (\mathbf{S} \times \mathbf{m})}{\lambda_\varphi^2}\right) = -\nabla \cdot \mathbf{J_S}$$

$\mathbf{J_S}$:
$$\mathbf{J_S} = -\frac{\mu_B}{e}\beta_\sigma\left(\mathbf{J_C} \otimes \mathbf{m} + \beta_D D_e \frac{e}{\mu_B}[(\nabla\mathbf{S})\mathbf{m}] \otimes \mathbf{m}\right)$$
$$- D_e \nabla \mathbf{S}$$

$\mathbf{J_C}$:
$$\mathbf{J_C} = \sigma\mathbf{E} - \beta_D D_e \frac{e}{\mu_B}[(\nabla\mathbf{S})\mathbf{m}]$$

TB :
$$\sigma(\theta) = \frac{\sigma_P + \sigma_{AP}}{2}\left(1 + \left(\frac{TMR}{2+TMR}\right)\cos\theta\right)$$

Fig.2 Coupled equations describing the magnetization dynamics $\mathbf{m}$ in the effective magnetic field $\mathbf{H}_{eff}$ and in presence of the electric current density $\mathbf{J_C}$. D_e is the diffusion coefficient, β_D and β_σ are the spin polarization parameters [12] in the ferromagnet, and σ_P and σ_{AP} are the TB conductivities in the parallel and anti-parallel magnetization configuration.

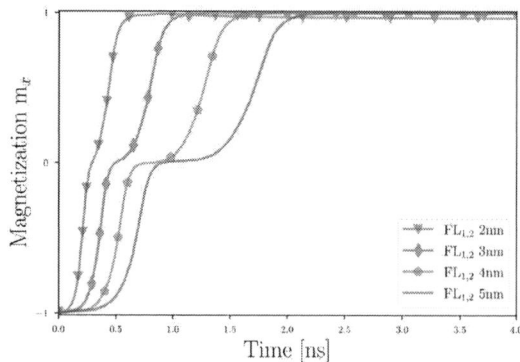

Fig.3 Switching of the ultra-scaled MRAM (Fig.1) from AP to P state, for bias of 1.5V and several lengths of FL$_1$=FL$_2$. The intermediary state at the total magnetization equals zero is observed for long FLs.

ferromagnetic parts FL$_1$ and FL$_2$ separated by TBs from the RL and between each other.

III. MULTI-LEVEL OPERATION IN ULTRA-SCALED MRAM

Fig.3 shows the FL magnetization switching of an ultra-scaled MRAM cell from anti-parallel (AP) to parallel (P) alignment of the composite FL with respect to the RL. The polarizations of the TBs (in Fig.1 from left to right) are equal to 0.6, 0.5, and 0.2, respectively. The bias voltage of 1.5V is applied to generate an electric current along negative OX direction. It this case the spin-polarized electrons impinge the FL$_1$ and create the torque forcing its magnetization to become parallel to that of RL. From the FL$_2$ side, the positive "holes" with the spin-polarization opposite to the magnetization of the FL$_2$ impinge the FL$_1$ and therefore create the torque in the same directions as the spins from the RL. As the two torques acting on the FL$_1$ from the RL and the FL$_2$ are additive, the FL$_1$ magnetization is rapidly inverted and becomes AP to the magnetization of the FL$_2$. The total FL magnetization is then zero as it is seen in Fig.3 [16]. As it takes a certain time to switch the magnetization of the remaining FL$_2$ part, the FL magnetization develops a plateau around $m_x = 0$. The plateau width is reduced by reducing the length of both FL$_1$ and FL$_2$, or of either of them as shown in Fig.4. Fig.4 also confirms the sequence during the AP to P switching. As the FL$_1$ part is inverted first, the magnetization in the intermediate state is positive if only the length of FL$_2$ is reduced and is negative if only the length of FL$_1$ is reduced, as the complementary part

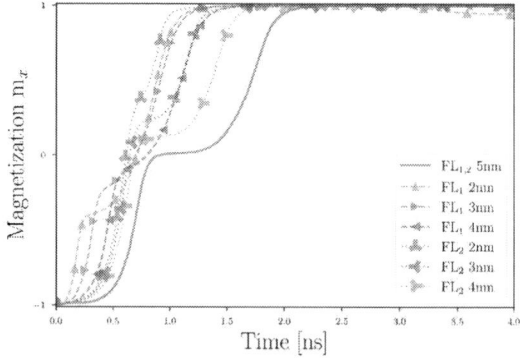

Fig.4. Switching from AP to P state, for bias of 1.5V, and several lengths FL$_1$ at FL$_2$=5nm and FL$_2$ at FL$_1$=5nm. The intermediary state at the total magnetization (FL$_2$−FL$_1$)/(FL$_2$+FL$_1$) is observed.

979-8-3503-1191-4/23 $31.00 © 2023 IEEE

2023 IEEE Latin American Electron Devices Conference (LAEDC)
Puebla, México, July 3-5, 2023

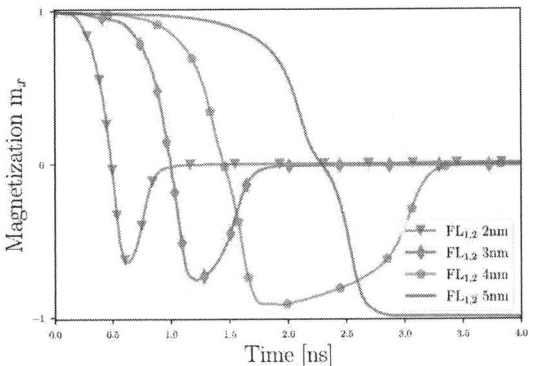

Fig.5 Magnetization trajectories for the switching from P to AP for several FL lengths in a symmetric configuration where both FL parts have the same length.

of the FL remains fixed at 5nm (Fig.4). While the switching at a reduced length of the FL becomes faster, the time the FL remains in the intermediary state is also reduced. However, the plateau of 0.5ns at $FL_1=FL_2=5nm$ is sufficient for resolving the intermediate magnetization state.

In Fig.5 the magnetization trajectories for the switching from parallel (P) to anti-parallel (AP) of the ultra-scaled MRAM are shown. To achieve the switching from P polarization shown in Fig.6, topmost panel, we now invert the current direction by applying the bias of −1.5V. The polarizations of the TBs are the same and equal to 0.6, 0.5, and 0.2, respectively. For $FL_1 = FL_2 = 5nm$ the magnetization reversal is again a sequential process [16] and it happens in an order inverse to that at AP to P switching. The middle part of the free layer FL_1 experiences the torques from RL and FL_2. However, in contrast to the AP to P switching, the torque acting from FL_2 is of an opposite sense to that of the RL. As the interface polarizations of the corresponding TBs are nearly equal, the corresponding torques nearly compensate each other, and the total torque acting on FL_1 is small. At the same time the torque acting from FL_1 on FL_2 is strong and it forces FL_2 to switch from P first (Fig.6, second panel). As soon as the magnetization of FL_2 is inverted, the torque acting from FL_2 on FL_1 changes sign. It then acts in the same sense as the

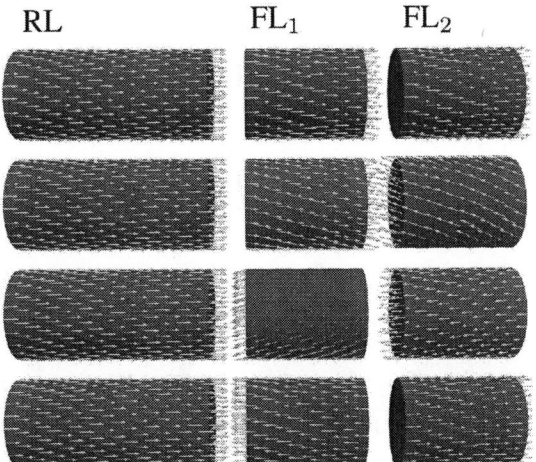

Fig.6 Sequential switching of the composite FL from P to AP. Initially FL_2 is reversed followed by the middle part FL_1 reversal. Bottom panel: novel FL configuration due to back-hopping effect of FL_2 [17].

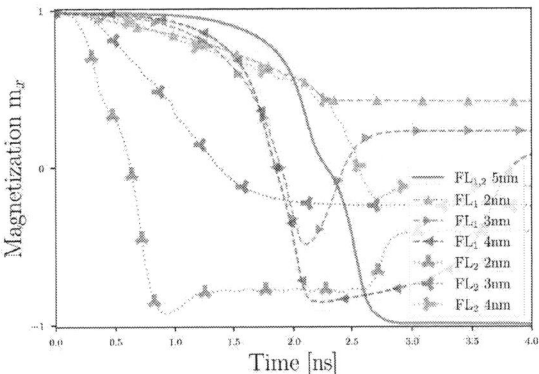

Fig.7 . Switching from P to AP state, for bias of −1.5V, and several lengths FL_1 at $FL_2=5nm$ and FL_2 at $FL_1=5nm$. The final state at the total magnetization $(FL_2−FL_1)/(FL_2+FL_1)$ is observed (except for $FL_1=FL_2=5nm$ shown by solid line).

torque from the RL. It inverts the magnetization of FL_1 fast, thus completing the switching from P to AP (Fig.6, third panel from top). Because FL_1 under the additive torques from the RL and FL_2 inverts the last, the intermediate state with zero magnetization is less pronounced compared to the case of AP to P switching [16].

Fig.5 displays that if the length of the FL is reduced, the total magnetization of the FL does not reach its AP value of −1. Instead, we observe what seems to be a switching failure resulting in the total FL magnetization returning to zero. This signifies that in the final states the magnetization of the parts FL_1 and FL_2 remain anti-parallel to each other. In order to establish their relative orientation in the final configuration, we chose the parts FL_1 and FL_2 of a different length (resulting in different partial magnetization). Results are shown in Fig.7. If the length of FL_2 is kept at 5nm and the length of FL_1 varies from 2nm to 4nm, the final total FL magnetization remains positive. It means that the longer part FL_2 becomes parallel to the RL. If, however, the length of FL_1 is set to 5nm and the length of FL_2 varies from 2nm to 4nm, the FL magnetization is less than zero (Fig.7). This signifies that the longer portion FL_1 is AP to the RL. It is only possible if FL_2 is parallel to RL, while FL_1 is anti-parallel to both. This configuration is shown in Fig.6, bottom panel. FL_2 flips due to the torque acting from FL_1. The flipping of FL_2 is similar to an effect of the reference layer back-hopping predicted in traditional MTJs [17]. The torque from the RL favors a certain (parallel) orientation of the FL. At the same time, the torque acting from the FL on the RL favors exactly the opposite (anti-parallel) orientation of the RL with respect to the FL. In practice this leads to an erroneous RL back-hopping and a writing failure [18] at high currents.

In the case of the composite FL the back-hopping of FL_2 results in a configuration (Fig.6, bottom) not previously present for P to AP or P to AP switching. It therefore represents an additional memory state and opens perspectives for a four-level memory operation in ultra-scaled STT-MRAM with composite FLs. However, to return from back to the P configuration (Fig.6, top) it seems necessary to invert the current and proceed from bottom to top, Fig.6, in inverse order. Is it possible to complete the switching cycle and proceed directly from Fig.6 bottom to top by only inverting FL_1?

2023 IEEE Latin American Electron Devices Conference (LAEDC)
Puebla, México, July 3-5, 2023

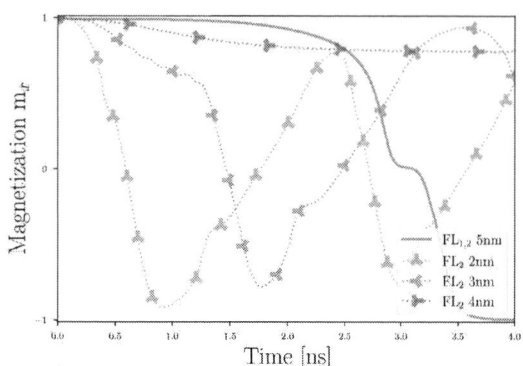

Fig.8 Magnetization trajectories for the switching from P to AP for different combinations of FL lengths (as in Fig.5). The polarizations of TBs are: 0.6, 0.9, 0.2.

The goal can be achieved if the torque acting on FL_1 from FL_2 is larger than that from the RL. Fig.8 shows the magnetization dependencies on time, when the spin polarization of the TB separating FL_1 and FL_2 is increased from 0.5 to 0.9. A clear cyclic magnetization behavior is observed for the lengths of FL_1=5nm and FL_2 equal to 3nm and 2nm. The last step in the cyclic switching shown in Fig.9 is achieved with the bias polarity typically employed for writing AP from P. At the same time, the intermediate state with a new magnetization configuration shown in Fig.9, top panel is also present in Fig.8. This configuration was not previously accessible [16]. The findings are promising for realizing a multi-level operation in ultra-scaled MRAM cells with a composite FL.

IV. CONCLUSION

The versatility of the spin and charge transport approach coupled to the magnetization dynamics to evaluate the spin-transfer torques and magnetization reversal of elongated metallic ferromagnetic layers separated by tunnel barriers is demonstrated by considering an ultra-scaled memory cell with a composite free layer. A cyclic switching through four distinct states of the composite free layer is shown, for the same current polarity. We benefit from the back-hopping effect, usually considered as parasitic, to demonstrate a multi-level operation in an ultra-scaled memory cell with a composite free layer.

ACKNOWLEDGMENT

Financial support by the Austrian Federal Ministry of Labour and Economy, the National Foundation for Research, Technology and Development and the Christian Doppler Research Association is gratefully acknowledged.

REFERENCES

[1] H. Honjo, K. Nishioka, S. Miura, H. Naganuma, T. Watanabe, T. Nasuno, T. Tanigawa, Y. Noguchi, H. Inoue, M. Yasuhira, S, Ikeda, T. Endoh, "25 nm iPMA-type Hexa-MTJ with Solder Reflow Capability and Endurance $> 10^7$ for eFlash-type MRAM", in Proc. IEDM Conf., pp. 10.3.1-10.3.4 (2022).

[2] T.Y. Lee, M. Lee, M.K. Kim, J.S. Oh, J.W. Lee, H.M. Jeong, P.H. Jang, M.K. Joo, K. Suh, S. H. Han, D.-E. Jeong, T. Kai, J. H. Jeong, J.-H. Park, J.H. Lee, Y.H. Park, E.B. Chang, Y.K. Park, H.J. Shin, Y.S. Ji, S.H. Hwang, K.T. Nam, B.S. Kwon, M.K. Cho, B.Y. Seo, Y.J. Song, G.H. Koh, K. Lee, J.-H. Lee, G. T. Jeong, "World-most Energy-Efficient MRAM Technology for Non-Volatile RAM Applications", in Proc. IEDM Conf., pp.10.7.1-10.7.4 (2022).

[3] S. Ikegawa, K. Nagel, F.B. Mancoff, S.M. Alam, M. Arora, M. DeHerrera, H.K. Lee, S. Mukherjee, G. Shimon, J.J. Sun, I. Rahman, F. Neumeyer, H.Y. Chou, Ch. Tan, A. Shah,

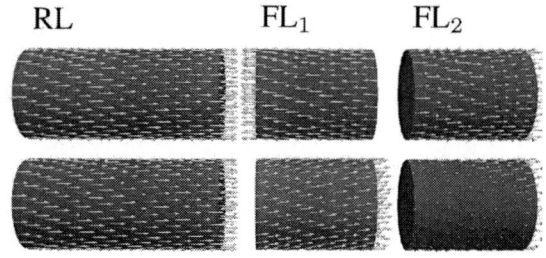

Fig.9 FL_1 reversal completing the switching cycle at bias −1.5V applied to the structure with the polarizations of TB 0.6 between RL and FL_1 and 0.9 between FL_1 and FL_2.

A. S. Aggarwal, "High-Speed (400MB/s) and Low-BER STT-MRAM Technology for Industrial Applications", in Proc. IEDM Conf., pp. 10.4.1-10.4.4 (2022).

[4] Y. Chih,, C. Chou, Y. Shih, C. Lee, W. Khwa, C. Wu, K. Shen, W. Chu, M. Chang, H. Chuang, T. Chang, "Design Challenges and Solutions of Emerging Nonvolatile Memory for Embedded Applications", in Proc. IEDM Conf., pp.2.4.1-2.4.4 (2021).

[5] G. Hu, C. Safranski, J. Sun, P. Hashemi, S. Brown, J. Bruley, L. Buzi, C. Demic, E. Galligan, M. Gottwald, O. Gunawan, J. Lee, S. Karimeddiny, P. Trouilloud, D. Worledge, "Double Spin-Torque Magnetic Tunnel Junction Devices for Last-Level Cache Applications", in Proc. IEDM Conf., pp. 10.2.1-10.2.4 (2022).

[6] I S. Ikeda, K. Miura, H. Yamamoto, K. Mizunuma, H. D. Gan, M. Endo, S. Kanai, J. Hayakawa, F. Matsukura, H. Ohno , "A Perpendicular-anisotropy CoFeB–MgO Magnetic Tunnel Junction", Nature Mater. 9, 721 (2010).

[7] H. Sato, M. Yamanouchi, S. Ikeda, S. Fukami, F. Matsukura, H. Ohno, "MgO/CoFeB/Ta/CoFeB/MgO Recording Structure in Magnetic Tunnel Junctions with Perpendicular Easy Axis", IEEE Transactions on Magnetics 49, 4437–4440 (2013).

[8] K. Nishioka, H. Honjo, S. Ikeda, T. Watanabe, S. Miura, H. Inoue, T. Tanigawa, Y. Noguchi, M. Yasuhira, H. Sato, T. Endoh, "Novel Quad Interface MTJ Technology and its First Demonstration with High Thermal Stability and Switching Efficiency for STT-MRAM beyond 2Xnm", in Symposium on VLSI Technology, pp.T120–T121 (2019).

[9] B. Jinnai, J. Igarashi, K. Watanabe, T. Funatsu, H. Sato, S. Fukami, H. Ohno, "High-Performance Shape-Anisotropy Magnetic Tunnel Junctions down to 2.3 nm", in Proc. IEDM Conf., pp.24.6.1–24.6.4 (2020).

[10] J. Slonczewski, "Current-Driven Excitation of Magnetic Multilayers", J. Magn. Magn. Mater. 159, L1-L7 (1996).

[11] L. Berger, "Emission of Spin Waves by a Magnetic Multilayer Traversed by a Current", Phys. Rev.B 54, 9353-9358 (1996).

[12] C. Abert, M. Ruggeri, F. Bruckner, C. Vogler, A. Manchon, D. Praetorius, D. Suess, "A Self-Consistent Spin-Diffusion Model for Micromagnetics", Scientific Reports 6, 16 (2016).

[13] S. Fiorentini, M. Bendra, J. Ender, R.L. Orio, W. Goes, S. Selberherr, V. Sverdlov, "Spin and Charge Drift-diffusion in Ultra-scaled MRAM Cells", Scientific Reports 12, 20958 (2022).

[14] CDL NovoMemLog, "ViennaSpinMag," last accessed on 30.03.2023. [Online]. Available: https://www.iue.tuwien.ac.at/viennaspinmag/

[15] J. Ender, M. Mohamedou, S. Fiorentini, R. Orio, S. Selberherr, W. Goes, V. Sverdlov, "Efficient Demagnetizing Field Calculation for Disconnected Complex Geometries in STT-MRAM Cells", in Proc. SISPAD Conf., pp. 213-2016 (2020).

[16] M. Bendra, S. Fiorentini, W. Goes, S. Selberherr, V. Sverdlov, "Interface Effects in Ultra-Scaled MRAM Cells", Solid-State Electron. 194, 108335 (2022).

[17] C. Abert, H. Sepehri-Amin, F. Bruckner, C. Vogler, M. Hayashi, D. Suess, "Back-Hopping in Spin-Transfer-Torque Devices: Possible Origin and Countermeasures", Phys. Rev. Applied 9, 054010 (2018).

[18] O. Golonzka, J. Alzate, U. Arslan, M. Bohr, P. Bai, J. Brockman, B. Buford, C. Connor, N. Das, B. Doyle, T. Ghani, F. Hamzaoglu, P. Heil, P. Hentges, R. Jahan, D. Kencke, B. Lin, M. Lu, M, Mainuddin, M. Meterelliyoz, P. Nguyen, D. Nikonov, K. O'brien, J.O Donnell, K. Oguz, D. Ouellette, J. Park, J. Pellegren, C. Puls, P. Quintero, T. Rahman, A. Romang, M. Sekhar, A. Selarka, M. Seth, A. J. Smith, A. K. Smith, L. Wei, C. Wiegand, Z. Zhang, K. Fischer, "MRAM as Embedded Non-Volatile Memory Solution for 22FFL FinFET Technology", in Proc. IEDM Conf., pp. 18.1.1-18.1.4 (2018).

979-8-3503-1191-4/23 $31.00 © 2023 IEEE

2023 IEEE Latin American Electron Devices Conference (LAEDC)
Puebla, México, July 3-5, 2023

Perylenediimide-based Acceptors for OPV Applications

Magaly Ramírez-Como*
Sección de Estudios de Posgrado e
Investigación, UPIITA
Instituto Politécnico Nacional
Mexico City, Mexico
ORCID: 0000-0002-9313-8337

Desiré Molina
Instituto de Bioingeniería
Universidad Miguel Hernández
Elche, Spain
ORCID: 0000-0002-1532-2032

Magali Estrada
Sección de Electrónica del Estado
Sólido, Ingeniería Eléctrica
CINVESTAV–IPN
Mexico City, Mexico
ORCID: 0000-0002-6244-6492

Luis Reséndiz
Sección de Estudios de Posgrado e
Investigación, UPIITA
Instituto Politécnico Nacional
Mexico City, Mexico
ORCID: 0000-0002-6272-9003

Josep Pallarès,
Department of Electric, Electronic and
Automatic Control
Universitat Rovira i Virgili
Tarragona, Spain
ORCID: 0000-0001-7221-5383

Ángela Sastre-Santos*
Instituto de Bioingeniería
Universidad Miguel Hernández
Elche, Spain
ORCID: 0000-0002-8835-2486

Lluis F. Marsal*
Department of Electric, Electronic and
Automatic Control
Universitat Rovira i Virgili
Tarragona, Spain
ORCID: 0000-0002-5976-1408

Abstract— Four new perylenediimides (PDIs) have been used as non-fullerene (NF) acceptors in solution-processed bulk heterojunction organic solar cells (OSC). Photovoltaic devices were fabricated using PTB7-Th as donor material. We investigated the effect of solvent vapor annealing (SVA) and thermal annealing on the performance of the fabricated inverted NF-OSCs. We found that solvent vapor annealing and thermal annealing of the active layer reduced the current density and therefore did not improve the performance of the fabricated solar cells. The highest performing cells are achieved using PTB7-Th:PDI-3 as an active blend with ZnO as the electron transport layer and MoO₃ as the hole transport layer. Most solar cells based in PDIs produced open circuit voltages in the same range as those typical solar cells made from PC₇₀BM.

Keywords— perylenediimide (PDI), non-fullerene acceptor, inverted organic solar cells, PTB7-th, polymer solar cells.

I. INTRODUCTION

Organic solar cells (OSCs) are considered a commercially viable alternative energy source due to their good processability, lightweight, low cost, and easy fabrication into large-area devices[1]–[3]. Fullerenes, such as PC₆₀BM and PC₇₀BM, which have high costs and poor light absorption properties in the visible region have dominated the field of the OSCs[4], [5]. Currently, the goal is to develop new materials that increase efficiency and reduce costs. In this sense, the

appearance of non-fullerene organic semiconductors as acceptor materials in active layers into OSCs have received great attention. Dramatic improvements in energy conversion efficiency (PCE) have been seen in non-fullerene OSC (NF-OSCs) achieving efficiencies approaching 20%[6]–[9]. Although polymers as acceptor materials have shown promising results initially, small-molecule non-fullerene acceptors have become the most studied entities. Among them, molecules built around the perylenediimide, subphtalocyanine, corannulene cores, have yielded the best results to date. The perylenediimide (PDI) family has been used as a cathode interlayer material in the conventional structure. PDIs are appropriate for working as cathode interlayers in OSCs due to their suitable energy levels, high electron mobility, high electron affinity, and facile modification[10]. The maximum PCE achieved using PDIs as cathode interlayers is 17.98% in ternary OSCs with D18-Cl:Y6:PC₇₁BM as active layer[11]. On the other hand, the perylenediimide family as acceptor material has shown balanced performance with several donor showing PCEs of over 5%[4], [12]. Moreover, PDI-based dimers have produced a PCE of more than 8%[13]. The initial problem that kind of materials presented was the formation of micrometer-sized acceptor domains, which are detrimental to the current density–voltage characteristics of the solar cells. This problem was solved by bridging two or more perylenediimide cores via their lateral ortho- or bay- positions which reduced the propensity for aggregation, as a result, obtained an active layer with crystallites of moderate size [14], [15].

In this contribution, we report four PDI-based molecules used as non-fullerene acceptor material in the active layer into inverted OSCs. Utilizing PTB7-Th as donor material, the fabricated iNF-OSCs have the following structure ITO/ZnO/PTB7-Th: non-fullerene acceptor/MoO₃/Ag. Solvent vapor annealing (SVA) and thermal annealing were used in order to improve the performance of solar cells.

This work was supported by the Spanish Ministerio de Ciencia e Innovación (MICINN/FEDER) under Grants PDI2021-128342OB-I00 and PID2020-117855 RB-I00; by the Agency for Management of University and Research Grants (AGAUR) ref. 2021-SGR-00739; by COST Action 20126 – NETPORE; by the Catalan Institution for Research and Advanced Studies (ICREA) under the ICREA Academia Award; by Diputació de Tarragona under Grant 2021CM14, 2022PGR-DIPTA-URV04, 2022/33; by the Generalitat Valenciana (CIPROM/2021/059) and the Advanced Materials program by MCIN with funding from European Union NextGenerationEU (PRTR-C17.I1) and Generalitat Valenciana (MFA/2022/028); by CONACYT program postdoctoral grants BP-PA-20220624083033039-2364083.

979-8-3503-1191-4/23 $31.00 © 2023 IEEE

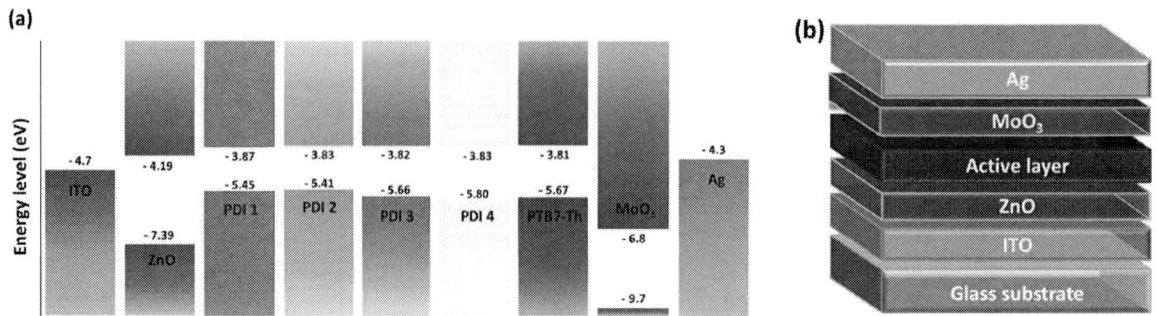

Fig. 1. (a) Energy band diagrams of iNF-OSCs. (b) The structure of the inverted bulk heterojunction NF-OSCs.

TABLE I.   PERFORMANCE PARAMETERS OF THE iNF-OSCs WITHOUT ANNEALING

Acceptor	J_{SC} (mA/cm²)	V_{OC} (mV)	FF (%)	R_S (Ω cm²)	R_{SH} (Ω cm²)	PCE (%)
PDI 1	5.08	510	39	4.2	227	1.02
PDI 2	3.64	770	33	10.3	392	0.92
PDI 3	4.38	710	34	7.7	304	1.07
PDI 4	3.84	740	36	5.4	369	1.02

TABLE II.   PERFORMANCE PARAMETERS OF THE iNF-OSCs WITH THERMAL ANNEALING (100 °C 10 MIN)

Acceptor	J_{SC} (mA/cm²)	V_{OC} (mV)	FF (%)	R_S (Ω cm²)	R_{SH} (Ω cm²)	PCE (%)
PDI 1	4.93	480	42	2.9	250	1.00
PDI 2	2.37	739	30	27.5	426	0.52
PDI 3	3.18	780	33	12.9	385	0.83
PDI 4	3.43	720	34	6.8	372	0.83

TABLE III.   PERFORMANCE PARAMETERS OF THE iNF-OSCs WITH SOLVENT VAPOR ANNEALING (SVA) IN CHLOROFORM FOR 10 MIN

Acceptor	J_{SC} (mA/cm²)	V_{OC} (mV)	FF (%)	R_S (Ω cm²)	R_{SH} (Ω cm²)	PCE (%)
PDI 1	5.01	550	36	5.8	202	1.00
PDI 2	2.68	710	35	13.0	431	0.67
PDI 3	2.28	720	35	42.2	534	0.58
PDI 4	3.34	760	36	6.1	407	0.92

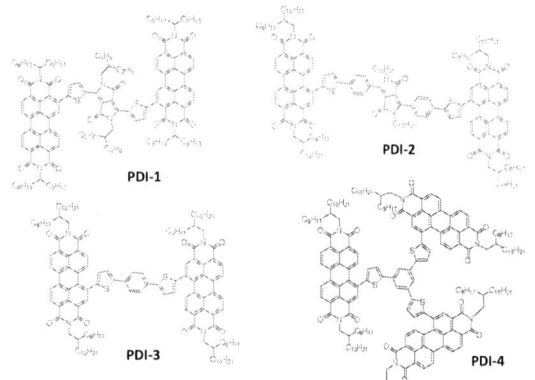

Fig. 2. Chemical structure of the studied PDIs

Comparative studies were carried out to research the effect of these annealing on the overall performance of fabricated solar cells.

II. EXPERIMENTAL SECTION

A. Materials

The substrate used was glass and a patterned of indium tin oxide (ITO) on glass. The resistivity of ITO is 10 Ω/sq purchased from Xin Yan Technology LTD. PTB7-Th donor polymer and PC₇₀BM acceptor fullerene was purchased from One-Material. Chloroform (anhydrous, ≥ 99%), Chlorobenzene (anhydrous, 99.8%), 1,8-Diiodooctane (98%), 2-methoxyethanol (99.9%), ethanolamine (99.5%), zinc acetate dihydrate (99.999%), molybdenum oxide (MoO₃, 99.97%) was obtained from Sigma-Aldrich. Silver (Ag, 99.99%) wire was purchased from Testbourne Ltd. PDI-DPP-PDI 1, PDI 2, PDI-3, and PDI-4 were synthesized by the group led by Ángela Sastre–Santos (Fig. 2).

B. Organic Solar cell fabrication

Inverted non-fullerene organic solar cells (iNF-OSCs) fabricated were based on the structure of ITO/ZnO/Active layer/MoO₃/Ag as shown in Fig. 1(b). ITO glass substrates were cleaned in a solution of deionized water and soap. Then, the substrates were put in an ultrasonic bath with acetone, methanol, and isopropanol high-purity solvents in a sequence for 10 min, respectively. Finally, the cleaned substrates were dried in at100 °C in an oven for 10 min. Before the first layer deposit, the substrates were treated with UV- Ozone for 30 min. The process for preparing the ZnO precursor solution is carried out as detailed in [16], [17]. The ZnO precursor solution was deposited on the ITO by spin-coated at 4000 rpm for 45 s and left for 1 h at 200 °C in air, obtaining a film with 30 nm thickness. The ZnO-coated ITO substrates were brought to a glove box for active layer deposition. The active blend was prepared in chloroform with the weight ratio of 1:1.5 of the PTB7-Th and the corresponding PDI, to get an overall concentration of 10 mg/ml. Then, the solution was continuously stirred for all the night at 40 °C and spin-coated at 1500 rpm for 60 s onto the ITO/ZnO substrates. Finally, hole transport layer, MoO₃ (10 nm), and metal contact, Ag (100 nm) was deposited by thermal evaporation at ≤ 1×10⁻⁶ mbar through a shadow mask leading solar cells with an area of 9 mm².

C. Device Characterization and Measurements

The current density – voltage (J–V) characteristics of the solar cells were conducted at room temperature with Keithley

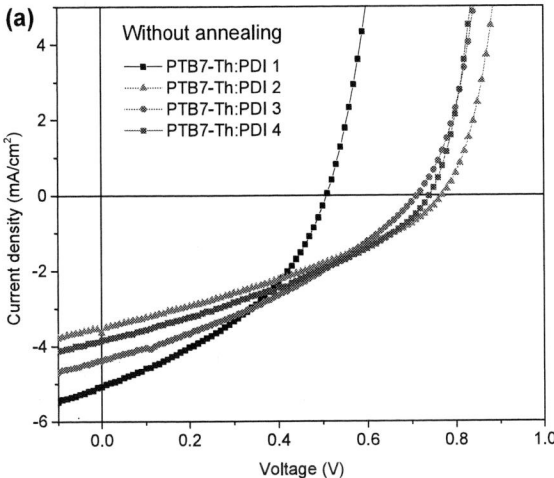

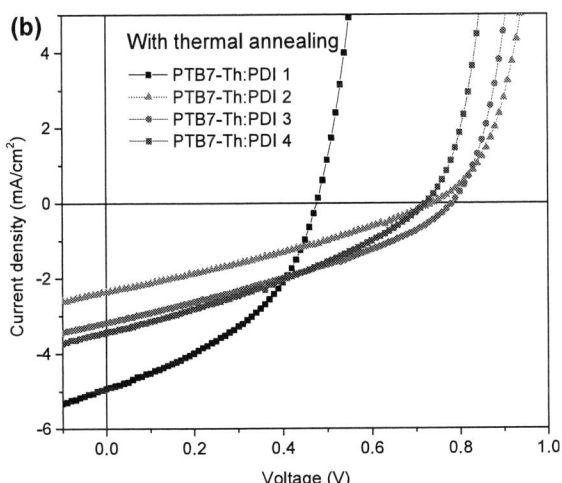

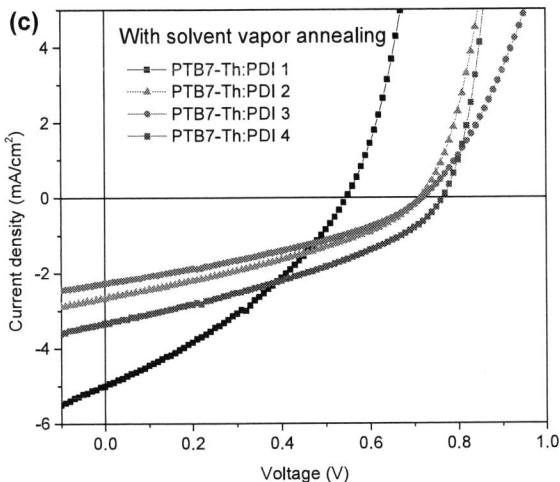

Fig. 3. J–V characteristic curves under AM 1.5 G illumination of the iNF-OSCs based in PDIs acceptors (a) without annealing, (b) with thermal annealing, and (c) whit solvent vapor annealing (SVA).

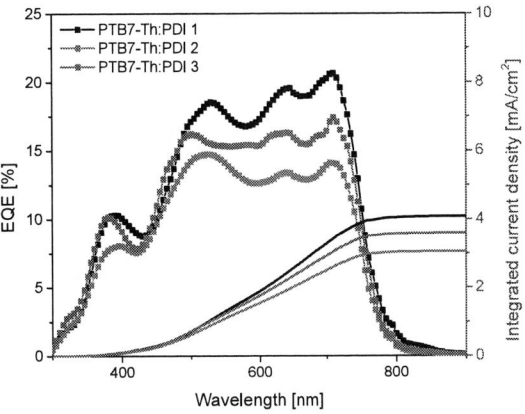

Fig. 4. External quantum efficiency (EQE) spectrum of iNF-OSCs without annealing.

2400 source-measure unit under light and dark conditions. The illumination was obtained from solar simulator (Abet Technologies model 11 000 class type A, Xenon arc). Calibration for the solar simulator intensity was operated with a Fraunhofer-certified photovoltaic cell to yield a 100 mW/cm² and AM1.5G spectrum. The external quantum efficiency (EQE) spectra of the solar cells were measured using Lasing IPCE-DC model equipment.

III. RESULTS AND DISCUSSION

In this research work, an iNF-OSCs structure of ITO/ZnO/Active layer/MoO₃/Ag as shown in Fig 1(b) with the schematic energy band in Fig. 1(a) were fabricated. the metals work functions and the energy positions of the band edges for the semiconductor were taken from [5], [17], [18] references. PTB7-Th was chosen as the polymer donor since the benzodithiophene–thienothiophenediyl copolymer family has shown the best results in devices fabricated with PDI acceptors[4], [5], [12]. The four PDIs were optimized for the donor/acceptor (D/A) ratio, annealing temperature, and solvent vapor annealing (SVA). Fig. 3 shown the current density–voltage characteristics (J–V) under light conditions of iNF-OSCs. It demonstrates the effect of different PDIs, in addition to the influence of the active layer SVA and the thermal annealing on the performance parameters of the fabricated solar cells.

Fig. 3(a) shown the effect of the PDIs molecules without SVA and thermal annealing. Table I shows the shows the performance parameters of the fabricated iNF-OSC: short circuit current density (J_{SC}), open circuit voltage (V_{OC}), power conversion efficiency (PCE), shunt (R_{SH}) and series (R_S) resistance. It can be noticed that the iNF-OSC with PDI 1 showed the highest J_{SC} and FF.

Fig. 3(b) showed the J–V of the solar cells due to active layer thermal annealing at 100 °C for 10 min. The observed results showed that J_{SC} of all fabricated solar cells decreased. It can be explained because commonly the thermal annealing decreases the thickness of the solar cell [19]. Therefore, when the thickness is decreased is expected to have less absorption as a result is obtained less current density. Additionally, it is convenient to mention that the iNF-OSC with PDI 1 still shows the highest J_{SC} and FF, as well as the lowest R_S, as illustrated in Table II.

Fig. 3(c) shows the *J–V* with the effect of the SVA for 10 min of chloroform. Initially, SVA was used in order to improve the morphology of the active layer and consequently facilitate current generation. However, it was obtained that SVA did not improve current density compared to non-annealed solar cells. On the other hand, iNF-OSC with PDI 1 maintained the highest J_{SC} and FF as listed in Table III.

Interestingly, the best performing devices were obtained when no SVA and thermal annealing was applied to the obtained the active layer. It was found that both SVA and thermal annealing do not have a positive effect on the efficiency of solar cells. The solar cells show modest PCEs, with unusually low FF and limited J_{SC}. However, in the solar cells with PDIs 2, 3 and 4 the V_{OC} is in the same range than those typical solar cells made from $PC_{70}BM$ [20]–[22]. The J_{SC} is higher with PDI 1 than with others PDIs while the V_{OC} is lower. Amongst the four PDIs, the PDI 3 was shown to produce the highest PCE without annealing, although that of PDI 1 was very similar and was better than the other PDIs under annealed conditions.

The J_{SC} trend is corroborated by the EQE spectrum of iNF-OSCs without annealing, as seen in Fig. 4. Importantly, the PDIs acceptors are seen to contribute significantly to photon conversion efficiency at lower wavelengths ($450 - 550$ nm).

IV. Conclusion

The inverted non-fullerene organic solar cells were fabricated using perylenediimide-based acceptor and PTB7-Th. It can be concluded the thermal annealing and solvent vapor annealing of the active layer showed no improved the performance parameters of solar cells. The best power conversion efficiency was enhanced for the solar cell with PDI 3. The solar cells exhibit a similar V_{OC} than PCBM-based solar cells. However, the density current–voltage characteristics leads a typically low FF and moderate J_{SC}. In order to identify which features of the active layer may play a key role in cell performance, an analysis of the morphological characteristics and photophysical characteristics should continue.

Acknowledgment

M. Ramírez–Como thanks Consejo Nacional de Ciencia y Tecnología (CONACYT) for supporting the postdoctoral scholarship under CVU 704226. This work was supported by the Spanish Ministerio de Ciencia e Innovación (MICINN/FEDER) PDI2021- 128342OB-I00, by the Agency for Management of University and Research Grants (AGAUR) ref. 2021- SGR-00739, COST Action 20126 - NETPORE and by the Catalan Institution for Research and Advanced Studies (ICREA) under the ICREA Academia Award and by the Diputació de Tarragona 2022/33.

References

[1] W. Hou, Y. Xiao, G. Han, and J.-Y. Lin, "The Applications of Polymers in Solar Cells: A Review," *Polymers (Basel)*, vol. 11, no. 1, p. 143, Jan. 2019, doi: 10.3390/polym11010143.

[2] O. a. Abdulrazzaq, V. Saini, S. Bourdo, E. Dervishi, and A. S. Biris, "Organic Solar Cells: A Review of Materials, Limitations, and Possibilities for Improvement," *Particulate Science and Technology*, vol. 31, no. 5, pp. 427–442, 2013, doi: 10.1080/02726351.2013.769470.

[3] A. Khalil, Z. Ahmed, F. Touati, and M. Masmoudi, "Review on organic solar cells," in *2016 13th International Multi-Conference on Systems, Signals & Devices (SSD)*, IEEE, Mar. 2016, pp. 342–353. doi: 10.1109/SSD.2016.7473760.

[4] T. Yu, W. He, M. Jafari, T. Guner, P. Li, M. Siaj, R. Izquierdo, B. Sun, G. C. Welch, A. Yurtsever, D. Ma, "3D Nanoscale Morphology Characterization of Ternary Organic Solar Cells," *Small Methods*, vol. 6, no. 1, p. 2100916, Jan. 2022, doi: 10.1002/smtd.202100916.

[5] C. Stenta *et al.*, "Diphenylphenoxy-Thiophene-PDI Dimers as Acceptors for OPV Applications with Open Circuit Voltage Approaching 1 Volt," *Nanomaterials*, vol. 8, no. 4, p. 211, Mar. 2018, doi: 10.3390/nano8040211.

[6] J. Lv, H. Tang, J. Huang, C. Yan, K. Liu, Q. Yang, D. Hu, R. Singh, J. Lee, S. Lu, G. Li, and Z. Kan, "Additive-induced miscibility regulation and hierarchical morphology enable 17.5% binary organic solar cells," *Energy Environ Sci*, vol. 14, no. 5, pp. 3044–3052, 2021, doi: 10.1039/D0EE04012F.

[7] M. Ramirez-Como, E. Moustafa, A. A. A. Torimtubun, J. G. Sanchez, J. Pallares, and L. F. Marsal, "Preliminary Study of the Degradation of PM6:Y7-based Solar Cells," in *2022 IEEE Latin American Electron Devices Conference (LAEDC)*, IEEE, Jul. 2022, pp. 1–5. doi: 10.1109/LAEDC54796.2022.9908202.

[8] E. Moustafa, M. Méndez, J. G. Sánchez, J. Pallarès, E. Palomares, and L. F. Marsal, "Thermal Activation of PEDOT:PSS/PM6:Y7 Based Films Leads to Unprecedent High Short-Circuit Current Density in Nonfullerene Organic Photovoltaics," *Adv Energy Mater*, vol. 13, no. 4, p. 2203241, Jan. 2023, doi: 10.1002/aenm.202203241.

[9] G. Li, L.-W. Feng, S. Mukherjee, L. O. Jones, R. M. Jacobberger, W. Huang, R. M. Young, R. M. Pankow, W. Zhu, N. Lu, K. L. Kohlstedt, V. K. Sangwan, M. R. Wasielewski, M. C. Hersam, G. C. Schatz, D. M. DeLongchamp, A. Facchetti, and T. J. Marks, "Non-fullerene acceptors with direct and indirect hexa-fluorination afford >17% efficiency in polymer solar cells," *Energy Environ Sci*, vol. 15, no. 2, pp. 645–659, 2022, doi: 10.1039/D1EE03225A.

[10] J. Yao, Q. Chen, C. Zhang, Z. Zhang, and Y. Li, "Perylene-diimide-based cathode interlayer materials for high performance organic solar cells," *SusMat*, vol. 2, no. 3, pp. 243–263, Jun. 2022, doi: 10.1002/sus2.50.

[11] M. Liu, Y. Jiang, D. Liu, J. Wang, Z. Ren, T. P. Russell, and Y. Liu, "Imidazole-Functionalized Imide Interlayers for High Performance Organic Solar Cells," *ACS Energy Lett*, vol. 6, no. 9, pp. 3228–3235, Sep. 2021, doi: 10.1021/acsenergylett.1c01482.

[12] S. M. McAfee, A.-J. Payne, A. D. Hendsbee, S. Xu, Y. Zou, and G. C. Welch, "Toward a Universally Compatible Non-Fullerene Acceptor: Multi-Gram Synthesis, Solvent Vapor Annealing Optimization, and BDT-Based Polymer Screening," *Solar RRL*, vol. 2, no. 9, p. 1800143, Sep. 2018, doi: 10.1002/solr.201800143.

[13] D. Meng, D. Sun, C. Zhong, T. Liu, B. Fan, L. Huo, Y. Li, W. Jiang, H. Choi, T. Kim, J. Y. Kim, Y. Sun, Z.-h Wang, and A. J. Heeger, "High-Performance Solution-Processed Non-Fullerene Organic Solar Cells Based on Selenophene-Containing Perylene Bisimide Acceptor," *J Am Chem Soc*, vol. 138, no. 1, pp. 375–380, Jan. 2016, doi: 10.1021/jacs.5b11149.

[14] D. Sun, D. Meng, Y. Cai, B. Fan, Y. Li, W. Jiang, L. Huo, Y. Sun, and Z. Wang, "Non-Fullerene-Acceptor-Based Bulk-Heterojunction Organic Solar Cells with Efficiency over 7%," *J Am Chem Soc*, vol. 137, no. 34, pp. 11156–11162, Sep. 2015, doi: 10.1021/jacs.5b06414.

[15] X. Zhang, C. Zhan, and J. Yao, "Non-fullerene organic solar cells with 6.1% efficiency through fine-tuning parameters of the film-forming process," *Chemistry of Materials*, vol. 27, no. 1, pp. 166–173, Jan. 2015, doi: 10.1021/cm504140c.

[16] M. Ramirez-Como, V. S. Balderrama, J. G. Sánchez, A. Sacramento, M. Estrada, J. Pallarés, and L. F. Marsal, "Solution-Processed Small Molecule Inverted Solar Cells: Impact of Electron Transport Layers," *IEEE Journal of the Electron Devices Society*, vol. 10, no. November 2021, pp. 435–442, 2022, doi: 10.1109/JEDS.2022.3165315.

[17] J. G. Sánchez, V. S. Balderrama, S. I. Garduño, E. Osorio., A. Viterisi, M. Estrada, J. Ferré-Borrull, J. Pallarès, and L. F. Marsal, "Impact of inkjet printed ZnO electron transport layer on the characteristics of polymer solar cells," *RSC Adv*, vol. 8, no. 24, pp. 1–9, 2018, doi: 10.1039/c8ra01481g.

[18] A. Sacramento, M. Ramírez-Como, V. S. Balderrama, J. G. Sánchez, J. Pallarés, L. F. Marsal, and M. Estrada, "Comparative degradation analysis of V_2O_5, MoO_3 and their stacks as hole

transport layers in high-efficiency inverted polymer solar cells," *J Mater Chem C Mater*, vol. 9, no. 20, pp. 6518–6527, 2021, doi: 10.1039/D1TC00219H.

[19] J. G. Sanchez, A. A. A. Torimtubun, V. S. Balderrama, M. Estrada, J. Pallares, and L. F. Marsal, "Effects of Annealing Temperature on the Performance of Organic Solar Cells Based on Polymer: Non-Fullerene Using V_2O_5 as HTL," *IEEE Journal of the Electron Devices Society*, vol. 8, pp. 421–428, 2020, doi: 10.1109/JEDS.2020.2964634.

[20] L. Reséndiz, V. S. Balderrama, G. Lastra, M. Ramírez, V. Cabrera, and M. Estrada, "Optimization of PFN thickness in inverted high-performance PTB7:PC_{70}BM solar cells," *Solid State Electron*, vol. 153, pp. 33–36, Mar. 2019, doi: 10.1016/j.sse.2018.12.013.

[21] M. Ramírez-Como, A. Sacramento, J. G. Sánchez, M. Estrada, J. Pallarès, V. S. Balderrama, L. F. Marsal, "Small molecule organic solar cells toward improved stability and performance for Indoor Light Harvesting Application," *Solar Energy Materials and Solar Cells*, vol. 230, no. June, p. 111265, Sep. 2021, doi: 10.1016/j.solmat.2021.111265.

[22] A. Sacramento, V. S. Balderrama, M. Ramirez-Como, J. G. Sanchez, M. Estrada, and L. F. Marsal, "Inverted Polymer Solar Cells Using V_2O_5/NiO as anode selective contact: Degradation Study," in *2020 IEEE Latin America Electron Devices Conference (LAEDC)*, San José: IEEE, Feb. 2020, pp. 1–4. doi: 10.1109/LAEDC49063.2020.9073402.

Effects of the compliance current in the electroforming process of HfO$_2$-based ReRAM devices

Silvana Guitarra
IMNE, Colegio de Ciencias e
Ingeniería
Universidad San Francisco de Quito
Quito, Ecuador
sguitarrra@usfq.edu.ec

Lionel Trojman
LISITE, ISEP,
Institut Supérieur dÉlectronique de
Paris,
Paris, France
lionel.trojman@isep.fr

Laurent Raymond
Aix-Marseille Université,
Université de Toulon, CNRS, CPT
Marseille, France
laurent.raymond@univ-amu.fr

Abstract— Electroforming is the activation process of ReRAM devices that initiates the resistive switching response by creating filamentary conduction paths in the oxide film. This step is fundamental for the proper operation of the device, and it is controlled by a compliance current (I$_c$) that prevents the hard dielectric breakdown of the ReRAMs. In this work, we study the influence of this current during the electroforming process and operation in HfO$_2$-based ReRAM devices of different areas. Some parameters from the IV curve were extracted and analyzed to understand how the process could affect the device's performance. We found a maximum value of I$_c$ needed to form the CF independent of the area device that does not provoke significant changes in the electrical behavior.

Keywords— *ReRAM, electroforming, compliance current, conductive filament, filamentary conduction*

I. INTRODUCTION

The ReRAM is a nonvolatile memory device studied due to the favorable characteristics that could allow it to replace flash memory technology [1]. These devices have a metal-insulator-metal structure whose operating principle relies on resistive switching controlled by an external voltage [2]. After fabrication, these devices are in a highly resistive state, called a pristine state. They need to be activated by an electroforming process that creates a conductive filament (CF) inside the insulator that connects the electrodes. During operation, the ReRAM can change between a high resistive state (HRS) and a low resistive state (LRS) due to the local changes inside the CF. Indeed, electroforming is a critical step for the functioning of the ReRAM devices because it determines the switching characteristics, especially the resistance, during the HRS and LRS.

For the electroforming, a positive voltage is applied to the top electrode, and a compliance current (I$_c$), controlled by the MOSFET transistor, is fixed to avoid the hard dielectric breakdown of the insulator layer during the forming and the set process [3]. In this work, we study the effect of the compliance current on the electroforming process of devices with nine different areas. Section II describes the samples and shows the current-voltage electroforming curves. Section III corresponds to the results and the discussion where some parameters are used to analyze the electroforming process and the answer during the operation of the devices. Finally, in section IV, we give the conclusions of this work.

II. ELECTROFORMING

A. Samples

The HfO$_2$-based ReRAM devices consist of TiN/HfO$_2$/Hf/TiN stacks with heights of 30nm/5nm/10nm/30nm, respectively. The samples have nine different areas 55x55nm^2, 65x65nm^2, 75x75nm^2, 85x85nm^2, 105x105nm^2, 135x135nm^2, 1x1μm^2, 3x3μm^2 and 5x5μm^2 [4] Due to the small size, the devices have a one-transistor one-resistor (1T1R) architecture to enable accurate electrical characterization (Fig. 1 (a)). The 1T1R array eliminates the sneak path current, decreasing energy consumption and causing higher reliability [5]. The metal-oxide-semiconductor field effect transistor (MOSFET) acts as an access device and allows the programming operation of a selected cell.

For electrical characterization, we used a four-probe station Cascade Alessi REL 4800 (Fig. 1 (b)). The measurements were programmed using the Keithley Interactive Test Environment (KITE).

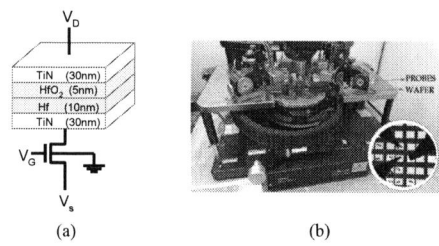

Fig. 1. (a) HfO2-based ReRAM devices consist of TiN/HfO$_2$/Hf/TiN stacks with heights of 30nm/5nm/10nm/30nm, respectively. (b) Cascade Alessi REL 4800 equipment and a wafer load probe station. The measurements were programmed using the Keithley Interactive Test Enviroment (KITE).

B. Electroforming process

During the electroforming step, the effect of compliance current, I$_c$, was studied under two conditions.. First, all samples described in section A were electroformed with I$_c$=5mA. Second, another group corresponding to the samples of 55x55nm^2, 65x65nm^2, 1x1μm^2, 3x3μm^2, and 5x5μm^2 were electroformed with I$_c$=10mA.

The IV characteristic curve allows us to recognize the point where the CF has been created inside the insulator due

979-8-3503-1191-4/23 $31.00 © 2023 IEEE

to an abrupt current increase (see Fig. 2). Initially, there is a low current answer due to the disconnection of the electrodes that maintain the system in a high resistive state. Thus, by applying a high voltage, the ReRAM switches into a LRS due to the formation of conductive paths [6]. We extract two parameters to quantify the process: the forming voltage (V_{form}) and current (I_{form}).

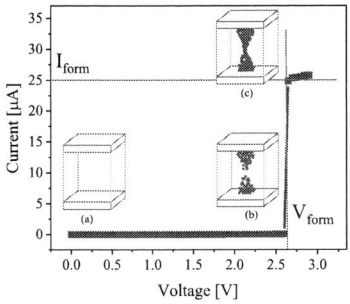

Fig. 2. Electroforming IV curve along with a schematic creation of the conductive filaments of oxygen vacancies during the process. The IV characteristic curve allows us to recognize the point where the CF has been created inside the insulator because of an abrupt current increase. The forming voltage (V_{form}) and the forming current (I_{form}) were extracted for all curves.

Comparing the IV electroforming curves under the two compliance currents, one can see that the influence of I_c is more significant in nm-size samples, where the I_{form} is larger with I_c=10mA. However, there is no apparent difference in um-size samples, as shown in Fig 3.

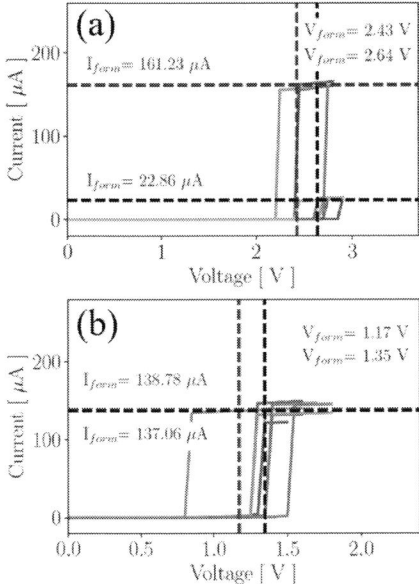

Fig. 3. IV curves during electroforming of two samples (a) 55x55nm^2 and (b) 5x5um^2. The results for I_c=5mA are in black, while for I_c=10mA, the results are in blue. One can see that the compliance current significantly affects the response of the nanometer size samples.

III. RESULTS AND DISCUSSION

The sudden current jump presented in the IV curve suggests the formation of the CF (see Fig. 2). In HfO$_2$-based ReRAMs, the electroforming is a field-driven process [7] where the electric field provides the required energy for moving the pre-existing vacancies. In addition, the field creates new oxygen vacancies due to the breaking of Hf-O bonds and allows the subsequent diffusion of oxygen ions [8]. It must be noted that all devices studied in this work have been fabricated under the same process, so all of them are expected to have similar pre-existing oxygen vacancy profiles [4]. However, we found that the response strongly depends on the sample area, as shown in Fig. 4, where the V_{form} and I_{form} as a function of the device area are presented.

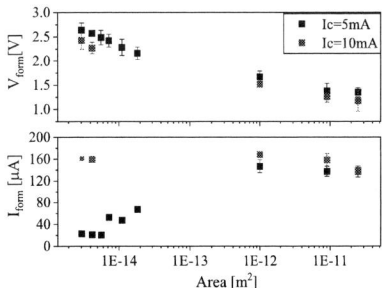

Fig. 4. Area scaling trends for (a) forming voltage (V_{form}) and (b) forming current (I_{form}), comparing the effect of I_c .

V_{form} decreases with the area when the samples are electroformed under both I_c, although V_{form} has a slightly smaller value when I_c is larger. This behavior has been reported before by Chen [9], and it is explained because a small area limits the locations to trigger electroforming. This V_{form} behavior could represent a problem from the point of view of memory scalability. Conversely, we found that I_{form} significantly differs between nm-size and um-size samples when I_c is 5mA, while it is almost independent when I_c is 10mA (see Fig. 2.6b). It could suggest that I_c=10mA could be a saturation value for this HfO$_2$-based ReRAM samples.

After the electroforming process, the CF is complete, and it is possible to consider that the dominant conduction mechanism is Ohmic, following the relationship $I_{form}=G_{form}V_{form}$. We determine the value of G_{form} and compare it along the area. As shown in Fig. 5, the tendency shows that the conductance increases with the area device. Further, the conductance value is more significant in all samples when the I_c=10mA. These G_{form} values would be related to the filament size, which is more significant in μm-size samples due to a higher concentration of available oxygen vacancies [10]. A higher compliance current provides more energy for CF formation.

2023 IEEE Latin America Electron Device Conference (LAEDC). Puebla, México, July 3-5, 2023

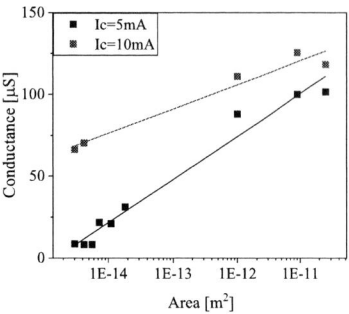

Fig. 5. Form conductance, G_{form}, as a function of the area. The tendency shows that the conductance increases with area device.

To analyze the effect of the electroforming process, under two values of compliance current, the I–V curves during the resistive switching of the two devices are presented in Fig. 6. The curves show the typical bipolar resistive switching behavior, and we cannot identify considerable differences in the LRS region. On the contrary, the curves tend to move away when the system is in HRS, especially under negative voltages. However, the variability during the HRS has already been reported before, and it is explained due to the conduction mechanism during this state [1,4,11]. According to the QPC model, once a CF is formed inside the HfO_2 switching layer, the tunnel barrier is the dominant factor that controls I-V characteristics with barrier thickness modification [12]. When the system is in LRS, the conduction through the CF is ideally ballistic, while in HRS is mainly due to the tunneling process.

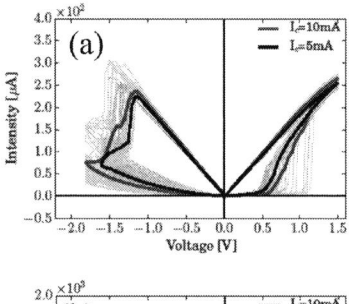

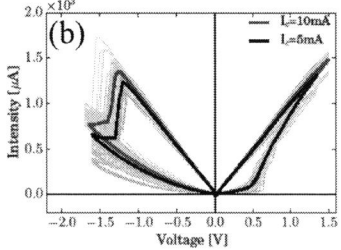

Fig. 6. I-V characteristic for a RRAM device of (a) $55 \times 55 nm^2$ and (b) $5 \times 5 \mu m^2$ during 50 cycles. The curves show the typical bipolar resistive switching behavior. We cannot identify considerable differences in the transition points, set and reset voltages, and in the LRS region.

During operation, in LRS the primary conduction mechanism is Ohmic. The continuous distribution of oxygen vacancies that creates the CF gives it a metallic behavior. Thus, we obtain the device's conductance to analyze how the

compliance current affects the ReRAM operation during this state. In Fig. 7, one can observe G as a function of the area

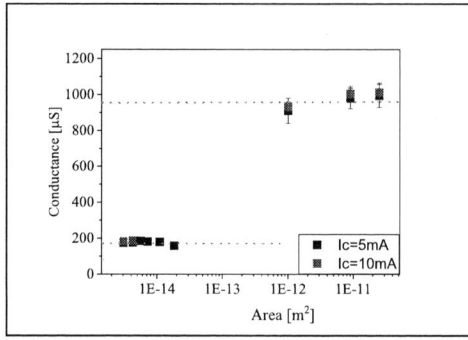

device.

Fig. 7. Conductance as a function of the area during operation of the ReRAM device in the LRS.

The main characteristic is the clear conductance difference between nm-size and um-size samples. The greater conductance suggests a wider CF due to a more significant number of conductive paths [10]. However, G has a similar value among samples whose areas are in the same order of magnitude, independently of the compliance current used during electroforming. A detailed analysis of electrical parameters associated with resistive switching can be found in [4].

IV. CONCLUSIONS

This paper analyzes the effect of the compliance current, I_c, during the activation process of Hf-based ReRAM devices of different areas. The results show that although there are some differences during the electroforming process, especially in nm-size samples, the devices' operation is unaffected during resistive switching. However, the conductance value of the devices during LRS, where the conduction could be considered as Ohmic, suggests that the size of the conductive filament in um-size samples is large.

REFERENCES

[1] D. Ielmini, "Resistive Switching Memories based on Metal Oxides: Mechanisms, Reliability and Scaling," Semicond. Sci. Technol. 31, 063002 (2016).

[2] H. -S. P. Wong *et al.*, "Metal–Oxide RRAM," in *Proceedings of the IEEE*, vol. 100, no. 6, pp. 1951-1970, June 2012.

[3] Zahoor, F., Azni Zulkifli, T.Z. & Khanday, F.A. Resistive Random Access Memory (RRAM): an Overview of Materials, Switching Mechanism, Performance, Multilevel Cell (mlc) Storage, Modeling, and Applications. *Nanoscale Res Lett* **15**, 90 (2020).

[4] S. Guitarra, L. Raymond and L. Trojman, "Stochastic multiscale model for HfO2 based resistive random access memories with 1T1R configuration," *Solid State Electronics*, 176, 2021.

[5] M. Mao, Y. Cao, S. Yu et al. "Optimizing latency, energy, and reliability of 1T1R ReRAM through appropriate voltage settings". In: *2015 33rd IEEE International Conference on Computer Design (ICCD)*. Oct. 2015, p. 359–366.

[6] Hu Q, Park MR, Abbas H, Kang TS, Yoon TS, Kang CJ (2018) Forming-free resistive switching characteristics in tantalum oxide and manganese oxide based crossbar array structure. Microelectron Eng 190:7–10.

[7] G. Bersuker, D. C. Gilmer, D. Veksler et al. "Metal oxide resistive memory switching mechanism based on conductive filament properties", In: *Journal of Applied Physics* 110.12 (2011), p. 124518.

979-8-3503-1191-4/23 $31.00 © 2023 IEEE

[8] J Joshua Yang, Feng Miao, Matthew D Pickett et al. "The mechanism of electroforming of metal oxide memristive switches", In: *Nanotechnology* .20.21 (2009), p. 215201.

[9] A. Chen. "Area and Thickness Scaling of Forming Voltage of Resistive Switching Memories". In : *IEEE Electron Device Letters* 35.1 (jan. 2014), p. 57–59. issn: 0741-3106

[10] Zhu, L.; Zhou, J.; Guo, Z.; Sun, Z. Synergistic Resistive Switching Mechanism of Oxygen Vacancies and Metal Interstitials in Ta2O5 . *J. Phys. Chem. C* **2016**, *120* (4), 2456–2463.

[11] Chen Wang, Huaqiang Wu, Bin Gao, Teng Zhang, Yuchao Yang, He Qian, Conduction mechanisms, dynamics and stability in ReRAMs, Microelectronic Engineering, Volumes 187–188, 2018, Pages 121-133.

[12] LM. Prócel, L. Trojman, J. Moreno, F. Crupi, V. Maccaronio, R. Degraeve, et al. "Experimental evidence of the quantum point contact theory in the conduction mechanism of bipolar HfO_2-based resistive random access memories," *J. Appl. Phys.* 2013;114:74509.

2023 IEEE Latin America Electron Devices
Conference (LAEDC)
Puebla, México, July 3-5, 2023

Development of a Biosensor for the Detection of Glucose Levels in Saliva Based on the Oxidation and Detection of Hydrogen Peroxide

Fernando Sánchez-Hernández
Departamento de Electrónica, Sistemas e Informatica
Instituto Tecnológico y de Estudios Superiores de Occidente (ITESO)
Tlaquepaque, Jalisco, México
fernando.sanchezh@iteso.mx

Esteban Martínez-Guerrero
Departamento de Electrónica, Sistemas e Informatica
Instituto Tecnológico y de Estudios Superiores de Occidente (ITESO)
Tlaquepaque, Jalisco, México
margres@iteso.mx

Elsie Evelyn Araujo-Palomo
Dispositivos y Materiales Avanzados, Departamento de Matématicas y Física
Instituto Tecnológico y de Estudios Superiores de Occidente (ITESO)
Tlaquepaque, Jalisco, México
elsie@iteso.mx

Cuauhtémoc R. Aguilera-Galicia
Departamento de Electrónica, Sistemas e Informatica
Instituto Tecnológico y de Estudios Superiores de Occidente (ITESO)
Tlaquepaque, Jalisco, México
cuauhtemoc@iteso.mx

José Luis Chávez-Hurtado
Departamento de Electrónica, Sistemas e Informatica
Instituto Tecnológico y de Estudios Superiores de Occidente (ITESO)
Tlaquepaque, Jalisco, México
josechavezh@iteso.mx

Patricia G. López-Cárdenas
Dispositivos y Materiales Avanzados, Departamento de Matématicas y Física
Instituto Tecnológico y de Estudios Superiores de Occidente (ITESO)
Tlaquepaque, Jalisco, México
plopez@iteso.mx

Jaime Ramírez-Angulo
Emeritus Professor New Mexico State University
Emeritus SNI III México
El Paso, Texas, USA.
jairamir@nmsu.edu

Abstract— This work presents the experimental results of cyclic voltammetry tests that correlate an electric current and the concentration of hydrogen peroxide within a controlled solution. As the concentration of hydrogen peroxide increases in the managed solution, so does the electric intensity. The goal of this data allows us to corroborate that the proposed biosensor is a reproducible and potential alternative, painless, and non-invasive method for monitoring glucose levels in patients who have diabetes. The transducer is a novel Ni Polycarbonate Track Etch nanowire sensor with a high sensitivity to detect low hydrogen peroxide in human fluids.

Keywords—Diabetes Mellitus, Hydrogen peroxide, Cycle Voltamperometry, Nano-Wire sensor

I. INTRODUCTION

Diabetes mellitus has emerged as a 21st-century pandemic, posing a significant global health threat. As reported by the WHO and the IDF, approximately 10.3% of adults worldwide have diabetes, with numbers predicted to rise to 643 million by 2030 and 783 million by 2045 [1]. The countries most affected are China, India, the United States, Brazil, Russia, and Mexico [2]. In Mexico, diabetes was one of the top four causes of death in 2020, causing over 151,000 fatalities [3]. Furthermore, as a public health concern, diabetes incurs high costs; in 2021, it accounted for an estimated USD 966 billion in global health expenditure in the U.S. alone [2]. In Mexico, per-patient costs are estimated to range from USD 700 to 3,200 annually, contributing 5% to 14% of health spending [2]. Fig. 1 shows a graphic summary of the number of deaths due to diabetes mellitus in each state within the Mexican Republic [4].

Deaths from Diabetes Mellitus by Federal Entity. México, INEGI 2020

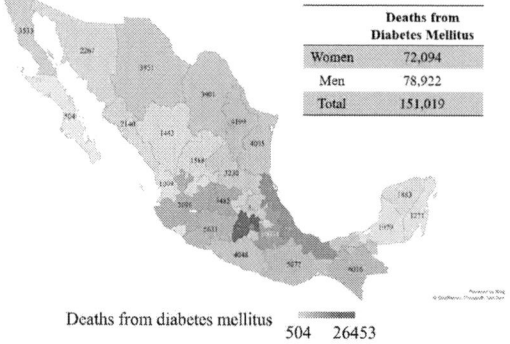

	Deaths from Diabetes Mellitus
Women	72,094
Men	78,922
Total	151,019

Deaths from diabetes mellitus 504 26453

Fig. 1. INEGI, Population Census 2020. Number of deaths in Mexico due to diabetes mellitus by federal entity and gender [5]. The states with the highest death concentration correspond to those with the highest population concentration.

Managing diabetes requires regular glucose monitoring and insulin dosing [2]. While blood glucose measurement is the standard method, daily monitoring devices can be expensive and cause discomfort from frequent finger pricking. Patients to overcome these challenges, alternative, predominantly non-invasive methods are being explored, such as using saliva, urine [5], tears [6], [7], or infrared light applied to the skin [8]. In addition, the composition of saliva, including glucose and other organic and inorganic components, has potential diagnostic value for various diseases [9]-[13].

This study proposes saliva as alternative medium for glucose level detection, using experimental results from H_2O_2 controlled solutions, and it is essential to mention that this study does not use saliva sample yet. Due to lower glucose concentrations in saliva compared to blood (see Table I) [14]-[17], a high sensitivity monitoring system is needed. Glucose

979-8-3503-1191-4/23 $31.00 © 2023 IEEE

in the saliva is oxidized, with byproducts monitored using glucose oxidase (Gox), an enzyme used for glucose electrooxidation [18].

TABLE I. SUMMARY OF GLUCOSE LEVELS COMPARISON BETWEEN BLOOD VERSUS SALIVA IN A HUMAN BODY

Glucose Levels	Blood (mMol/L)	Saliva (mMol/L)	Reference
Normal	4.9 − 5.3	0.1 − 1.5	[14]-[17]
Prediabetes	5.4 − 6.18	1.6 − 2.37	[14]-[17]
Diabetes	> 7.0	> 3.0	[14]-[17]

II. GLUCOSE MONITORING METHOD

Biosensors use enzymatic and non-enzymatic methods to detect substances in samples. Enzymatic methods leverage specific enzymes to catalyze reactions with the target analyte. In contrast, non-enzymatic methods rely on the physical or chemical properties of the analytes to react with a particular reagent. This research proposes indirectly monitoring glucose levels by measuring the concentration of hydrogen peroxide (H_2O_2), a byproduct of the reaction between glucose oxidase (GOx) and glucose. Both enzymatic and non-enzymatic methods can detect H_2O_2. Enzymatic methods involve the enzyme peroxidase, which catalyzes a reaction with a substrate like luminol to produce a detectable signal. Non-enzymatic methods use a reagent that reacts with H_2O_2 to generate a measurable signal [19]-[21]. This research employs a non-enzymatic electrochemical method to indirectly determine glucose levels by monitoring H_2O_2, a byproduct of β-D-glucose oxidation, catalyzed by glucose oxidase (GOx). The process can be summarized in two steps: firstly, glucose oxidase catalyzes the conversion of β-D-glucose to D-glucose-δ-lactone, creating H_2O_2, with molecular oxygen acting as the electron acceptor (1), and secondly, the lactone non-enzymatically hydrolyses into D-gluconic acid (2). As the concentration of H_2O_2 is proportional to the amount of oxidized glucose, it indirectly measures the glucose levels in the sample [22].

A summary of various H_2O_2 sensors from multiple studies detailing their sensor type, transducer base, detection range, and sensitivity, as shown in Table II. However, specific sensors' techniques, materials, and structures may obstruct their potential for miniaturization into compact devices due to factors like requiring extensive and expensive spectroscopy equipment or using less accessible materials like graphene. Alternatively, the proposed sensor in this study uses affordable, washable, reusable nickel nanostructures requiring minimal maintenance, presenting a more viable option for small-scale integration.

$$β\text{-}D\text{-}glucose + O_2 \rightarrow D\text{-}glucone\text{-}δ\text{-}lactone + H_2O_2 \quad (1)$$

$$D\text{-}glucone\text{-}δ\text{-}lactone + H_2O \rightarrow D\text{-}gluconic \quad (2)$$

TABLE II. SUMMARY TABLE TO COMPARE THE DIFFERENT PROPOSALS FOR HYDROGEN PEROXIDE SENSORS, CLASSIFIED BY METHOD, TRANSDUCER, AND DETECTION RANGE

Method	Transducer	Detection Range (mM)	Sensitivity (µM)	Ref.
Electrochemical	Ni and Co	5e-3 − 9.0	0.5	[23]
Spectroscopy	Co & Mn oxides	1e-4 − 1.27e3	0.0020	[24]
Spectroscopy	Mn oxide	0.01 − 0.1	0.5	[25]
Electrochemical	Ni	0 − 4.73	1,530	Our

III. DESCRIPTION OF THE EXPERIMENT AND TEST BENCH

A. Novel Nano-Wire Sensor

This proposal introduces a novel non-enzymatic sensor design based on Ni Polycarbonate Track Etch (PCTE) nanowires. These structures can increase their sensitivity by altering their surface-to-volume ratio concerning the analyte being tested by adjusting the nanowire length or sensor area, thus changing the nanowire number or density. These

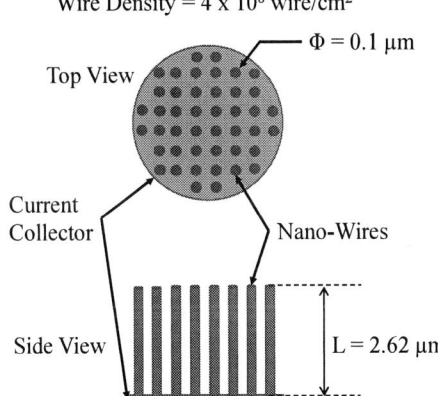

Fig. 2. Simplified diagram of the Ni Polycarbonate Track Etch (PCTE) nano-wire sensor [5], [26], and [27]. The figure shows an image of the top and side views. The length of the nanowire is 2.62 µm, nanowire diameter is 0.1 µm, and nanowire density is equal to 4 x 10⁸ wire/cm².

parameters make the sensor's sensitivity adaptable to specific requirements. Furthermore, this sensor is unique due to its optimization for high sensitivity to H_2O_2 fluctuations, which is desirable for measuring changes in low-glucose human fluids like saliva. More on the nanowire sensor design can be found in references [5], [26], and [27]. A simplified diagram is depicted in Fig. 2.

B. Potentiostat and Test Bench

The study involved lab testing and experimentation using the Potentiostat/Galvanostat Stanford Research Systems (SRS) EC301, focusing on cyclic voltammetry (CV) tests with H_2O_2 solutions at varying concentrations, depicted in Fig. 3 electrical schematic diagram. It was vital to ensure the equidistant placement of the working electrode (WE), counter electrode (CE), and reference electrode (RE) within the beaker. The distance between electrodes is 1 cm.

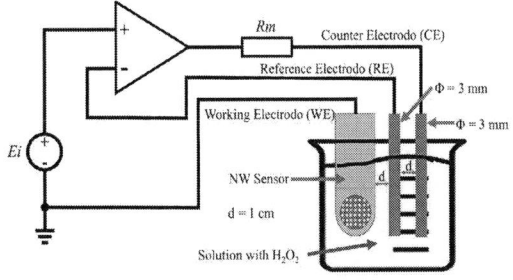

Fig. 3. CV Test bench diagram. All electrodes must be at the same distance between them and the same space at the bottom of the beaker.

In this experiment, the reference electrode was made of Ag/AgCl, the counter electrode from Pt, both having a 3 mm diameter, while the working electrode was the nanowire sensor.

IV. CYCLIC VOLTAMMETRY MEASUREMENTS RESULTS

The experimental findings in Fig. 4 showcase six characteristic CV test curves corresponding to H_2O_2 in discrete concentrations of 0.00, 0.34, 1.27, 2.54, 3.81, and 4.73 mM/L. These curves overlap for easier comparison among the concentrations. The analysis focuses on four areas: two non-faradaic currents segments, reduction peak current, and oxidation peak current. However, this study discusses the non-faradaic current segment 1 and the reduction peak current. These areas demonstrate variations in electrical variables corresponding to changes in concentration, particularly noticeable as an increased electric current in the non-faradaic current segment 1 with rising H_2O_2 concentration, and similar changes in the reduction peak current area.

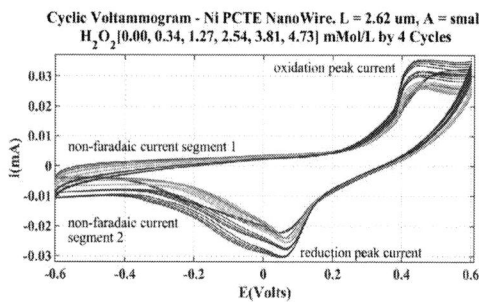

Fig. 4. Experimental results of cyclic voltammetry tests were obtained using controlled solutions of H_2O_2 at various concentrations ranging from 0.00 to 4.73 mMol/L. The tests revealed four distinct areas of interest: non-faradaic current segments 1 and 2, reduction peak current, and oxidation peak current. The Ni PCTE nano-wire sensor was utilized for the experiment.

A. Non-Faradaic currents segment 1

The research demonstrates an incremental progression in the electric current with increasing concentrations of H_2O_2, as illustrated in Fig. 5. The minimum current curve corresponds to zero H_2O_2 concentration, establishing a baseline current level. Each successive curve shows a corresponding uptick in the current level, mirroring the H_2O_2 concentration up to 4.73 mMol/L. Notably, shifts in the current's magnitude are tied to alterations in concentration, not voltage variations within a range of -600 mV to -500 mV, with a bias voltage of 0 volts. The data suggest a direct correlation between the measured current and H_2O_2 concentration. Throughout the specified potential range, the electric current values maintain near constancy. In Fig. 6, these findings point to a linear relationship between H_2O_2 concentration and the magnitude of electric current.

B. Peak Reduction Current Analysis

The reduction peak current data analysis, shown in Fig. 4, entails averaging the maximum current values from each test cycle. The results are graphed in Fig. 7, revealing a non-linear pattern. The linear model boasts an R-square (R^2) value of 0.93, which is respectable. However, a model closely conforming to experimental data is needed for medical applications to minimize error and mitigate any potential negative impact on patient health. It is typically reflected in an

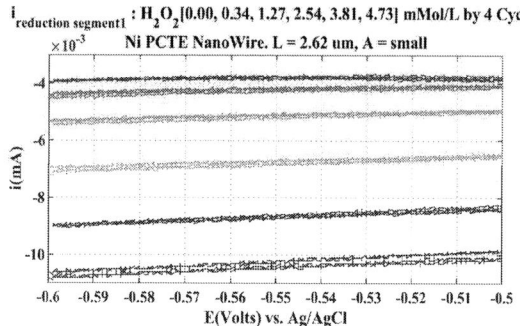

Fig. 5. Magnification are of non-Faradaic current segment 1. The range of the site is limited from -600 mV to -500 mV. The electrical intensity change through the input potential is constant, and the magnitude of the electrical current is affected by changes in the concentration of the controlled solution of H_2O_2. As a result, a discretized behavior can be observed in the electrical current.

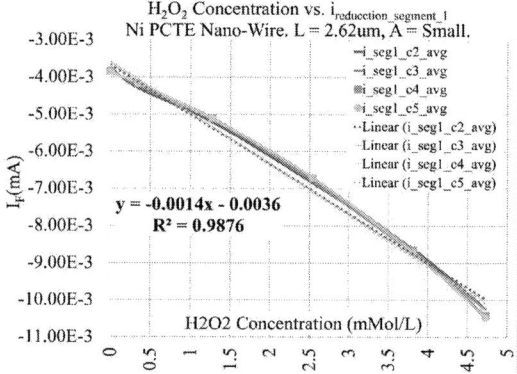

Fig. 6. The figure displays the average results of the curves from Fig. 5 for each test cycle, labeled as C_2, C_3, C_4, and C_5, where, for instance, C_2 corresponds to data from cycle two. The curves reveal a linear trend, indicating a direct correlation between H_2O_2 concentration and the magnitude of the electric current: as the former increases, so does the latter. The attained model demonstrates a high coefficient of determination (R^2) of at least 0.98, suggesting a reliable linear approximation.

R^2 value approaching unity. For glucose detection applications, it is recommended to aim for an R^2 greater than 0.95, as the goal is to provide a reliable monitoring method for patients with diabetes mellitus. The combined data analysis from Fig. 6 and Fig. 7 allows precisely estimating H_2O_2 concentration in the test solution. This is because monitoring the current at two points reduces the error due to some noise source.

V. CONCLUSIONS AND FUTURE WORK

The experimental results presented in this research show a high correlation between measured electrical current and the H_2O_2 concentration, a byproduct of glucose oxidation due to a reaction with the glucose oxidase enzyme. So, these results are promising as an electrochemical method to measure glucose in fluids with low glucose concentrations, such as human saliva. However, it is necessary to continue the research and the analysis of measured data. Therefore, two next steps are proposed; the first is to identify the more linear data region to make the indirect glucose measurements under constant potential. The second is to design an electronic device to amplify and process the nano sensor signal to show the measure in a display.

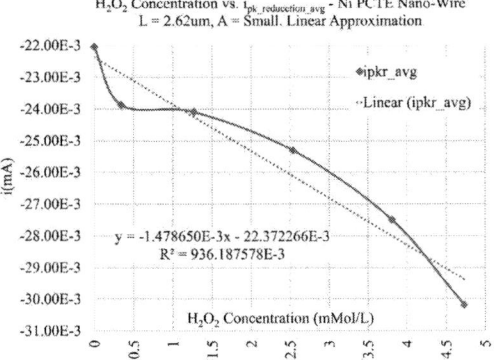

Fig. 7. The correlation between H_2O_2 concentration and the reduction peak current is not linear, as per the average values derived from all test cycles from C2 to C5. The linear model does not fully align with the experimental data, with an R^2 value of 0.93. An approach to enhance this nonlinear behavior is proposed in [26].

ACKNOWLEDGMENT

INTEL DE MEXICO and CONACYT supported this work. Charles-Suárez Oscar Luis, for their invaluable support and contributions at the laboratory assistance.

REFERENCES

[1] World Health Organization. (2019, May 13). Diabetes. Retrieved from https://www.who.int/health-topics/diabetes#tab_1

[2] Hernández-Ávila, M., Gutiérrez, J. a. T., & Reynoso-Noverón, N. (2013, March 4). "Diabetes mellitus en México. El estado de la epidemia". Salud Publica De Mexico, 55, 129. https://doi.org/10.21149/spm.v55s2.5108

[3] PMFarma. Los números de la diabetes en México. Retrieved February 27, 2023, from https://www.pmfarma.com.mx/noticias/1359-los-numeros-de-la-diabetes-en-mexico.html

[4] Instituto Nacional de Estadística y Geografía. (2021). Defunciones por diabetes mellitus por entidad federativa de residencia habitual de la persona fallecida y grupo quinquenal de edad según sexo, serie anual de 2010 a 2021. Retrieved February 27, 2023, from https://www.inegi.org.mx/app/tabulados/interactivos/?pxq=Mortalidad_Mortalidad_04_aa50eec5-79fe-45a7-aee2-a2d7ec0feeba&idrt=127&opc=t

[5] Lopez-Cardenas, P. G., Alcala, E., Sanchez-Torres, J. D., & Araujo, E. (2021, July 28). "Enhancing the sensitivity of a class of sensors: A data-based engineering approach". In 2021 IEEE 21st International Conference on Nanotechnology (NANO) (pp. 1-4). IEEE. https://doi.org/10.1109/nano51122.2021.9514352

[6] Google contact lens could help diabetics track glucose. (2014, January 18). CBC. Retrieved from https://www.cbc.ca/news/science/google-contact-lens-could-help-diabetics-track-glucose-1.2500274

[7] Control, D. I. (2015, October 1). Next Generation Google Glucose Monitoring Devices in the Making. Diabetes In Control. Retrieved from http://www.diabetesincontrol.com/next-generation-google-glucose-monitoring-devices-in-the-making-2/

[8] Zhou X., & Pero Y. (2015, October 20). Noninvasive glucometer (Patent No. CN105232055A). Retrieved from https://patents.google.com/patent/CN105232055A/en

[9] Young, J., & Schneyer, C. A. (1981, February). "Composition of saliva in mammalia". Australian Journal of Experimental Biology and Medical Science, 59, 1-53. https://doi.org/10.1038/icb.1981.1

[10] Lenander-Lumikari, M., & Loimaranta, V. (2000)." Saliva and Dental Caries". Advances in Dental Research, 14(1), 40–47. https://doi.org/10.1177/08959374000140010601

[11] Kidd, E. (2005, May). Essentials of Dental Caries: The Disease and Its Management, 3rd ed. Oxford University Press, ISBN: 978-1-118-93582-8.

[12] Castagnola, M., Picciotti, P. M., Messana, I., Fanali, C., Fiorita, A., Cabras, T., Calò, L., Pisano, E., Passali, G. C., Iavarone, F., Paludetti, G., & Scarano, E. (2011, December). Potential applications of human saliva as diagnostic fluid. Acta Otorhinolaryngologica Italica, 31(6), 347-357.

[13] Jung, D. W., Jung, D., & Kong, S. H. (2017, November). "A Lab-on-a-Chip-Based Non-Invasive Optical Sensor for Measuring Glucose in Saliva". Sensors, 17(11), 2607. https://doi.org/10.3390/s17112607

[14] De Vasconcelos, A. T. R., Soares, M. J. G. O., Almeida, P. S., & Soares, T. C. (2010). "Comparative study of the concentration of salivary and blood glucose in type 2 diabetic patients". Journal of Oral Science, 52(2), 293–298. https://doi.org/10.2334/josnusd.52.293

[15] Uthayasankar, D., Jayaraj, G., & R3, G. (2020). "Saliva in diabetes - A review". European Journal of Molecular & Clinical Medicine, 7(1), ISSN 2515-8260. https://ejmcm.com/article_2309.html

[16] Fares, S., Said, M., Ibrahim, W. H., Amin, T. T., & Saad, N. E. S. (2019, March). "Accuracy of salivary glucose assessment in diagnosis of diabetes and prediabetes". Diabetes and Metabolic Syndrome: Clinical Research and Reviews, 13(2), 1543–1547. https://doi.org/10.1016/j.dsx.2019.03.010

[17] Indunil Dharmakeerthi, K., Priyashan Ponweera, M., Hasanjana Moragoda, E., Sandaruwan Galgamuwa, L., Jayasekara, K., Kaluarachchi, V., & Bulugahapitiya, U. (2021, April). "Correlation between blood glucose and salivary glucose in type 2 diabetes mellitus patients". Malaysian Journal of Medicine and Health Sciences, 17, eISSN 2636-9346. https://medic.upm.edu.my/upload/dokumen/2021040613022706_MJMHS_0021.pdf

[18] Heller, A., & Feldman, B. (2008, May). "Electrochemical Glucose Sensors and Their Applications in Diabetes Management". Chemical Reviews, 108(7), 2482–2505. https://doi.org/10.1021/cr068069y

[19] Guo, J., Cui, H., Zhou, W., & Wang, W. (2008, January). "Ag nanoparticle-catalyzed chemiluminescent reaction between luminol and hydrogen peroxide". Journal of Photochemistry and Photobiology A-Chemistry, 193(2–3), 89–96. https://doi.org/10.1016/j.jphotochem.2007.04.034

[20] Sakura, S. (1992, June). "Electrochemiluminescence of hydrogen peroxide-luminol at a carbon electrode". Analytica Chimica Acta, 262(1), 49–57. https://doi.org/10.1016/0003-2670(92)80007-t

[21] Tarladgis, B. G., Pearson, A. M., & Dugan, L. R. (1962, January 1). "The chemistry of the 2-thiobarbituric acid test for the determination of oxidative rancidity in foods". I. Some important side reactions. J Am Oil Chem Soc, 39, 34–39. https://doi.org/10.1007/BF02633347

[22] Kornecki, J. F., Carballares, D., Tardioli, P. W., Fernandez-Lafuente, R., Berenguer-Murcia, Á., Alcántara, A. R., & Fernandez-Lafuente, R. (2020, July 14). "Enzyme production of d-gluconic acid and glucose oxidase: successful tales of cascade reactions". Catalysis Science & Technology, 10(17), 5740–5771. https://doi.org/10.1039/d0cy00819b

[23] Hussain, M. M., Nisar, A., Qian, L., Karim, S., Maaz, K., Liu, Y., Sun, H., & Ahmad, M. (2021, February 24). "Ni and Co synergy in bimetallic nanowires for the electrochemical detection of hydrogen peroxide". Nanotechnology, 32(20), 205501. https://doi.org/10.1088/1361-6528/abe4fb

[24] Wu, Z., Li, C., Zhu, F., Liao, S., Yu, J., Yang, H. G., & Chen, X. (2017, December). "Binary cobalt and manganese oxides: Amperometric sensing of hydrogen peroxide". Sensors and Actuators B-Chemical, 253, 949–957. https://doi.org/10.1016/j.snb.2017.07.012

[25] Di, W., Zhang, X., & Qin, W. (2017, April). "Single-layer MnO2 nanosheets for sensitive and selective detection of glutathione by a colorimetric method". Applied Surface Science, 400, 200–205. https://doi.org/10.1016/j.apsusc.2016.12.204

[26] Lopez-Cardenas, P. G., Alcala, E., Sanchez-Torres, J. D., and Araujo, E. "A Resampling Approach for the Data-Based Optimization of Nanosensors," 2021 18th International Conference on Electrical Engineering, Computing Science and Automatic Control (CCE); (2021, November), https://doi.org/10.1109/cce53527.2021.9633114

[27] Lopez-Cardenas, P. G., Alcala, E., Sanchez-Torres, J. D., & Araujo, E. (2023, April 4). "Improving self-supported nanowire arrays using response surface methodology for synthesis of H_2O_2 nanostructured sensor," 2023 Materials Chemistry and Physics. https://doi.org/10.1016/j.matchemphys.2023.127729

2023 IEEE Latin American Electron Devices Conference (LAEDC)
Puebla, México, July 3-5, 2023

"Accelerating Engineering Education and Workforce Development in Automation & Control for the Semiconductor Industry Based on Cognitive Neuroscience"

Luis Fernando Cruz
Vicerrectoría de Investigaciones
Universidad Manuela Beltran
Bogota, Colombia
luis.cruz@docentes.umb.edu.co

Luis Miguel Quevedo M.
Member IEEE
JWG S.A.S
Tampa,FL, USA
lquevedo@ieee.org

Wilfrido Alejandro Moreno
Electrical Engineering Department
University of South Florida
Tampa,FL, USA
wmoreno@usf.edu

Abstract—The rapid development of the semiconductor industry represents a global challenge to meet the needs of Science, Technology, and Innovation in the 21st century. Advancements in semiconductors optimize computer technology with innovations, among others, in nanotechnology, quantum technology, cognitive computing, Cyber-Physical Human Systems, Artificial Intelligence, and Robotics. The rapid development of semiconductors demands an imperative acceleration of training and the development of a highly skilled workforce. There is a gap between the knowledge and skills of students in the current education system and the fast development of new knowledge. To accelerate comprehensive training in the semiconductor workforce we must start at an early stage, this paper proposes that STEM education must be pedagogically grounded in the advances of cognitive neuroscience and learning. By understanding how the brain functions and how it learns, the process can be enhanced, the education of an Integral Engineer needs to be addressed by forming professionals equipped with adaptability, complex problem-solving, assertive communication, emotional intelligence, interdisciplinary teamwork skills, systems, analytical and critical thinking.

This paper proposes curricular transformation processes, ethical formation, pedagogical foundations of interdisciplinary research laboratories, enhanced by an international network of remote laboratories, and a TRUE system of cultural transformation.

Keywords—controls systems, workforce, integral engineer, neuroscience, semiconductors industry.

I. INTRODUCTION

The development of the semiconductor industry represents a global challenge to respond to the Science, Technology and Innovation needs of the 21st century. Advances in semiconductors greatly impact technology trends in areas such as information technology with full-stack innovation, nanotechnology, clean energy, quantum technology, bioengineering, cognitive computing, ubiquitous robotics, Cyber-Physical Human Systems (CPHS), Artificial Intelligence (AI), Complex Adaptive Systems and new control driven architectures. To strengthen R&D and the workforce of the future, billions of dollars are been invested in the semiconductor industry via the CHIPS and Science Act in the

USA[1], and also in the private sector in flagship companies such as Micron, Qualcomm, Intel, TSMC, Samsung, and SkyWater among others. The rapid development of the semiconductor industry in the USA urgently requires accelerating the training and development of a talented workforce which is expected that more than 50,000 new engineering jobs will be created in the semiconductor sector.

II. OVERVIEW OF THE SEMICONDUCTOR INDUSTRY AND THE NEED FOR THE INTEGRATION OF AUTOMATION AND CONTROL STRATEGIES

The semiconductor industry presents inherent complexity not only from a technological standpoint but also due to the need for safeguarding proprietary processes and product data. Consequently, this complexity imposes high levels of intricacy on the overall engineering design process which in turn will require a highly specialized workforce[2]. The semiconductor industry operates on trade secrets, patents, and cutting-edge technological advancements driven by applied research and engineering. This specialization poses challenges for academic institutions, as they must not only provide standard curriculum that should emphasize comprehensive understanding of semiconductor processing and technologies but also offer hands-on training through specialized laboratories. To address these challenges, academic institutions must establish strategic partnerships with semiconductor industry stakeholders, enabling workforce development through in-company internships and hands-on training at industry partner locations. These collaborations are expected to expedite students' skill acquisition while familiarizing them with the special process and fab equipment employed by the industry.

By observing the technology trends within leading semiconductor companies like Advanced Semiconductor Materials Lithography (ASML), NVIDIA, INTEL, and SAMSUNG, several technological challenges can be identified. For instance, ASML faces challenges with the development and implementation of Extreme Ultraviolet (EUV) Lithography Technology, NVIDIA's challenge is to ensure technologies roadmap alignment, while INTEL and SAMSUNG continue to be challenged with the complex task

979-8-3503-1191-4/23 $31.00 © 2023 IEEE

of maintaining precise control over 3σ while reducing device features size to as small as 2 nm (20Å) [3].

In addition, it is critical to emphasize that the semiconductor industry must align itself also with technology roadmaps that define future semiconductor needs driven by Industry 5.0 and Advanced Manufacturing, including the development of flexible fabs. This alignment ensures that the semiconductor industry keeps pace with new developing technological needs, enabling the realization of advanced manufacturing processes and applications. By considering these examples and acknowledging the need for technology roadmap alignment, the semiconductor industry can effectively address the challenges posed by rapid technological advancements and remain at the forefront of innovation and competitiveness.

In the International Roadmap for Devices and Systems 2022 Edition Factory Integration, it was presented the potential use of 200 mm wafers for specific development projects [4]. This focus presents an opportunity to reduce prototyping costs significantly. Moreover, it paves the way for research & development of new technologies aimed at cost reduction in production processes. Interestingly, the possibility arises for a fabrication facility to incorporate both 200 mm and 450 mm wafers, thereby creating additional challenges for engineers to address logistical complexities within such a facility.

In addition to virtual environments, the integration of advanced distributed and autonomous control strategies is imperative across all processes within a semiconductor fabrication facility (fab). This integration is necessary to handle Metrology-acquired process data from a data-centric perspective [4]. The acquired data encompasses a range of vital information from key fab support infrastructure, including measurements of air quality, water purity, environmental temperature, as well as calibration data for equipment domains such as lasers used for position control, material deposition, chemical mechanical planarization, wafer handling robots, and many others.

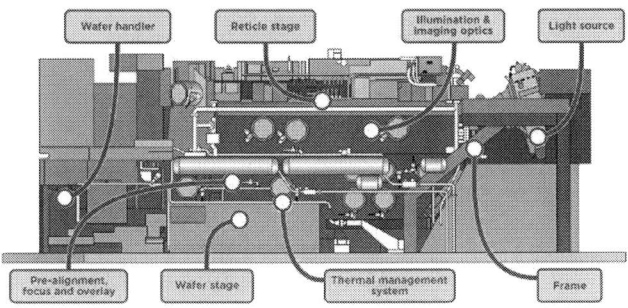

Fig. 1. Subsystems of a Photolithography machine[5].

This integration is facilitated by technologies such as Digital Twins, Run to Run (R2R), Virtual Reality (VR), and other digital solutions [3]. These technologies enable decision-making support and the emulation of scenarios across individual fab processes, coupled processes, or even the entire fab. Furthermore, manufacturers must implement

factory integration to swiftly respond to production requirements. This integration relies on the utilization of big data, which originates from the implementation of Industrial Internet of Things (IIoT) and aligns with Industry 4.0 and 5.0 initiatives. However, implementing these technologies pose challenges for control systems, particularly in the semiconductor industry, where data protection becomes paramount, waiting for the implementation of in-situ control mechanisms [4].

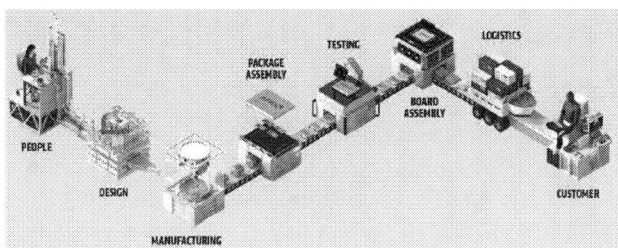

Fig. 2. Cycle from Chip Design to Consumer Product[6].

III. NEED TO ACCELERATE LEARNING FOR WORKFORCE DEVELOPMENT FOR THE SEMICONDUCTORS INDUSTRY

The semiconductor industry is expected to undergo significant changes in the next decade, requiring a prepared and adaptable workforce. Academic institutions are facing the task of dynamically adjusting their curricula to align with these forthcoming changes [7]. Additionally, faculty members must capitalize on the latest advancements in neurosciences and incorporate the findings on effective teaching and learning strategies. This approach ensures that students are not only equipped with fundamental knowledge but also develop essential soft skills such as critical thinking, analytical mindset, problem resolution, broad vision, teamwork & dynamics, conflict management, and emotional intelligence, all within the framework of personalized learning.

The adoption of new technological tools and personalized learning assistants will play a pivotal role in maintaining students' curiosity and engagement throughout their educational experience. By leveraging these tools, academic institutions can foster a dynamic learning environment that meets the evolving demands of the semiconductor industry. The challenges of training a workforce for the semiconductor industry in this era of rapid technological advancement are just as critical and unique as the industry itself. Thus, it is vital for educational institutions to adapt, innovate, and prepare students to excel in this dynamic and specialized field.

In the broader context of engineering workforce development, it is crucial to equip professionals with modern digital engineering frameworks, such as Model-based Systems Engineering (MBSE), that can be applied throughout the entire product life cycle. This encompasses activities ranging from design and development to verification and validation of complex systems. As these frameworks become increasingly essential, it presents a significant opportunity for educational institutions and workforce development organizations to proactively accelerate the development and implementation of programs that cater to these emerging needs.

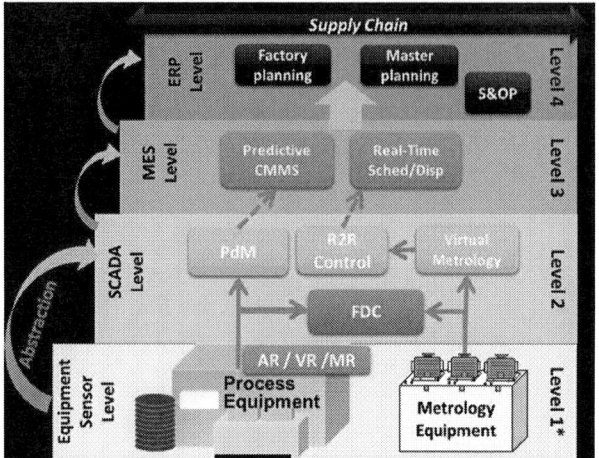

Fig. 3. Digital Twin representation from the perspective of the International Society of Automation (ISA-95) Levels*[4].

The objective of this paper is to show the fundamentals of neuro-learning that allow for accelerating comprehensive STEM education and semiconductor workforce in the short term. Knowing how the brain works and how it is empowered facilitates the education of a comprehensive engineer to develop skills in systems thinking, analytical thinking, critical thinking, creativity & innovation processes, complex problem solving, adaptability, assertive communication, emotional intelligence, and interdisciplinary teamwork among others.

To align with initiatives such as Chips for America, which prioritize education and training for the semiconductor industry, it is essential to establish a comprehensive educational pathway starting from K-12 schools and extending through career technical education programs offered by community colleges and universities [7]. Creating an environment that allows students to reinforce their knowledge through hands-on experiences becomes crucial in stimulating their neuro structure and enhancing their ability to grasp complex concepts.

Collaboration among educational institutions, training providers, and the semiconductor industry plays a pivotal role at this stage. It enables the development of programs that effectively prepare students to tackle critical and emergent situations within semiconductor fabs, empowering them to find real-time solutions [8]. By bridging the gap between academia and industry, students gain practical insights and a deeper understanding of the challenges faced in the field, fostering their ability to contribute meaningfully upon entering the workforce.

The learning acceleration process is the result of activating educability, i.e., student learning potential, through teachability, i.e., pedagogical mediations that stimulate the development of educability in each disciplinary context. Therefore, to accelerate learning, the educator must design effective pedagogical mediations or teaching-learning scenarios that facilitate the activation and strengthening of

cognitive functions. However, teachability is generally not in line with the educability of the learner.

The basis of educability is neurobiological. The structure of the neural networks and the connectivity of the teacher's brain is different from that of the student due to experience, skills and knowledge built over time. The student must create his/her own networks and strengthen them, that is the challenge of learning and the limitation of traditional STEM education.

Learning is a complex process given by a system of systems that connects in the brain between 80-100,000 million neurons in a network of thousands of billions of synapses or connections regulated by neuronal nodes or specialized centers, i.e., cerebral cortex, hippocampus, cerebral amygdala, etc. It is an electrical and biochemical neuromodulator system that facilitates signal transmission between neurons, synaptic plasticity, modulates attention, memory consolidation and meaningful learning. Among the main neurotransmitters that regulate learning are acetylcholine, associated with synaptic plasticity, memory consolidation and plays an important role in the formation of new memories; serotonin, facilitates the modulation of synaptic plasticity and influences memory and emotional regulation necessary for the learning process; glutamate, facilitates the ability to modify and strengthen neural connections during learning; dopamine regulates the formation of associations between stimuli and rewards. In summary, it integrates motivation with reward and reinforcement to optimize learning; norepinephrine is associated with alertness, attention and memory consolidation. In addition to the classical neurotransmitters, there are other groups of neurotransmitters such as peptides that include somatostatin, substance p and cholecystokinin that regulate synaptic transmission and memory. This network generates patterns that modulate learning. Neural networks activate in an integrated way basic and higher cognitive function in different parts of the brain, i.e., perception, memory, attention, language, metacognition, creativity, problem solving, decision making, etc. [9].

The brain connectivity is dynamic and changing. Factors such as exercise, learning and sleep facilitate epigenesis or the formation of new neurons in the hippocampus and a permanent rotation of synapses that generates brain plasticity[10],[11]. Learning, therefore changes the structure of the brain and permanently updates it.

To activate neural networks efficiently in the brain, two forms of learning must be energized. A focused mode, i.e., conscious mode, and diffuse mode, i.e., an unconscious mode. The focused mode requires intentional concentration. It is modulated by the prefrontal cortex and from there connects with information zones of long-term memories, mediated by neuromodulators such as acetylcholine. This mode connects with the learner's prior knowledge and facilitates learning new concepts or skills [12]. It activates working memory. This memory, although effective for new learning, is limited and has the risk of losing information if it suffers interference from other stimuli because natural memory dissipators intervene and tend to erase the information[13].

The first mistake of an educator is to load the student with information, i.e., formulas, figures, etc., for long periods without understanding that the neural structures created are different and that there is cognitive overload for the student. Overlearning makes the neural system inefficient, confuses and demotivates the student.

In the learning process, error must be tolerated, which is necessary to make cognitive adjustments and build adequately. If the teacher demands everything without error from the beginning, he/she induces the student to work by heart, without analysis, and does not facilitate meaningful learning. Therefore, it does not accelerate learning, it slows it down and makes it ineffective. An efficient focused mode subsequently favors creativity, innovation and complex problem solving[15],[16]. In the focused mode, neurobiologically speaking, there are previously created structures, established paths and established search routes.

In this sense, it is necessary to periodically reflect on the curricular structure. Rigid curricular processes do not allow for the renewal of content, teaching-learning strategies and evaluation processes that facilitate learning. This has several implications: first, it does not take into account that the cognitive structure of digital natives requires different teachability dynamics from traditional curricula, designed for digital migrants in terms of content and time management, teaching strategies and evaluation processes. Secondly, it limits the entry of new subjects and contents that facilitate the student to develop skills from the university and not wait until the industry, for example, the inclusion of physics and other subjects such as materials science in laboratories, creation and design of manufacturing, desktop simulations, virtual augmented realities for simulated factories, connection with the industry from the beginning of the program, permanent practices by nanotechnology and semiconductor centers, remote laboratories in the teaching of semiconductors, interdisciplinary teamwork, etc.

The second mode of learning is the diffuse mode. This mode is related to a neural rest and is modulated by neurotransmitters such as dopamine, serotonin etc. It has to do with concepts not thought of before, there is no previous pattern. It modulates a new way of thinking. Neurobiologically it involves new neuronal connections and new pathways. It does not require a fixed concentration like the focused mode. Students can see things in a broader way. It occurs in moments of relaxation, exercise, walks in nature, rest moments, during sleep, etc. To activate it, neural rest spaces must be provided in different places and activities. It is important to emphasize that the diffuse mode is not effective if an adequate neural structure has not been previously created through the focused mode. In this unconscious mode of learning, sleep plays a preponderant role in order to deactivate the conscious prefrontal cortex and allow other areas of the brain to talk to each other for solutions to emerge. Sleep allows to eliminate toxic substances, to reinforce the most important part of the memories and to erase the less important ones [17],[18],[19].

The fuzzy mode practices over and over again the most difficult part of learning by repeating neural patterns and increases the ability to solve problems provided that structures are well established through the focused mode. In order to assimilate knowledge, develop new and challenging skills, solve complex problems, develop creative processes and accelerate learning, the learning modes must be alternated. It is not possible to be in focused and diffuse modes at the same time[20].

Accelerating neural connections during the learning process and skill development requires building and strengthening patterns through repeated use and practice in real contexts from the beginning. The more abstract the concepts, as in the case of mathematics, the more practice is required because the neural patterns are weak due to the abstract nature of the discipline[19],[21]. If a practical application is not seen and felt emotionally, the synapses are weak. Concrete neural patterns are needed through practical immersion from the beginning of training in the semiconductor industry so that the cognitive structure integrates the practical, and generate meaningful learning that lasts, helps to solve real-life problems and facilitates processes of creativity and innovation. The hands-on needs to be integrated concurrently with the theory, otherwise the learning process and its effectiveness can be delayed.

The student starts by understanding something with focused attention and fixing the concept, solving a problem, repeatedly re-solving the problem without seeing the solutions, even getting it wrong several times until the student deepen a pattern of thinking that will become permanent. Accelerated learning happens when students personally engage with and resolve their understanding of the material they believe they comprehend. Even if the student is in practical immersion from the beginning, it is essential to give time to let the neural cement dry in the cognitive system, while this happens there may be errors. This should be kept in mind by the educator.

Interaction with other people in laboratories or class teamwork favors areas of the brain such as the hippocampus with the birth of new neurons and enhancing memory and learning[22]. Teamwork with more capable people or an effective accompaniment of the teacher favors the Zone of Proximal Development (ZDP), a zone to which a student can reach to achieve his or her greatest potential, as opposed to doing it alone, in which case he or she would not achieve the same potential. Interaction in interdisciplinary teams also favors the transfer of knowledge from other disciplines and enhances learning because although the disciplinary concepts are different, the neuronal patterns and schemes are similar[14],[16]. The pattern to solve a problem in medicine can be applied in engineering and vice versa.

In the focused mode the prefrontal cortex is connected by a neural link to different areas of the brain and creates compact packets of information with the use of analogy to converting a heavy computer file into a ZIP file. As the information is packaged, the memory and learning capacity of the neural system is released[23],[24]. Scattered pieces of information are linked together through meaning and context.

These pieces of information are composed of a complex neural network activity that fires together synchronously to create an idea, a concept or an action that the mind can easily access at any given moment. Hyperlinks are created from the prefrontal cortex that connect to the system and make it efficient. This process creates memory traces that connect to other related memory traces. The construction and use of these packets of information promotes learning and creativity [25]. The packets become the larger, simplifying picture of what is being studied and accelerates learning. Repetitions, practice and alternation with the fuzzy mode through disruptive pedagogical processes favors the development of these neural patterns.

The neural nanofragments built up through experience, interaction and practice with the material being studied grow larger and more complex. The larger and more robust the networks, the more creative interpretations and more complex problems can be drawn upon at any given time as the material under study is processed, without getting lost in ineffective detail[17]. However, if basic concepts are missing in conceptualization and practice (e.g., in the case of physics or materials science), the structure will be incomplete. The fragments must be adjusted to a broader conceptual vision of what is being learned, and here the teacher plays an essential pedagogical role.

Once created and strengthened, the fragments can be modulated by the brain unconsciously to enhance cognitive functions and accelerate learning. However, it is not enough to create the patterns, it is necessary to know when to use them and when not to use them, this generates neural pathways for decision making such as those that the student will require in his professional life in the industry. For this, it is necessary to practice with related and unrelated problems and to interact permanently with interdisciplinary teams[24]. In other words, to develop creativity and innovation, one must learn to connect the cerebral cortex to various neural networks distributed in the brain from different paths, provided that a solid foundation has been previously built.

At a more advanced stage of learning the neural packages must be combined in new and original ways[9],[14],[16]. This is the basis for innovations. This is done by updating and applying knowledge in the semiconductors industry with permanent cognitive challenges. In short, a well-structured and up-to-date library of conceptual fragments must be maintained to apply knowledge practically and solve problems permanently and not sporadically to take an exam. From this baseline, combinations of neural networks are made from which new ideas emerge and possible solutions to complex problems are proposed in various possible scenarios (as in professional life) and transfers are made from other fields of knowledge in a flexible and effective way.

The focused and diffuse modes are influenced by emotional processes. Brain centers such as the amygdala and limbic system regulate emotional responses such as excessive stress, anger, depression and fear that affect learning and motivation[26]. The brain has a diffuse connection system of neuromodulators that connect the cerebral cortex with the emotional regulation system. Therefore, cognition and emotion are intertwined and feedback on each other. Perception, attention and emotions are involved with learning

and memory. The amygdala as part of the limbic system together with the hippocampus, a center specialized in memory and learning, influences not only the consolidation of conceptual neural networks but also higher cognitive processes such as decision making, innovation and complex problem solving[9],[14],[16]. Learning processes are accelerated if intrinsic and extrinsic motivation is optimized through pedagogical mediations that facilitate learning modes and decrease the activation of negative emotional responses.

The Fig. 4. depicts a Student's Learning Process in the brain that shows the levels at which the system receives information through the senses and how it receives information internally through neuroendocrine pathways.

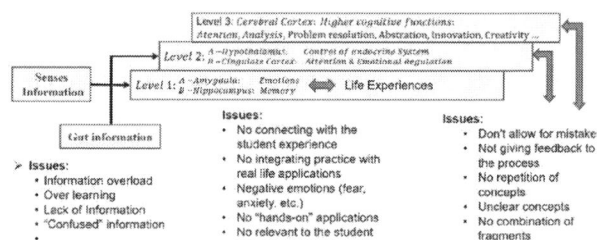

Fig. 4. Student's Learning Process in the Brain – "A Simplified Model"

At this point there may be information overload, over-learning, lack of information, or very confusing information on the part of the educator. The information must be clear and concrete. The system has different levels of processing that work in a distributed and almost simultaneous way. It integrates emotional circuits of the limbic system such as the cerebral amygdala and hippocampus with memory. These circuits are connected to the student's life experience. The problems that hinder processing are no connecting with the student experience, no integrating practice with real life applications, negative emotions, i.e., fear, anxiety, etc., no "hands-on" applications, no relevant to the student. Accelerated workforce training in semiconductors requires practice in real-world environments from the beginning in collaboration with industry. The information is processed simultaneously at other intermediate levels such as the hypothalamus that dynamizes processes of attention, emotional regulation, neuroendocrine control, and connection with the cerebral cortex where higher functions such as analysis, synthesis, complex problem solving, abstraction, innovation, creativity etc. are processed. The issues that hinder processing at this level are, unclear concepts, not proper creation of fragments, no combination of fragments, no repetition spaced of concepts. not giving feedback to the learning process and don't allow for mistakes.

Accelerated learning requires two paths. On the one hand, a bottom-up pathway that involves building, strengthening and accessing increasingly broad and new robust networks and neural patterns when required from different pathways. On the other hand, a top-down pathway that allows the student to see what is being learned and how it can be adjusted. It involves a process of metacognition[14]. That is, the ability to realize how to modulate and optimize one's own cognitive functions and learning, modulating its educability. The educator can facilitate both paths or can block them with inadequate and rigid teachability processes.

IV. CONCLUSIONS

Learning is a complex and adaptive system of information processing. The system must be properly modulated to achieve agile and efficient learning. Although the system is regulated by the dynamic interaction of multiple factors that must be considered such as genetics, environment, culture, diet, health conditions, sleep and physical activity, among others, it is the way in which information is entered and processed, through adequate pedagogical mediations, which will be decisive for meaningful learning. The information that is received and processed is conscious and unconscious. 21st century tools must be used in order to accelerate the teaching of STEM knowledge & skills, bringing new elements and concepts to faculty without discarding traditional methodologies. The paper emphasizes the opportunity that the technology provides, such as personalized learning in order to enhance the classroom experience by leveraging with the latest advances on cognitive neurosciences to accelerate train the required semiconductor workforce.

ACKNOWLEDGMENT

The authors would like to extend their sincere appreciation to all those who have contributed to the successful realization of this research endeavor. First and foremost, we express our deepest gratitude to the Electrical Engineering Department at the University of South Florida (USF) for their unwavering support in facilitating transdisciplinary Research, Development, and Education (RD&E). Over the past decade, their continuous assistance has been instrumental in the progress and achievements of our macro-project. We would also like to thank the Nanotechnology Research & Education Center (NREC) at USF for their invaluable support in semiconductor processing and metrology education. The authors also thank the organizers of the IEEE Latin American Electron Devices Conference.

REFERENCES

[1] National Institute of Standards and Technology. (2023). A Vision and Strategy for the NSTC [PDF]. Retrieved from https://www.nist.gov/system/files/documents/2023/04/27/A%20Visio n%20and%20Strategy%20for%20the%20NSTC.pdf

[2] J. Moyne and J. Iskandar, "Big Data Analytics for Smart Manufacturing: Case Studies in Semiconductor Manufacturing," Processes Journal, Vol. 5, No. 3, July 2017. Available on-line: http://www.mdpi.com/2227- 9717/5/3/39/htm

[3] International Roadmap For Devices And Systems, 2022 Edition Factory Integration. Available Online: https://irds.ieee.org/images/files/pdf/2022/2022IRDS_FAC.pdf

[4] Lopez, F., Moyne, J., Barton, K., and Tilbury, D., "Process capability-aware scheduling/dispatching in wafer fabs," Advanced Process Control Conference XXIX, October 2017. Available via: http://apcconference.com.

[5] National Institute of Standards and Technology. (2023). Vision For Success: Facilities for Semiconductor Materials and Manufacturing Equipment [PDF]. Available online: https://www.nist.gov/system/files/documents/2023/06/23/Vision%20f or%20Success-Facilities%20for%20Semiconductor%20Materials%20and%20Mfr%2 0Equipment.pdf

[6] AMD, Introduction to semiconductors June 2023. Available online: https://www.amd.com/en/technologies/introduction-to-semiconductors

[7] National Institute of Standards and Technology. (2023). Workforce Development Planning Guide Guidance for CHIPS Incentives Applicants [PDF]. Available online: www.nist.gov/system/files/documents/2023/03/30/CHIPS%20Workfo rce%20Development%20Planning%20Guide%20%281%29.pdf

[8] Control for Societal-scale Challenges: Road Map 2030, Eds. A. M. Annaswamy, K. H. Johansson, and G. J. Pappas, IEEE Control Systems Society Publication, 2023, https://ieeecss.org/control-societal-scale-challenges-roadmap-2030 (p. 217)

[9] Poeppel, D, Mangun,G, Gazzaniga M. "The Cognitive Neurosciences" Sixth Edition,Publisher :The MIT Press.2020

[10] Van Ooyen, A., & Butz-Ostendorf, M. (Eds.). (2017). The rewiring brain: a computational approach to structural plasticity in the adult brain. Academic Press.

[11] Hua Tang, Mitchell R. Riley, Balbir Singh, Xue-Lian Qi, David T. Blake, Christos Constantinidis, "Prefrontal cortical plasticity during learning of cognitive tasks", Nature Communications, 13, 1, (2022).

[12] Sarafyazd M, Jazayeri M. "Hierarchical reasoning by neural circuits in the frontal cortex". Science. 2019 May 17;364(6441):eaav8911

[13] Miller EK, Lundqvist M, Bastos AM. "Working Memory 2.0". Neuron. 2018 Oct 24;100(2):463-475.

[14] Keith Sawyer . "The Cambridge Handbook of the Learning Science". Cambridge University Press 3rd edition 2022.

[15] Zeraati, R., Shi, YL., Steinmetz, N.A. et al. " Intrinsic timescales in the visual cortex change with selective attention and reflect spatial connectivity". Nat Commun 14, 1858 (2023).

[16] Charles Lang, George Siemens, Alyssa Friend Wise, Dragan Gašević, Agathe Merceron (Eds.). 2022. "Handbook of Learning Analytics"(2nd. ed.). SoLAR, Vancouver, BC.

[17] Cruz, L., Moreno,W. Neurobiological Computation and Neural Networks. In M. Puri., Y. Pathak., V. Sutariya., S. Tipparaju., W. Moreno, Artificial Network for Drug Delivery and Disposition. (pp. 103-120). Waltham: ELSEVIER. Publication: Science Direct (2016)

[18] Inmordino-Yang,M,H.,A.Christoulou,and V.Singh. Rest Is Not Idleness:Implications of the Brain's Default Mode for Human Development and Education. " Perspectives on Psychological Science 7, N°.4 (2012):352-64.

[19] Takeuchi,H., Y. Taki, H. Hashizume, Y. Sassa,T. Nagase, R. Nouchi, and R. Kawashima. "The Association Between Resting Functional Connectiviy and Creativity" Cerebral Cortex 22,N° 12 (Jan 10 2012): 2921-29.

[20] Lundqvist, M., Brincat, S.L., Rose, J. et al. " Working memory control dynamics follow principles of spatial computing". Nat Commun 14, 1429 (2023).

[21] Jacob A. Miller, Arielle Tambini, Anastasia Kiyonaga, Mark D'Esposito, "Long-term learning transforms prefrontal cortex representations during working memory", Neuron, 110, 22, (3805-3819.e6), (2022).

[22] Lee C, Lee BH, Jung H, Lee C, Sung Y, Kim H, Kim J, Shim JY, Kim JI, Choi DI, Park HY, Kaang BK. Hippocampal engram networks for fear memory recruit new synapses and modify pre-existing synapses in vivo. Curr Biol. Feb 6;33(3):507-516.e3. 2023

[23] Feng-Kuei Chiang, Joni D. Wallis, Erin L. Rich, "Cognitive strategies shift information from single neurons to populations in prefrontal cortex", Neuron, 110, 4, (709-721.e4), (2022).

[24] Tang, H., Qi, X. L., Riley, M. R. & Constantinidis, C. Working memory capacity is enhanced by distributed prefrontal activation and invariant temporal dynamics. Proc. Natl Acad. Sci. USA 116, 7095–7100 (2019)

[25] Park JC, Bae JW, Kim J, Jung MW. Dynamically changing neuronal activity supporting working memory for predictable and unpredictable durations. Sci Rep. 2019 Oct 29;9(1):15512.

[26] Choi DI, Kim J, Lee H, Kim JI, Sung Y, Choi JE, Venkat SJ, Park P, Jung H, Kaang BK. "Synaptic correlates of associative fear memory in the lateral amygdala". Neuron. Sep 1;109(17):2717-2726.e3. 2021.

2023 IEEE Latin American Electron Devices Conference (LAEDC)
Puebla, México, July 3-5, 2023

Applications: Photonic Devices   Confirmation Number: xxxx

High-Performance Germanium P-I-N Photodiodes for High-Speed, Hard X-Ray Imaging

Ziang Guo[1], Sergei Mistyuk[1], Arthur Carpenter[2], and Charles E. Hunt[1]

[1]Department of Electrical and Computer Engineering, University of California – Davis, USA

[2]National Ignition Facility, Lawrence Livermore National Lab, USA

Abstract

A P-I-N hard X-ray photodetector made using epitaxial germanium has been designed, fabricated, and tested. The device is tested from 6keV to 28 keV synchrotron radiation to verify quantum efficiency and the temporal response of epitaxial germanium.

Background

High-speed x-ray direct imaging, especially in the higher (> 20keV) energies has proved challenging for silicon photodetectors. Existing silicon-based devices require the use of thick depletion layers to achieve sufficient external quantum efficiency (QE) for imaging. Germanium, with a higher atomic weight and higher energy K-shell transition, proves superior at stopping high energy photons and allows the use of thinner depletion regions for improved temporal and spatial resolution [1]. Compared to competing materials such as gallium arsenide and lead telluride, germanium has superior hole mobilities, an indirect bandgap to eliminate photon recycling effects, and, in the case of lead telluride, a larger bandgap for reduced

dark currents [2]. We demonstrate here a germanium-based P-I-N photodetector capable of high-speed, high-QE direct X-ray imaging above 12 keV.

Device Description

The device design is a vertical P-I-N photodiode to be operated in reverse-bias, forming the absorption region. The device structure as shown in Figure 2 was formed epitaxially on a highly-doped n-type germanium wafer. The intrinsic regions of the P-I-N structures have varying widths (tested: 10μm, 30μm, 60μm, and 245μm.)   On the n-type surface, a 5-nanometer aluminum oxide layer and a 70-nanometer Ni layer alleviates Fermi-level pinning and provides near-ohmic behavior [3]. The p-type surface is passivated with 19 micrometers of polyimide [4] and has a metallic contact consisting of 200 nanometers of palladium and 200 nanometers of gold. The device is bonded to a SubMiniature version A (SMA) connector using silver conductive adhesive epoxy, as shown in Figure 1.

Experimental Procedure

The P-I-N photodiodes were fabricated at the Center for Nano-MicroManufacturing (CNM2). High-energy X-ray testing was conducted at the Advanced Light Source (ALS) (Lawrence Berkeley National Laboratory) using synchrotron radiation from 6 keV

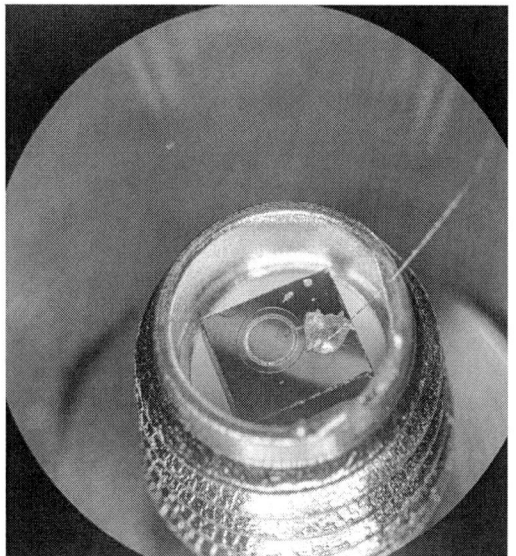

Figure 1 – Packaged germanium ringed anode photodiode device with a 1.39mm diameter.

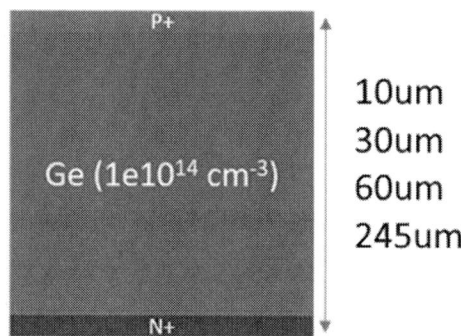

Figure 2 – Epitaxial structure of photodiodes.

979-8-3503-1191-4/23 $31.00 © 2023 IEEE   168

to 28 keV. Packets of electrons accelerated to 1.9 GeV were run through a superconducting magnet to generate magneto-bremsstrahlung radiation. The emitted x-ray beam passed through a monochromator for energy selection, followed by an ion chamber for intensity measurements, then into a collimator, and finally incident on the active region of the device under test (DUT).

QE measurements were performed on 28 germanium P-I-N diodes with all listed thicknesses as well as a reference silicon AXUVHS5 from the Opto Diode Corporation. DUTs were connected to a Keithley 2400 source measure unit (SMU) and a shutter was used to collect alternating bright and dark current data. Devices were left unbiased to prevent impact ionization from skewing results. Energies were swept in steps of 2 keV from 6 keV to 28 keV, and the resultant current and x-ray intensity were recorded. Figure 3 demonstrates the experimental setup for QE testing.

The temporal testing consisted of testing seven 10μm devices as well as the same silicon photodiode from the QE testing. The DUT was connected to the DC+AC port on a 13GHz Picosecond Pulse Labs bias tee. The inductively coupled port was connected to a power supply for reverse bias, the capacitively coupled port was connected to a Microcircuits 3GHz low-pass filter and then a low noise amplifier (LNA). The output of the amplifier was connected to a Tektronix 694C 3GHz oscilloscope for data acquisition. Unfortunately, the LNA was destroyed part way through experimental testing. The device was biased at -45V to be in the reverse-saturation region of the diode. Monochromatic X-ray photons were incident on the active region of the device and traces were collected with the oscilloscope. Due to the monochromator being broken day of testing, devices were only tested with 17 keV X-rays. Figure 4 shows the experimental setup for temporal response testing.

Results

Processed external QE data are shown in Figure 5. As can be seen, epitaxial germanium external QE agrees well with calculated internal QE data using mass-attenuation coefficients for germanium measured by the National Institute of Standard and Technology (NIST) [5]. This indicates that epitaxial germanium shares mass-attenuation coefficients with bulk germanium, and that our germanium P-I-N photodiodes were performing as expected.

The temporal responses of germanium and silicon photodiodes are shown in Figure 6. The average rise time of germanium devices is ~140ps, showing excellent temporal response properties, while the silicon rise time is ~180ps. Both show a full-width half-max (FWHM) of ~1.09ns. Data indicates that a charge-evacuation time below the targeted 2ns from the intrinsic region is possible.

Conclusion

Single-pixel germanium P-I-N photodiode devices intended for hard X-ray imaging have been designed and fabricated. Quantum efficiency and temporal data have been collected at ALS to demonstrate the superior photon-electron conversion efficiency and improved temporal resolution of germanium-based photodetectors compared to existing silicon-based devices.

References

[1] C. E. Hunt *et al.*, "p-i-n High-Speed Photodiodes for X-Ray and Infrared Imagers Fabricated by In Situ-Doped APCVD Germanium Homoepitaxy," *IEEE Trans. Electron Devices*, vol. 67, no. 8, pp. 3235–3241, Aug. 2020.

[2] K. Gupta, M. Chakraverty, Vinay B, A. R. Khan and V. Meshram, "Non-classical scaling approaches for ultra deep sub micron technology," *2013 International Conference on Emerging Trends in Communication, Control, Signal Processing and Computing Applications (C2SPCA)*, Bangalore, India, 2013, pp. 1-6, doi: 10.1109/C2SPCA.2013.6749377.

[3] S. Mistyuk, Z. Guo, A. Kumar, Q. Shao, C. E. Hunt, A. Carpenter, "Interlayers on Homoepitaxial n-Ge for Fermi Level Depinning," In preparation.

[4] S. Mistyuk, Z. Guo, A. Garafalo, Q. Shao, C. E. Hunt, A. Carpenter, "Surface Cleaning and Passivation of Epitaxially Grown Germanium," In preparation.

[5] J.H. Hubbell, S.M. Seltzer, "X-Ray Mass Attenuation Coefficients". NIST Standard Reference Database 126, 1994.

2023 IEEE Latin American Electron Devices Conference (LAEDC)
Puebla, México, July 3-5, 2023

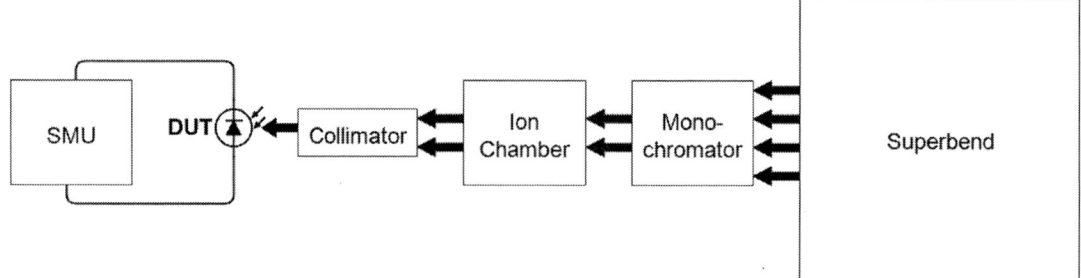

Figure 3 – Schematic representation of QE experimental setup.

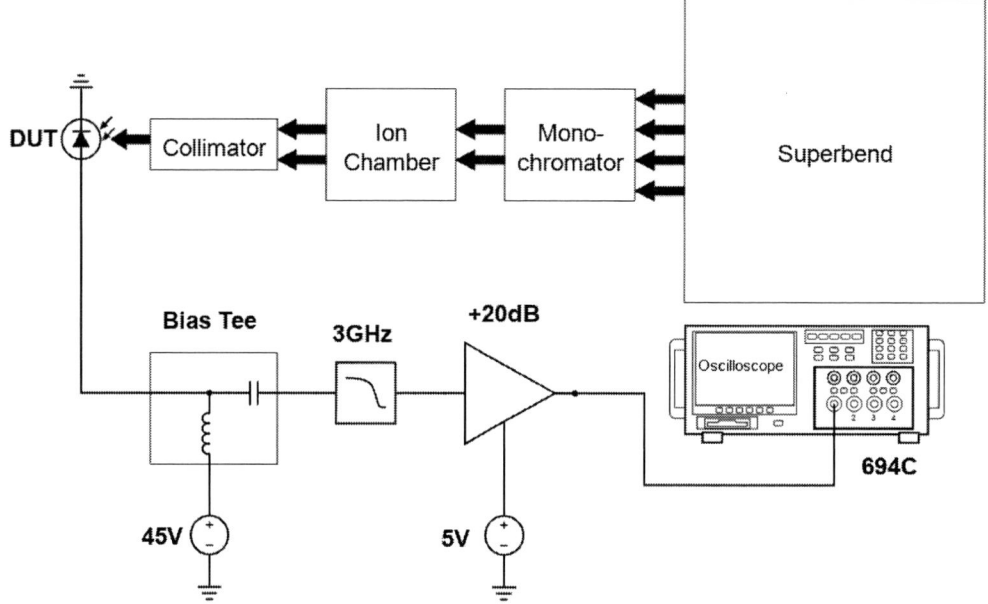

Figure 4 – Experimental setup for temporal testing.

979-8-3503-1191-4/23 $31.00 © 2023 IEEE   170

2023 IEEE Latin American Electron Devices Conference (LAEDC)
Puebla, México, July 3-5, 2023

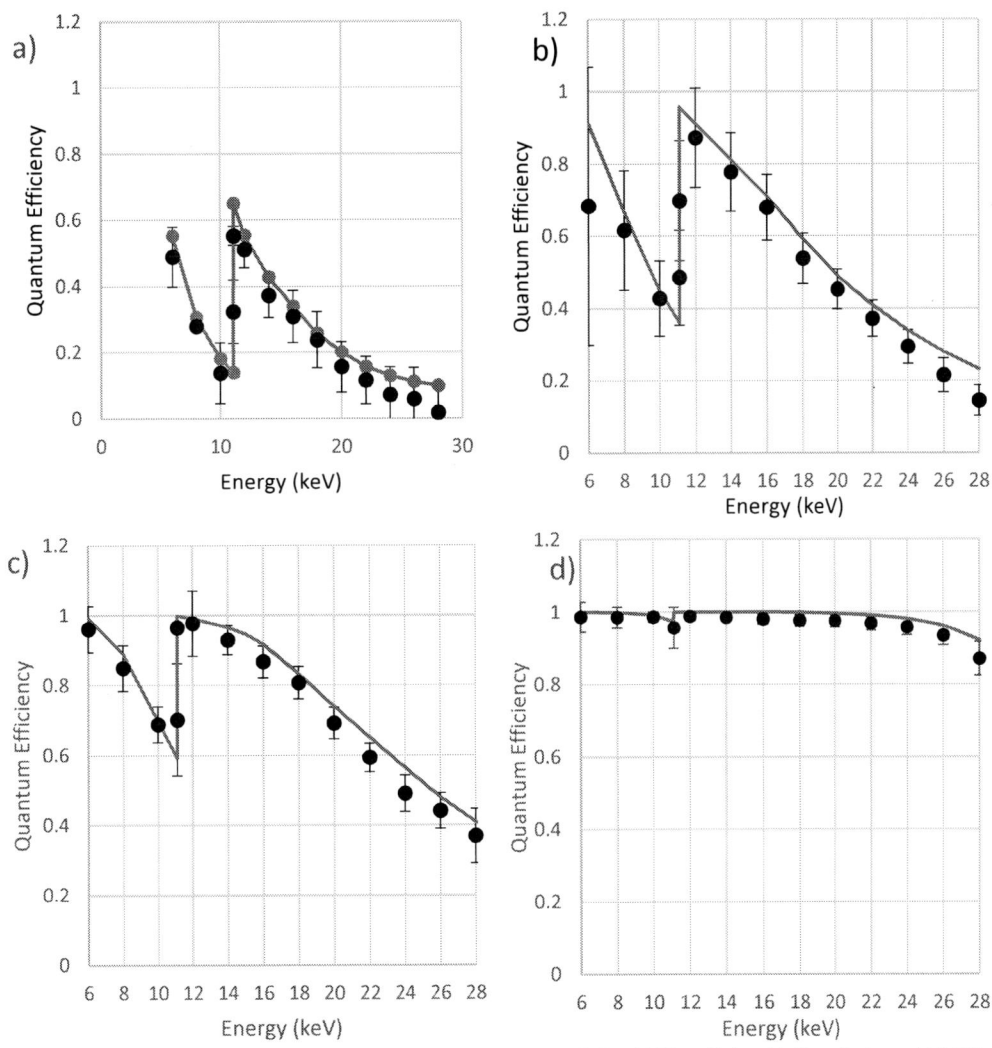

Figure 5 – Processed germanium quantum efficiency from ALS of (a) 10μm diodes, (b) 30μm diodes, (c) 60μm diodes, and (d) 245μm diodes.

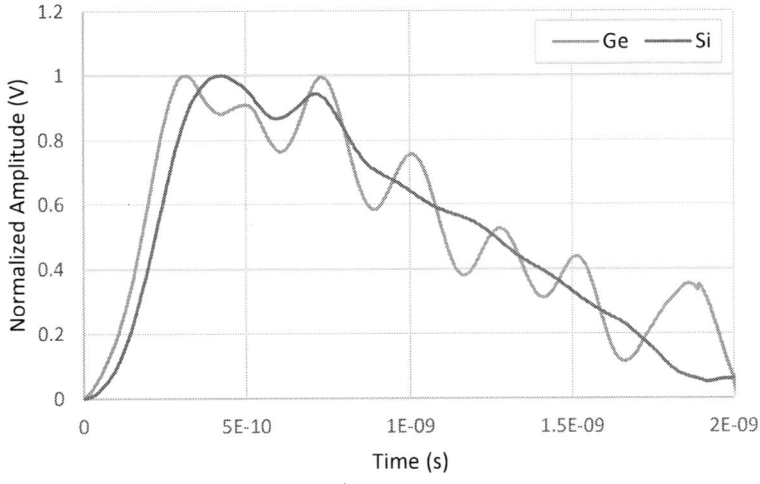

Figure 6 – Averaged and normalized oscilloscope traces.

2023 IEEE Latin American Electron Devices Conference (LAEDC)
Puebla, México, July 3-5, 2023

Development of a turbine spirometer prototype and signal digitalization

Heyul Chavez
Direccion de Investigacion
Universidad Peruana de Ciencias
Aplicadas
Lima, Peru
u812426@upc.edu.pe

Carlos Herrera
Direccion de Investigacion
Universidad Peruana de Ciencias
Aplicadas
Lima, Peru
carlos.herrera.trujillo@gmail.com

Felix Llanos Tejada
Instituto de Investigacion en Ciencias
Biomedicas
Universidad Ricardo Palma
Lima, Peru
felix.llanos.tejada@gmail.com

Jorge Bazán Mayra
Dirección Regional de Salud de
Cajamarca
(DIRESA-Cajamarca)
Cajamarca, Peru
jbazanm@diresacajamarca.gob.pe

Javier M. Moguerza
Escuela Superior de Ingenieria
Informatica
Universidad Rey Juan Carlos
Madrid, Spain
javier.moguerza@urjc.es

Carlos Raymundo
Direccion de Investigacion
Universidad Peruana de Ciencias
Aplicadas
Lima, Peru
carlos.raymundo@upc.edu.pe

Abstract— The present research is part of the project called PHUKUY, which is a low-cost portable digital spirometer prototype designed for remote monitoring of patients with COPD and COVID-19. It also includes additional sensors for improved patient monitoring. This article presents the development of the spirometry module. A turbine spirometer was chosen for its small and lightweight size, easy cleaning and maintenance, and low-cost implementation, which allows for portability. The airflow was successfully digitized, and its direction was detected using a microcontroller, a pair of IR receivers and emitters. This information was then sent via Bluetooth to be displayed in real-time on a mobile phone. For a 3-liter calibration, a maximum error of ±0.115 liters and a standard deviation of 0.051 were achieved.

Keywords—Spirometer, airflow, COPD, COVID-19, respiratory disease.

I. INTRODUCTION

Spirometry is used to measure the volume of air that can be exhaled, it is one of the simplest, most reliable and most accessible techniques to measure the mechanics of breathing.

With this technique, different diseases or anomalies can be detected, such as small lungs, either due to a lung disease or by birth, since large lungs are capable of receiving a greater amount of air than small ones, in addition, it can be determined if the bronchi are or are not obstructed, among other diseases [1].

The medical use of spirometry was made possible by the invention of John Hutchinson in 1844, who built the first spirometer and was the first to use the term expiratory vital capacity and developed normal standards based on measurements made on approximately 200 people [2].

This research is part of the project called PHUKUY [3] (Fig. 1), which is a prototype of a portable spirometer, particularly suitable for home and personal use. It is composed of a system, a software and a method for obtaining, storing and displaying the results of spirometry tests, a sensor for the nose hook, an infrared body temperature sensor and an ambient temperature sensor, an oximetry sensor for measurement of arterial oxygen saturation level (SpO2), heart rate, and an ambient pressure sensor; linked to a web and mobile technology platform that stores the information in the cloud. This device integrates all the measurements carried out by the users in the technological platform, allowing from the analysis of their data to provide an on-demand telemedicine service both at the level of control and monitoring, as well as in the process of rehabilitation of patients.

In the present work, the spirometry module will be explained, the method that was used to detect the air flow and the direction in which it is directed, as well as its digitization.

Fig. 1. PHUKUY – Portable spirometer (design)

II. STATE OF THE ART

Currently on the market, there are different types of spirometers, from incentive spirometers that are the cheapest with an approximate value of USD 16 to specialized spirometers that can cost more than USD 2000, this price difference is due to the functionalities that each one has. spirometer and the technology they use, there are different types such as: volumetric spirometer, water or bell spirometer, piston spirometer, bellows spirometer, Fleisch-type pneumotachograph, Lilly-type pneumotachograph, turbine spirometer, hot-wire spirometer, ultrasound, among others [4].

III. CONTRIBUTION

After studying the technologies used for the development of a spirometer, we chose to develop a turbine-based spirometer. This choice was made because it occupies minimal space and has a lightweight design, making it portable. Additionally, this technology allows for easy cleaning and maintenance, and it is cost-effective to implement.

One important characteristic of the turbine spirometer is its ability to measure the velocity and direction of airflow. The working principle is based on the airflow passing through a

979-8-3503-1191-4/23 $31.00 © 2023 IEEE

turbine that contains a propeller inside. The propeller rotates at a speed proportional to the airflow (see Fig. 2). As the propeller spins, its blades interrupt a beam of light emitted by two infrared sensors located at each end of the headpiece (Fig. 3). This interruption allows for the measurement of the interruption time, and through a transformation of this variable, the airflow can be obtained.

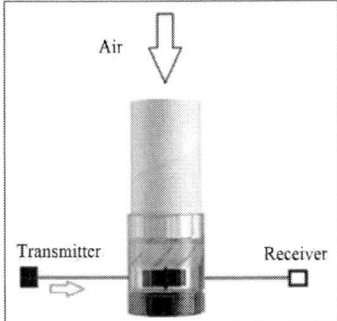

Fig. 2. Operating principle

The rotation detection is done by "Sensor 1" (see Fig. 3), which is intended to obtain the interruption time, and "Sensor 2" (see Fig. 3) is intended to determine the state of the blade to detect the direction of rotation.

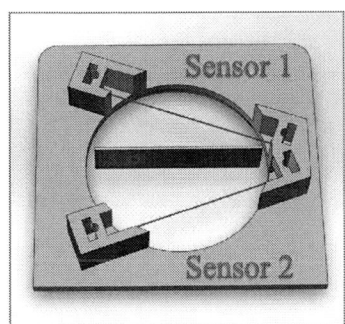

Fig. 3. IR sensors

Fig. 4 shows the printed support and the infrared emitters and receivers that are located in the upper part of the proposed spirometer.

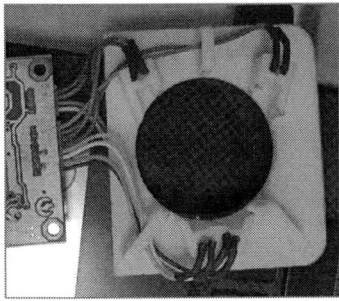

Fig. 4. Sensor support

For example, in Fig. 5, you can see the two possible interruptions in "Sensor 1" clockwise (blue color) or counterclockwise (green color), the microprocessor calculates the time that the blade interrupts the sensor since it is proportional to the RPM with which the blade rotates.

Fig. 5. Possible interruptions

Then, to obtain the direction in which the blade rotates, the status of "Sensor 2" is read, when the interruption of "Sensor 1" occurs. When the blade rotates clockwise (light blue blade), and the interruption occurs in "Sensor 1", "Sensor 2" is also interrupted (see Fig. 6), that is, the microcontroller obtains the value "0". of "Sensor 2", however, when the blade is rotating counterclockwise (green blade) and interrupts "Sensor 1", "Sensor 2" is not interrupted (see Fig. 6), and the microcontroller obtains a "1", in this way it is possible to obtain the direction of rotation of the blade.

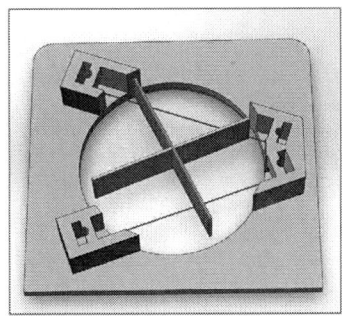

Fig. 6. Rotation detection

Fig. 7 shows a conceptual map that summarizes the connection of the proposal.

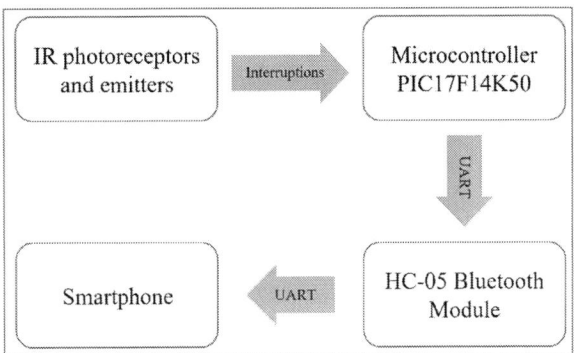

Fig. 7. Proposal concept map

In Fig. 8 you can see the tests with the obstacle sensor modules (consisting of the pair of photoreceptors and IR emitters) in order to validate the correct count of the interruption time of the disposable turbine by means of a module implemented with the PIC17F14K50 microcontroller. (Fig. 9).

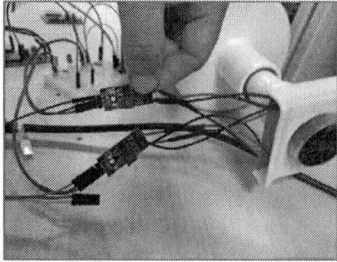

Fig. 8. Pair of photoreceptors and IR emitters

Subsequently, the connection and configuration of the HC-05 bluetooth module and the PIC17F14K50 microcontroller (Fig. 10) were made in order to send the information to a cell phone application via bluetooth (Fig. 11).

The information frames are sent every 50 ms, they include information on the packet number and the time in microseconds of the interruption of the sensors.

With this information, the calibration was carried out, with the help of the SFM3300-D air flow sensor (Fig. 12), in which a constant air flow was injected to obtain the flow and with the same flow it was injected into the turbine to obtain the relationship between air flow and the RPM of the turbine.

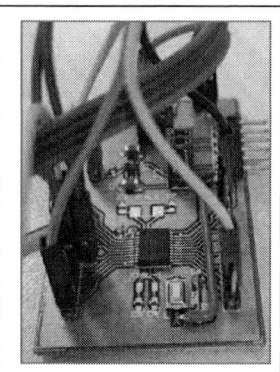

Fig. 9. PIC17F14K50

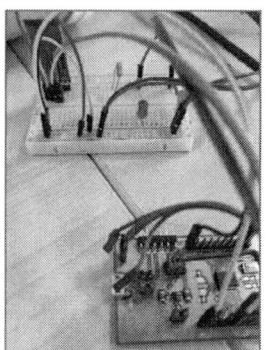

Fig. 10. Microcontroller connection and Bluetooth module HC-05

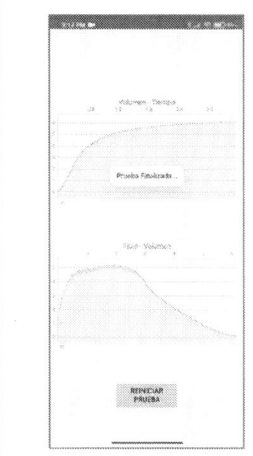

Fig. 11. Mobile application

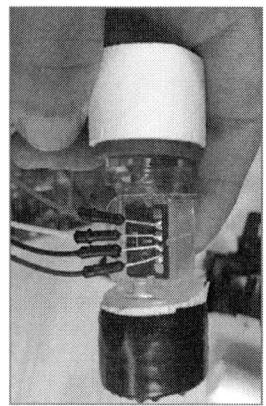

Fig. 12. SFM3300-D Air Flow Sensor

Table 1 shows the relationship between the turns per minute made by the turbine and the air flow.

TABLE I.   RELATIONSHIP RPM VS FLOW (L/s)

RPM	Flow (L/s)
173	0.057
215	0.068
257	0.077
353	0.102
440	0.120
540	0.147
673	0.187
783	0.223
883	0.252
1050	0.297
1210	0.353
1370	0.413
1525	0.465
1690	0.510
1800	0.570
2100	0.673
2359	0.770
2920	0.917
3750	1.167
4570	1.433

With this information, the relationship between the time it takes the turbine to turn around and the air flow obtained with the SFM3300-D sensor can be obtained. Fig. 13 shows the relationship obtained, as well as its equation.

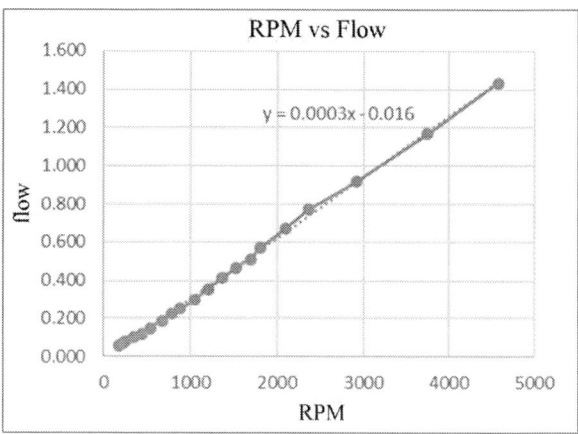

Fig. 13. Relationship between RPM and Flow

By averaging and obtaining a trend line, the following equation was obtained:

$$y = 0.0003\,x - 0.016 \qquad (1)$$

Where:

x: RPM

y: Flow in L/s.

With the equation obtained, it is possible to obtain the air flow from the loop interruption time (this is transformed into RPM) generated by the air blown by the patient.

This transformation is carried out in the cell phone, since it has great processing capacity and in this way the microcontroller is not overloaded, this processed information is shown in the cell phone application.

IV. RESULTS

The validation of the spirometer was carried out with the syringe "Hans Rudolph Calibration Syringe 3 liters - Series 5530" (Fig. 13), which has the following characteristics:

- Total volume: 3L
- Minimum graduation (ml): 100
- Stroke volume (cm): 37.47
- Cylinder diameter (cm): 10.16
- Weight (grams): 2268

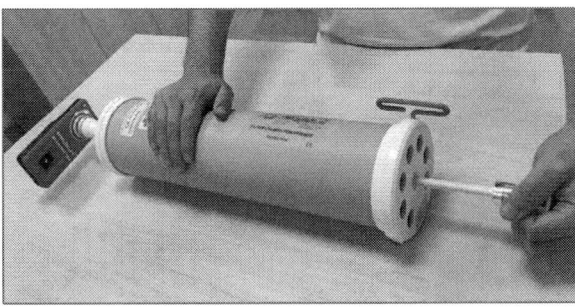

Fig. 14. Tests with the calibration syringe

The calibration syringe has 3 Liters of oxygen, the results were obtained by performing different tests and are shown in Fig. 15 where the y axis represents the Liters and x the tests performed.

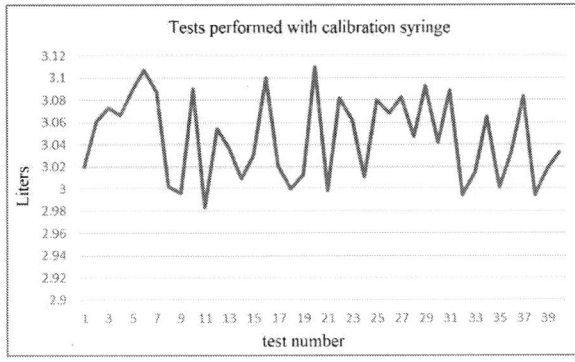

Fig. 15. Calibration Syringe Tests

Table 2 shows the summary table of the statistics of the tests performed.

The results show a variation between ±0.115 Liters and an average of 2.96 Liters, which is very close to 3 Liters.

TABLE II. TEST STATISTICS WITH CALIBRATION SYRINGE

Mean	2.9608
Typical error	0.0081
Median	2.9437
Standard deviation	0.0514
Sample variance	0.0026
Kurtosis	1.0684
Asymmetry coefficient	1.0173
Range	0.2151
Minimum	2.9006
Maximum	3.1157
Account	40

V. CONCLUSIONS

With the help of the flow sensor it was possible to obtain a relationship between the rotation of the turbine and the air flow. For a 3-liter calibration, a maximum error of ±0.115 liters and a standard deviation of 0.051 were achieved.

The frame is sent every 50 milliseconds in order to obtain an image in real time in the application.

With this technique it is possible to obtain the air flow that passes through the turbine and the direction in which it goes (inspiration or expiration), however, it is not enough to obtain the patient's lung capacity, since the flow is not constant. and when air stops flowing through the turbine, it continues to rotate because there is inertia in the turbine, thus generating a non-existent air flow.

ACKNOWLEDGMENT

To the Research Directorate of the Universidad Peruana de Ciencias Aplicadas for the support provided to carry out this research work through the UPC-EXPOST-2023-1 incentive.

To CONCYTEC, which through its executing unit FONDECYT financed the "Leaders in Innovation Fellowships 2020" program with contract No. 008-2021 of the financial scheme E009-2020-01-RAENG.

To CONCYTEC, which through its executing unit FONDECYT with contract number 043-2020-FONDECYT financed the development of the project through the call E067-2020-01-02 Special Projects: Response to COVID-19.

REFERENCES

[1] Juan Carlos Vazquez Garcia and Rogelio Perez Padilla (2018) "Manual de espirometría, tercera edición".

[2] John Hutchinson (1852) "The spirometer, the stethoscope, & scale-balance; their use in discriminating diseases of the chest [&c.]" Número 109 de The spirometer, the stethoscope, & scale-balance; their use in discriminating diseases of the chest [&c.].

[3] Chavez, H., Ronceros, J., Salas, S., et al. (2021). Design of a low-cost digital spirometer for remote monitoring of COPD and COVID patients. In: Waldemar Karwowski, Tareq Ahram, Mario Milicevic, Darko Etinger and Krunoslav Zubrinic (eds) Human Systems Engineering and Design (IHSED2021): Future Trends and Applications. AHFE (2021) International Conference. AHFE Open Access, vol 21. AHFE International, USA. http://doi.org/10.54941/ahfe1001198.

[4] Rivero-Yeverino, Daniela. (2019). Espirometría: conceptos básicos. Revista alergia México, 66(1), 76-84. https://doi.org/10.29262/ram.v66i1.5

AUTHOR INDEX

Aguilera-Galicia, Cuauhtémoc R. 158
Almonacid, Florencia.. 17
Alta, Roxana Pastrana.. 55
Alvarez, Ana Beatriz.. 50
Álvarez-Botero, Germán A. 58
Álvarez-Botero, Germán...................................... 75
Amaya, Marco 1
Angst, Lucas... 50
Ansari, Md. Hasan Raza 101
Araujo-Palomo, Elsie Evelyn.............................. 158
Ascencio-Blancarte, Jorge E. 71
Bastos-Filho, T 31
Bendra, Mario 145
Bestelink, Eva.. 137
Blanco-Díaz, C. F.. 31
Bravo, J. Juan Avilés.. 93
Brown, Ian P. 141
Bucher, Matthias.. 36
Bulhosa, Lucas Dos Santos.................................. 118
Cáceres, Fernanda De Lourdes 62
Calle, Luis.. 1
Calvo-López, B.. 67
Carpenter, Arthur 168
Castel, Jordi 1
Chavez, Heyul 172
Chávez-Hurtado, José Luis.................................. 158
Chejne, Farid 26
Chevas, Loukas.. 36
Comesaña, Enrique 9, 17
Côrrea, Arthur Fonseca 118
Costa, Fernando J... 105
Cruz, Luis Fernando 162
Díaz, Camilo A. R.. 31
Doria, Rodrigo T. 105
Estrada, Magali 132, 149
Fernández, Eduardo F. 17
Fernández, Julián G. 9
Ferrusca-Rodríguez, Daniel 79
Fiorentini, Simone 145
Florez, Ruben... 50
Franco, Ernesto.. 97
Galhardo, Marcos André Barros 118
Garcia-Barrientos, Abel 145
García-Loureiro, Antonio J. 9
García-Loureiro, Antonio..................................... 17
Garduño, Salvador Ivan 132
Goes, Wolfgang 145
Golec, Patryk 137

Gonzalez-Azuara, Arlen....................................... 58
González-Cely, A. X. 31
Goyal, Priyanshi.. 83, 88
Guitarra, Silvana 154
Guo, Ziang 168
Gutiérrez, E. A. 13
Henao-Bravo, Elkin Edilberto 40
Henao-Bravo, Elkin.. 26
Heredia-Rios, Manuel J.. 45
Hernández, Luis 123
Hernandez-Matinez, Luis 45
Herrera, Carlos 172
Herrera-Guerra, Alcides Amado........................... 40
Huerta, L. Palacios 93
Hunt, Charles E. 168
Iñiguez, Benjamin 137
Jiménez, David... 21
Kaur, Harsupreet 83, 88
Linares-Aranda, Monico 45
Lobato-Morales, Humberto 58, 75
Longoria-Gandara, Omar 71
Lopez, Walter Estrada 55
López-Araiza, Karla G. 71
López-Cárdenas, Patricia G. 158
Lozano, Javier F. 17
Macêdo, Wilson Negrão.. 118
Machado, João Paulo De Andrade 118
Maggi, Luis 50
Makris, Nikolaos 36
Marsal, Lluis F. 149
Martínez, Alisson Waleska 62
Martínez, Kevin Fabricio 62
Martínez-Guerrero, Esteban 158
Mavredakis, Nikolaos 21
Mayra, Jorge Bazán 172
Méndez-Jerónimo, Gabriela 58, 75
Méndez-V, J. 13
Misra, D. 105
Mistyuk, Sergei 168
Moguerza, Javier M. 172
Molina, Desiré.. 149
Molina-Reyes, Joel 79, 113
Morales, Alfredo 123
Moreno, Luis Aiquipa 55
Moreno, M. Moreno 93
Moreno, Mario 97, 123
Moreno, Wilfrido Alejandro 162
Moreno-Moreno, Mario 45

Muñoz, Lilia 128
Navlakha, Nupur 101
Pacheco-Sanchez, Anibal 21
Paixão, Thuanne 50
Palomino-Quispe, Facundo 50
Pavanello, Marcelo A. 5
Pavanello, Marcelo 109
Povolotskyi, Michael 5
Quevedo, M. Luis Miguel 162
Raleva, K. 13
Ramírez-Angulo, Jaime 158
Ramírez-Como, Magaly 132, 149
Ramírez-García, Eloy 21
Rangel-Patiño, Francisco E. 71
Rayas-Sánchez, José E. 71
Raymond, Laurent 154
Raymundo, Carlos 172
Reis, Felipe Cabral 118
Reséndiz, Luis 149
Reyes-Valderrama, María Isabel 132
Rios, J. Federico Ramirez 93
Rocha-Aguilera, Daniel 79, 113
Rodríguez-Lugo, Ventura 132
Rossetto, Alan 109
Sacramento-Orduño, Angel 132
Sánchez, A. Morales 93
Sánchez-Hernández, Fernando 158
Sanin-Villa, Daniel 26
Sanz-Pascual, M. T. 67
Sastre-Santos, Ángela 149
Selberherr, Siegfried 145
Seoane, Natalia 9, 17
Serrano, Victoria 128
Serrano-Reyes, A. 67
Shen, Z. John 141
Soares, Caroline S. 5
Soares, Caroline 109
Sporea, Radu A. 137
Sverdlov, Viktor 145
Tejada, Felix Llanos 172
Torres, Alfonso 97, 123
Trevisoli, Renan 105
Trojman, Lionel 154
Tsakalis, Konstantinos 128
Valdez-Sandoval, Leslie M. 21
Vasileska, D. 13
Vasileska, Dragica 5, 109
Vega-Ochoa, Edgar A. 71
Velandia, Oscar 123
Villanueva, Jairo Mendez 5
Villarreal, Vladimir 128

Villegas-Ceballos, Juan Pablo 26, 40
Wang, Wendi 141
Wang, Ziyi 5, 109
Wirth, Gilson 5, 109
Zavala, Ricardo 123
Zeinati, Aseel 105

IEEE
445 Hoes Lane
Piscataway, NJ 08854-4141

ISBN 979-8-3503-1191-4